GRAPHS AND QUESTIONNAIRES

NORTH-HOLLAND MATHEMATICS STUDIES

32

Graphs and Questionnaires

Claude François PICARD

1980

NORTH-HOLLAND PUBLISHING COMPANY – AMSTERDAM • NEW YORK • OXFORD

© *North-Holland Publishing Company, 1980*

All rights reserved. No part of this publication may be reproduced, stored in a retrieval system, or transmitted, in any form or by any means, electronic, mechanical, photocopying, recording or otherwise, without the prior permission of the copyright owner.

ISBN: 0 444 85239 5

Translation and revised edition of:
Graphes et Questionnaires (Tome I and Tome II)
©Bordas (Gauthier-Villars), 1972

Translated by:
J. C. Ault, M. Lobenberg, and N. L. Aggarwal

Publishers:
NORTH-HOLLAND PUBLISHING COMPANY
AMSTERDAM • NEW YORK • OXFORD

Sole distributors for the U.S.A. and Canada:
ELSEVIER NORTH-HOLLAND, INC.
52 VANDERBILT AVENUE, NEW YORK, N.Y. 10017

Library of Congress Cataloging in Publication Data

Picard, Claude François, 1926-
   Graphs and questionnaires.

   (North-Holland mathematics studies ; 32)
   Abridged translation of Grapes et questionnaires.
   Bibliography: p.
   Includes index.
   1. Graph theory. I. Title.
QA166.P5213   511'.5   80-10162
ISBN 0-444-85239-5

PRINTED IN THE NETHERLANDS

PREFACE

It gives me great pleasure to see the publication now of the English version of my work on graphs and questionnaires. As there are already a great many execllent texts on graph theory in the English language, it has seemed appropriate to shift the emphasis more firmly onto the subject of questionnaires by reducing the graph theoretical content from the first volume of the French edition to three chapters. There we set out those elements of graph theory essential for the development of the mathematical techniques used in the theory of questionnaires. Chapter 3 is devoted to a discussion of several operations on graphs which, although they have appeared in many separate publications, have not yet been given a systematic treatment in any other book. Chapter 1 corresponds to chapters 1, 2 and 3 of the French edition, chapter 2 to chapters 4 and 5 and chapter 3 to chapter 6 and chapter 7. The reduction has been achieved by leaving out certain sections for which no summary is given. The remaining topics are carried over without modification.

Chapters 4 to 10 constitute the translation of the second volume of the French edition.

Following the bibliography arranged by chapter, a supplementary list of papers on the subject of questionnaires rounds off the work.

Except as indicated by the above remarks, the account of the contents given in the Preface to the French edition naturally still applies.

The topics dealt with in this book have aroused the interest of several authors who have been able to make original contributions to the theory and to guide young postgraduate students into this line of research.

This translation is leaving the presses at just the right time and I hope that a favourable reception by research workers, engineers and technicians will facilitate further progress to new extensions and applications.

I should like to thank the translators who have often contributed appreciable improvements as compared with the French text and have made every effort to spot any errors.

I am also grateful to the North-Holland Publishing Company for the care they have taken over the preparation and presentation of the book.

C.F. Picard

PREFACE TO THE FRENCH EDITION

Graphs are mathematical entities whose theory facilitates the discussion of the relationships between the elements of a set. Introduced about 1870 by the pioneers of combinatorics who were then still known as geometers, graphs have recently been recognized as the most adaptable tools for certain organizational problems. After first being used in operations research, graphs have now been introduced into information theory where they have proved to be of use in many different areas.

Questionnaires are defined to be valuated graphs intended to elaborate choice or decision models and to process certain information. The discussion of problems in questionnaires is based on certain types of graph to be more specially treated in the first volume and involves the language of probability theory. So once the general definitions and the principal construction techniques have been developed, a chapter is devoted exclusively to information theory for discrete sources. Good text books on graphs are available to French readers at various levels, but there are scarcely any devoted to information theory. So it has been thought necessary to introduce it in axiomatic form in a setting providing easy study.

Based mainly on collective research carried out at the "Centre National de la Recherche Scientifique" in collaboration with various foreign institutions especially in the United States and East Germany, the results presented here are, or have been, the subject of courses delivered to young research workers (Diplôme d'Etudes Approfondies) or to students of the "Maîtrises d'Informatique et de Mathématiques et Applications Fondamentales" at Parisian Universities and also to those already engaged in applied mathematics in France and abroad (for example, Toronto, Prague, Barcelona and Rome).

The first volume of this book is devoted to graphs and chapter 1 presents most of the basic definitions and properties. I have used the concepts of latticoid and lower quasi-strong connectivity introduced by Claude Berge (1958) and Bernard Roy (1961) and I work systematically with the concept of a graph in the sense of what is often called a directed graph (Berge, 1958) or 1-graph (Berge 1970) but nevertheless without ruling out symmetry, reflexivity or multiple arcs; this method which combines rigour with generality often leads to original and direct proofs. Chapter 2 which deals with the representation of graphs, offers the opportunity to satisfy the reader's curiosity to manipulate properly a certain number of graphs: after a few informal examples (§2.8)[1] I give indications for the construction of graphs possessing

---

[1] In the references to paragraphs and subparagraphs the first number indicates the chapter, here chapter 2, paragraph 8.

certain properties, (§2.9). Diagrams and figures are often the necessary support of intuition, and the reader will understand that a drawing may be used to represent a graph, but that a graph does not have to be a drawing; the reader will therefore be able to personalize the examples by constructing his own graphs. In chapters 3 and 4, devoted mainly to paths, circuits and circuitless graphs, the general theory of graphs is developed and the basic tools for dealing with questionnaires are constructed. Therein will be found on the one hand classical results due principally to Berge's school with Camion, Ghouila-Houri and Roy, and on the other a certain number of new results often unpublished. Chapter 5 is a transition to the valuation of graphs and gives some examples of fundamental algorithms, sources of applications in operations research through data processing techniques. The last two chapters of volume 1 are devoted to operational aspects of the theory. Presented algebraically, these two chapters are essential for the construction, transformation and generation of graphs and include results sometimes very recent which are not completely available in book form. In the second volume, these results are supposed to be known.

The second volume is devoted to questionnaires. My initial intention was to offer a new edition of my book, printed by Gauthiers-Villars in 1965, with a review of all principal results obtained since then. But the study of latticoid questionnaires undertaken essentially since 1967, the realization of questionnaires studied more particularly by F. Dubail and E. Lüdde, the pseudoquestionnaires of D. Chenais and M. Terrenoire, and the adjunction of a cost function (G. and S. Petolla and G. T. Duncan) enriched the problematics of questionnaires to such an extent that the theory could no longer be satisfactorily formulated without a basis extending from graphs to probability spaces. The reader acquainted with questionnaires will find that definitions and the more usual terms and symbols have been somewhat modified (for example the notion of an answer is substituted for that of an eventuality the set of questions is denoted by F ...) but these modifications should not be unacceptable as they give to the theory a more unified character and a generalization in a setting which is both wider and more exact. The first chapter of the second volume (chapter 8) is devoted to the general properties of questionnaires and provides a link between valuations a priori of certain graphs and probability spaces. Certain aspects which are trivial for arborescent questionnaires become delicate in the case of latticoids for which various choices of representations and interpretations are possible. Chapter 9 is devoted to transformations, operations and constructions of questionnaires. There, the probability distribution of the answers plays the most important part; nevertheless its suppression introduces new problems for graphs. In chapter 10 the necessary properties and sufficient conditions for questionnaires of minimum routing-length have been collected; this gives a new proof of Huffman's algorithm, deduced from the more general construction elaborated by S. Petolla when the cost function is not uniform. This fundamental type of algorithm, known as a downstream to upstream process is the basis of an extensive area of research whose thread is pursued all through the volume. This fundamental type of algorithm appears somewhat as the complementary part of the algorithms adopted by dynamic programming and must be called for when constraints oppose the free realization of the questionnaire with given answers. An almost complete exposition of information theory for discrete sources is undertaken in chapter 11. I then have recourse to Shannon and Renyi's theory of probabilistic information, but I attempt to show also that an information related to a measurable space which is not probabilizable, of the Kampe, Fériet and Forte type can be useful for the discussion of processing methods. In chapters 11 and 12, some classical theorems on information theory are reformulated or made more precise with the aid of questionnaire theory and the role of optimal agent played by a certain number of questionnaires is pointed out; information appears as a measure of the richness of a set of answers wherein the routing-length gives the minimum cost of acquisition. Chapters 13 and 14 give an extension of the theory by introducing economic functions, utility of answers, questions' cost or formulation and probability of the accuracy of answers. Both aspects of a question are precisely presented; it is a vertex with which is associated a partition of a set of answers. Applications of questionnaires to programming, data-processing and thence to disciplines ranging from human sciences to medicine, to economic and business studies are proposed in these last two chapters where I have attempted to preserve the rigorous aspect of the book by indicating some wide perspectives and exceptionally also some details, without any pretention to exhaustivity.

The 14 chapters nearly all include some quite simple exercises of application and each volume ends with a few problems, the detailed solutions of which, provide complements and precisions. For each volume a bibliography relative to each chapter provides a means of referring to sources, of deepening the understanding of certain questions, or openings to new modes of expression; only a few repetitions are made. A bibliography of collected works will also be found. The authors of theorems or propositions are generally named in the text and the reader will find the reference in the bibliography. However I have omitted the author's name when the result is already classical, mainly in graph theory, or ... yet unpublished. Each volume ends with an index of definitions and a table of the principal symbols with reference to the paragraph numbers where they are used for the first time.

I wish to express my thanks to all research workers and technicians of the research team "Structures de l'Information" of the C.N.R.S. with whom most original results have been found, established and discussed. I equally thank Louis Nolin for accepting this book in the "Programmation" collection which I am pleased to see printed with the care and quality for which the firm Gauthiers-Villars is reknowned.

Sougy, August 1971

CONTENTS

Preface		v
Preface to the French edition		vi

Chapter I Fundamental properties of graphs 1

1.1	Ordered pairs and product sets	1
1.2	The graph concept	2
1.3	Elementary operations and transitive closures	8
1.4	Connectivity, equivalence and preorder	11
1.5	Graph representations	17
1.5.1	Various definitions of graphs	17
1.5.2	Graph isomorphisms	18
1.5.3	Adjacency matrices	19
1.5.4	The incidence matrix of a graph	20
1.5.5	Computer representation of graphs	20
1.5.6	Valuations	21
1.5.7	Coding	21
1.6	Paths, circuits and cocircuits	22
1.6.1	Chains and concatenation	22
1.6.2	Cocircuits, cocycles and cycles	24
Exercises		26

Chapter II Latticoids and arborescences 29

2.1	Circuitless graphs	29
2.2	Arborescences and trees	33
2.3	Arborescences and data processing	41
2.3.1	Simplexes and arborescences	41
2.3.2	Monoids and arborescences	43
2.3.3	Chains, paths and arborescent procedures	44
2.3.4	Coding	45
2.3.5	Finite and infinite graphs	47
2.4	Transportation networks	48
Exercises		50

Chapter III Operations on graphs 51

3.1	General definitions	51
3.2	Unary operations	52
3.3	Transformations	53
3.4	Cartesian operations	58
3.4.1	Product and sum	58
3.4.2	Classes of vertices	63
3.4.3	Connectivities	66
3.4.4	Valuations	72
3.5	Latticoid operations	75
Exercises		77

Chapter IV		General properties of questionnaires	81
	4.1	Preliminaries	81
	4.2	The concept of a questionnaire	82
	4.2.1	Axioms and definitions	82
	4.2.2	Cutsets of a questionnaire	85
	4.2.3	Partitions of the answers	88
	4.2.4	Probabilities in an arborescent questionnaire	91
	4.3	Routing	93
	4.3.1	The arborescence of paths in a latticoid questionnaire	93
	4.3.2	Compatible arborescent questionnaires	96
	4.4	Probabilities in a latticoid questionnaire	101
	4.4.1	Probabilities of the vertices	101
	4.4.2	Probabilities of the arcs and conditioning	103
	4.4.3	Example of a semantic	107
	4.4.4	A restriction of the theory	109
	4.5	Routing length	111
	Exercises		114
Chapter V		The construction of questionnaires	117
	5.1	Operations on questionnaires	117
	5.1.1	Definitions	117
	5.1.2	Operations and routing length	122
	5.2	Valuations on the answers and the arcs	126
	5.3	L-optimal supports	130
	5.3.1	Homogeneous questionnaires	130
	5.3.1.1	a-1 is a divisor of N-1	131
	5.3.1.2	a-1 is not a divisor of N-1	134
	5.3.2	Heterogeneous questionnaires	136
	5.4	Properties of arborescent questionnaires	137
	5.4.1	The number of vertices and notation	137
	5.4.2	Arborescences of minimal height	139
	5.4.3	Questionnaires with balances support	142
	5.4.4	Arborescences and questionnaires of maximal height	143
	5.4.5	Extremal properties of the supports	145
	Exercises		147
Chapter VI		Optimal routing	149
	6.1	Determination of an L-optimal questionnaire	149
	6.2	Necessary conditions for L-optimality	149
	6.2.1	Substitutions of arcs	149
	6.2.2	Transfers of arborescences	150
	6.2.3	Sub-questionnaires	153
	6.3	A sufficient condition for L-optimality	155
	6.4	Huffman's algorithm	157
	6.5	Questionnaires and coding	162
	6.6	Equiprobable polychotomic questionnaires	165

6.6.1	A characteristic property of homogeneous balanced arborescences	165
6.6.2	Equiprobable dichotomic questionnaires	168
6.6.2.1	Optimal questionnaires	168
6.6.2.2	Routing in a dichotomic questionnaire which is not optimal	169
Exercises		172

Chapter VII Informational study of questionnaires — 175

7.1	Introduction to information	175
7.1.1	Hartley and Shannon's forms	175
7.1.2	Questionnaires in the sense of Shannon	179
7.2	Axiomatics of information	182
7.2.1	Faddeev's axioms	182
7.2.2	Some axiom systems	185
7.3	Properties of information	191
7.3.1	Convexity and concavity	191
7.3.2	Independence and dependence	194
7.4	Processed information and transmitted information	201
7.5	Other definitions of information	207
7.5.1	Information for incomplete distributions	207
7.5.2	Measure, probability and information	209
7.5.2.1	Probability and information	209
7.5.2.2	Information for measure spaces	210
7.5.2.3	Non-probabilistic questionnaires	213
Exercises		215

Chapter VIII Information and routing length — 217

8.1	Information and routing in questionnaires	217
8.1.1	Inefficiency and noise	217
8.1.2	L-optimal questionnaires ($L_H \geqslant I$)	221
8.1.3	Questionnaire product of two polychotomic questionnaires	227
8.1.4	Heterogeneous questionnaires	235
8.2	Contribution of information	241
8.2.1	Maximization of processed information and Shannon-Fano's algorithm	241
8.2.2	Partitions in equiprobable dichotomic questionnaires and choice	242
8.2.3	Minimization of the contributed information and Huffman's algorithm	244
8.2.3.1	Dichotomic questionnaires	244
8.2.3.2	Polychotomic questionnaires in the strict sense	244
8.2.3.3	Heterogeneous questionnaires	245
8.2.3.4	Polychotomic questionnaires in the broad sense	247
8.2.3.5	Informational interpretation of Huffman's algorithm	247
8.2.4	Heterogeneous information and acquisition	248
8.3	Quasi-questionnaires	252
8.3.1	Quasi-answers and quasi-questions	252

8.3.2	Instantaneous codes	255
8.3.3	Upper bounds for L-optimal questionnaires	259
Exercises		267

Chapter IX   Conditioning of the questions and answers   269

9.1	Limitations and extensions	269
9.2	Utilities of the answers	269
9.2.1	Useful length	269
9.2.2	Useful information	272
9.3	Cost of the questions	275
9.3.1	Costs and expenses	275
9.3.2	Free costs	277
9.3.3	Binding of the costs to the bases	280
9.3.4	Logarithmic costs	283
9.4	Questionnaires in the sense of Campbell	284
9.4.1	Questionnaires and Rényi's information	284
9.4.2	Charges and expenses	287
9.5	Questionnaires in the broad sense	288
9.5.1	Infinite questionnaires	288
9.5.2	Questionnaires with circuits	290
9.5.3	Flow charts and circuits	292
9.6	Realizable questionnaires	294
9.6.1	Constraints in questionnaires	294
9.6.2	A detection problem	295
9.6.3	Partitions and formable questions	300
9.6.4	Arborescent realizable L-optimal questionnaires	303
9.6.5	The equivalence of constraints and costs	305
9.7	Questionnaires in practice	306
9.7.1	The dynamic aspect of interrogation	306
9.7.2	A random experiment	307
9.7.3	Absorption tests	309
9.7.4	Weighings	310

Chapter X   Interrogations, comparisons, sortings   313

10.1	Indirect interrogations and pseudo-questionnaires	313
10.1.1	Direct and indirect interrogations	313
10.1.2	Pseudoquestionnaires	313
10.1.3	Routing, information, convergence	317
10.1.4	Applications to pattern recognition	321
10.1.4.1	Diagnosis aid	321
10.1.4.2	Segmentation in a population	323
10.1.4.3	Word recognition	323
10.2	Comparisons and questions	325
10.2.1	Compatible realizable latticoids and arborescences	325
10.2.2	Products of arborescent questionnaires	328
10.2.3	Sequential questionnaires	331
10.3	Questionnaires for sorting	333
10.3.1	L-optimal sorting	333
10.3.2	Realizable sortings	338

Problems 343
Solutions to Problems 355
Tables 397
Bibliography 401
Index 421
Main Symbols 429

Chapter One

FUNDAMENTAL PROPERTIES OF GRAPHS

We start by recalling the concepts of *ordered pair* and *product set*. The concept of a graph will be introduced through the idea of relations defined on a set and we will thereby elucidate some of its fundamental aspects, in particular the properties of symmetry and connectivity.

1.1 Ordered pairs and product sets

Let X and Y be sets, x an element of X and y an element of Y.

Definition 1. To say "z is an *ordered pair*" means

$$\exists x \; \exists y; \; z = (x,y) \tag{1}$$

The element x is referred to as the *first projection* of z and y is referred to as the *second projection* of z. We write

$$x = pr_1 z \quad \text{and} \quad y = pr_2 z \tag{2}$$

The presentation of a particular ordered pair $z = (a,b)$ is equivalent to the statement

$$\exists x \; \exists y; \; z = (x,y) \quad \text{and} \quad x = a, \; y = b \tag{3}$$

The uniqueness of an ordered pair is described by the axiom

$$\forall x \; \forall x' \; \forall y \; \forall y' \, ((x,y) = (x',y')) \Rightarrow (x=x' \text{ and } y=y') \tag{4}$$

Any (binary) relation R between the elements x and y can be interpreted as a property of the ordered pairs (x,y). We will also write xRy to mean x is related to y by R.

It is essential to maintain the distinction between the projections of an ordered pair. That is, the first projection (source object) must always be written in front of the second projection (target object).

Definition 2. The *direct product* XY of sets X and Y is the set of all ordered pairs

$$z = (x,y) \text{ in which } x \in X \text{ and } y \in Y \tag{5}$$

X is referred to as the *first factor* and Y as the *second factor* of XY.

The direct product of the set $Z = XY$ with another set T consists of the ordered pairs

$$u = (z,t) \quad \text{in which} \quad z \in XY \quad \text{and} \quad t \in T,$$

that is of all

$$u = ((x,y),t) \quad \text{in which} \quad x \in X, \ y \in Y \ \text{and} \ t \in T \qquad (6)$$

Definition 3. To say "u is an ordered triple" means

$$\exists x \ \exists y \ \exists t \ ; \ u = (x,y,t).$$

Property 1. Up to isomorphism the direct product of sets is associative.

For, the typical element of $(XY)T$ is $u = ((x,y),t)$, where $x \in X$, $y \in Y$ and $t \in T$ and, by axiom (4), this may be uniquely associated with the ordered triple (x,y,t).

Similarly the typical element of $X(YT)$ is $v = (x,(y,t))$, where $x \in X$, $y \in Y$ and $t \in T$ and again by axiom (4), this may be uniquely associated with the ordered triple (x,y,t).

Thus there is an isomorphism between the sets $(XY)T$ and $X(YT)$.

We may write

$$(XY)\ T = X(YT) = XYT \qquad (7)$$

In general $XY \neq YX$.

When $Y = X$, we will write $XX = X^2$.

The elements of a product of n sets are known as *ordered n-triples*. Ordered pairs and triples are special instances of this.

Definition 4. A *(binary) relation* R on a set X may be given as a subset Γ of X^2, where an ordered pair $(x_1, x_2) \in \Gamma$ if and only if $x_1 R x_2$.

1.2 The Graph Concept

Definition 5. A *graph* is an ordered pair

$$G = (X, \Gamma)$$

in which X is a set and Γ is a subset of X^2.

Terminology: For the sake of simplicity we will refer to Γ also as a relation on X and we will freely use the symbol Γ to denote the relation, that is, we will sometimes write $x\Gamma y$ to mean $(x,y) \in \Gamma$. This conforms with the usual terminology used in "graph theory" from Silvester (1877) or Koenig (1913) to Berge, Zytkov and Sachs (1970), though Bourbaki calls "graph" what we call "relation" and "correspondence" what we call "graph, G". Note that "graph" as defined here is also known as a "directed graph".

Definition 6. Let $G = (X, \Gamma)$ be a graph. Then each element of X is known as a *vertex* of G and each element of Γ is called an *arc* of G.

Thus G consists of vertices and arcs and Γ is said to be the *set of arcs of* G.

"Ordered pair" will still be used to denote a general element of X^2.

The arc $u = (a,b)$ is said to *join* the vertices a and b (or b and a) or to *join* a *to* b and we write

$$a\Gamma b \quad \text{or} \quad (a,b) \in \Gamma \tag{8}$$

Analogously we will also write $a\cancel{\Gamma}c$ or $(a,c) \notin \Gamma$ to mean a is not joined to c by an arc in the graph.

The projections x and y of the ordered pair z are called the *extremities* of the arc z; x is its origin or *initial extremity* (or *vertex*) and y is its *terminal extremity* (or *vertex*). The arc (x,y) will also be said to *leave from* (or *originate at*) x and to *enter into* y.

The set of all the terminal extremities of arcs leaving from the vertex x will be denoted by Γx and the set of all initial extremities of arcs entering into x will be denoted by $\Gamma^{-1}x$. Similarly if $A \subseteq X$, ΓA will denote the set of all terminal extremities of arcs originating in A and $\Gamma^{-1}A$ will denote the set of all initial extremities of arcs entering A, that is, entering into vertices in A.

An arc $u = (a,b)$ is said to be *incident* (*outwardly*) with a and (*inwardly*) with b; a and b are said to be *adjacent*; the elements of Γx are adjacent to x.

Definition 7. If in an arc z, $pr_1 z = pr_2 z$, then z is said to be a *loop*. Any arc which is not a loop is said to be a *proper arc*.

Note that $|\Gamma x|$ and $|\Gamma^{-1}x|$ may take any values including infinity and zero. If $|\Gamma x| = |\Gamma^{-1}x| = 0$, x is said to be *isolated*.

In diagrams, the vertices of a graph will be represented by points and the arcs by lines (possibly curved) marked by arrows to indicate the direction from the initial to the terminal vertices (Figure 1.1).

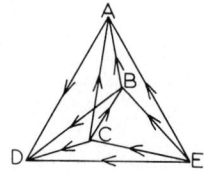

Figure 1.1

If $a\Gamma b$, then $b\Gamma^{-1}a$. The graph $G' = (X,\Gamma')$ obtained from $G = (X,\Gamma)$ by reversal of the projections so that $b\Gamma'a \Leftrightarrow a\Gamma b$ has as its set of arcs the set Γ' which is also denoted by Γ^{-1}, and is often called the *inverse relation* of Γ.

Definition 8. A *path* in a graph G is a set of arcs in G ordered in such a way that the initial vertex of any one (except possibly the first) is equal to the terminal vertex of the one preceding it.

If the set of arcs is finite, then the path is said to be *finite*. Otherwise it is *infinite*.

Any finite path must have an *origin* a and a terminal vertex b, being respectively the initial vertex of the first arc and the terminal vertex of the last arc of the path. The path denoted by [ab] will consist of ℓ arcs

$$(a_1,a_2), (a_2,a_3), (a_3,a_4), \ldots, (a_k,a_\ell), (a_\ell,b) \quad (9)$$

The path [ab] is said to *link* a and b (or a to b) and it may be exhibited more explicitly by using the alternative notation [a a_2 a_3 ... a_ℓ b]. We will also say that the path goes from a to b.

The path formed by the consecutive arcs u, v,....., z will also be denoted by [uv...z] or [uz]. This is an ordered n-tuple (u,v,...,z) in which the projections are elements of X^2.

We distinguish between the arcs and vertices which make up a path by saying that the arcs *belong to* the path and the vertices *lie on* the path.

The graph

$$G = (\{a,a_2,\ldots,a_\ell,b\}, \{(a,a_2), (a_2,a_3),\ldots,(a_\ell,b)\}),$$

in which the arcs are the arcs of a path and the vertices are the projections of those arcs, is said to be a *walk*. The set of vertices in a walk is known as its *itinerary*.

A path which does not use any arc more than once is said to be *simple*. Otherwise a path which may use an arc more than once is said to be *non-simple*.

A path is *elementary* if the initial vertices of its arcs are all distinct.

If a,...,f are vertices and $u_1, u_2, \ldots$ arcs such that $pr_1v_1 = pr_1u_3$, then

$$[abcdef] \quad \text{and} \quad [u_1v_1w_1u_2v_2]$$

are elementary paths,

$$[abcdebf] \quad \text{and} \quad [u_1v_1w_1u_2w_2u_3]$$

are simple but not elementary paths and

$$[abcdebcf] \quad \text{and} \quad [u_1v_1w_1u_2w_2v_1w_3]$$

are non-simple paths.

Every elementary path is a simple path according to these definitions.

The vertices of an elementary path are *totally ordered*.

DEFINITION 9. A *circuit* is a path [aa] in which the terminal vertex coincides with the origin.

A circuit (or a path) is determined by the different arcs of which it is composed or by the initial vertices of these arcs (and by the terminal vertex of the last arc).

A circuit is therefore an ordered set of arcs in which the terminal vertex of any one coincides with the initial vertex of the one following it and the terminal vertex of the last arc coincides with the initial vertex of the first.

A *non-simple* circuit is one in which the arcs may not all be distinct. Any simple circuit can be made into a non-simple one merely by going round twice.

An *elementary* circuit is an elementary path in which the initial vertex of the first arc coincides with the terminal vertex of the last.

Any cyclic permutation of the list of initial vertices of the consecutive arcs (or of the ordered list of arcs) produces the same circuit. Thus *the order in a circuit is not unique*.

DEFINITION 10. The graph $G = (X,\Gamma)$ is said to be *symmetric* whenever the relation Γ is symmetric, that is, when

$$x\Gamma y \Rightarrow y\Gamma x \tag{10}$$

Note that a symmetric graph may have loops because (10) allows Γ to be reflexive.

In order to indicate the existence of *both arcs* (x,y) and (y,x) when x and y are distinct in a symmetric graph we will speak of *the edge* $)x,y(.$ Thus a symmetric graph without loops has twice as many arcs as distinct edges. Such graphs are sometimes referred to as *simple graphs* (Berge, Sachs).

DEFINITION 11. A graph $G = (X,\Gamma)$ is said to be *anti-symmetric* whenever the relation Γ is anti-symmetric, that is, when

$$(x,y) \in \Gamma \Rightarrow (y,x) \notin \Gamma \tag{11}$$

It follows from (11) that an anti-symmetric graph cannot have loops.

From time to time we will come across the problem of

constructing an anti-symmetric graph by suppressing all the loops and half of the proper arcs of a given symmetric graph. This problem which consists of choosing exactly one of the ordered pairs (a,b) and (b,a) for each edge)a,b(of the symmetric graph is known as *orientation of a symmetric graph*. Conversely it is possible to generate a symmetric graph G_S from any given graph $G = (X,\Gamma)$. Here $G_S = (X,\Gamma_S)$, where $\Gamma_S = \Gamma \cup \Gamma'$ and

$$\Gamma' = \{z' \mid z' = (pr_2 z, pr_1 z) \text{ and } z \in \Gamma\},$$

is clearly symmetric. We say that G_S is the *symmetric graph based on* G.

Definition 12. A *chain* in a symmetric graph G is a path in G and a *cycle* in G is a circuit of G.[1]

Definition 13. We use the terms *edges*, *chains* and *cycles* of a graph G to mean the edges, chains and cycles respectively of the symmetric graph G_S based on G, except for those cycles which consist of the two arcs (a,b) and (b,a) of an edge in G_S when one of them is not in G.

We distinguish the chains formed by consecutive edges u,v,w and w,v,u by saying that they have opposite *directions of flow*.

Consequently a chain in a graph G is a set of arcs in which the initial vertex of any one is the same as the initial or terminal vertex of the preceding or following one. Every vertex except the extremal ones belongs to two consecutive ordered pairs.

A chain in a non-symmetric graph is a path if and only if it satisfies the stronger condition given in Definition 8 (cf. Property 1.9).

Chains and cycles are said to be non-simple or elementary when the corresponding paths and circuits are non-simple or elementary. That is when there may be repeated arcs or when the vertices are all distinct (except for the ends).

The edge corresponding to the arcs (a,b) and (b,a) is denoted by)a,b(and a chain with extremities a and b is denoted by]ab[. We will also write)a,b($\in \Gamma$ instead of the more correct form (a,b) $\in \Gamma_S$ and it means that aΓb or bΓa.

To construct a chain in a graph it is not actually necessary to refer to the symmetric graph constructed on G. For example the chain]abcd[can be described in any graph $G = (X,\Gamma)$ by the statement

---

[1] We say that the two arcs (paths) corresponding to the same edge (chain) have opposite orientation. The concepts introduced here are for a simple graph.

$(a,b) \lor (b,a) \in \Gamma$, $(b,c) \lor (c,b) \in \Gamma$, $(c,d) \lor (d,c) \in \Gamma$,

where the Boolean symbol $\lor$ means "or" (and allows the possibility of both). For, to say that]abcd[is a chain means that the edges)a,b(,)b,c(and)c,d(all belong to the subset $\Gamma_s \subset X^2$ in the symmetric graph based on G. This graph then also has a chain]dcba[.

Remark. In a symmetric graph, for any arc (a,b) we can define both the circuit [aba] and the edge)a,b(. We will use the word "circuit" in this context when we wish to indicate that both arcs are present and "edge" when we have at least one of the two.

Note that]abcda[and]adcba[are distinct cycles in a symmetric graph but that, for example]abcda[,]adcba[and]bcdab[all represent the same cycle in an antisymmetric graph.

Definition 14. An *s-graph* is an (s+1)-tuple

$$\Pi = (X, \Gamma_1, \Gamma_2, \ldots, \Gamma_s) \quad (12)$$

in which the first element is a set X and each of the other s elements Γ_i is a subset of X^2 determined by some relation R_i on X. (i = 1,2,...,n).

A 1-graph is what we have already called a graph.

In an s-graph, many arcs having the same extremities ($pr_1 z$, $pr_2 z$) can be distinct. That is, for any element of X^2 any or all of the s-relations $R_1, \ldots, R_n$ may be valid.

We say that u is an arc of Π if there is an i for which $u \in \Gamma_i$. If $u = (x,y)$, $u \in \Gamma_i$ and $v = (x,y)$ but $v \neq u$, then there is a $j \neq i$ such that $v \in \Gamma_j$. The indices i of the relations are sometimes also called *labels*.

Symmetry, reflexivity and antisymmetry can all be defined in s-graphs (for each set of arcs Γ_i). The various definitions of paths and chains extend to s-graphs: a path is a set of arcs belonging to the sets Γ_i (not necessarily all from the same one). A symmetric s-graph is usually referred to as a *multigraph* (Berge, 1970).

Definition 15. A *subgraph* S of a graph $G = (X,\Gamma)$ is a graph $S = (X',\Gamma')$, where $X' \subset X$ and

$$\Gamma' = \{z \mid z \in \Gamma \text{ and } pr_1 z \in X', pr_2 z \in X'\}$$

that is, $\Gamma' = \Gamma \cap X'^2$.

We say that S is *generated by* X' and Γ' is the *restriction* of the relation Γ to the subset X'. We also say that S is obtained from G by *suppression of the vertices* belonging to

X-X'.

Definition 16. A *partial graph* of a graph $G = (X,\Gamma)$ is a graph $P = (X,\Gamma'')$ in which the set of arcs Γ'' is a subset of Γ, that is, $\Gamma'' \subset \Gamma$.

We say that P is obtained from G by *suppression of the arcs* in $\Gamma - \Gamma''$.

A partial subgraph is a partial graph (X',Γ'') of a subgraph (X',Γ').

Note that in this case $(X',\Gamma'') \subset (X',\Gamma') \subset (X,\Gamma)$, where inclusion applies both to the vertices and the arcs.

A walk in G is an example of a partial subgraph of G.

1.3 Elementary Operations and Transitive Closures

Definition 17. An *elementary operation* θ on two graphs G_1 and G_2 is any operation by which we may define a third graph G_3 by composing the elements of G_1 and G_2 according to a law (denoted θ) acting on the vertices and the same law acting on the arcs. Thus if $G_1 = (X_1,\Gamma_1)$ and $G_2 = (X_2,\Gamma_2)$, then

$$G_3 = (X_3, \Gamma_3), \text{ where } X_3 = X_1 \theta X_2 \text{ and } \Gamma_3 = \Gamma_1 \theta \Gamma_2.$$

We may also write $G_3 = G_1 \theta G_2$.

Two examples of such elementary operations are:

Union of two graphs

$$G_1 \cup G_2 = (X_1 \cup X_2, \Gamma_1 \cup \Gamma_2).$$

Intersection of two graphs

$$G_1 \cap G_2 = (X_1 \cap X_2, \Gamma_1 \cap \Gamma_2).$$

Both these operations are associative and commutative because the union and intersection of sets are associative and commutative.

If (a,b) is an arc, we say that a is an *antecedent* of b and b is a *successor* of a. If $[acb]$ is a path, we say that c is the *predecessor* of b on this path.

The relative positions of two vertices in a graph can be made more precise by means of other types of relation (such as equivalence, pre-order or order) which are more general than that of belonging to the same ordered pair. However the corresponding information is already included in the specification of the graph.

The set Γa of all terminal vertices of arcs having a as initial vertex is referred to as the *set of successors* of a.

Similarly the terminal vertices of all paths having a as origin are known as the *proper descendants* of a while the origins of all paths having terminal vertex b are the *proper ancestors* of b.

Let $A(x)$ and $\mu(x)$ be the sets of proper ancestors and proper descendants of x.

Property 2. The following three properties are equivalent:

(i) $x \in A(x)$,
(ii) $x \in \mu(x)$,
(iii) There exists a circuit passing through x.

For, if $x \in A(y)$, then there exists a path $[xy]$ which in the case when $y = x$ becomes a circuit passing through x. Similarly if $x \in \mu(y)$, then there exists a path $[yx]$ which in the case $y = x$ is again a circuit passing through x. Finally the existence of a circuit $[x\ x]$ is equivalent to the existence of a path in which x is both the origin and the terminal vertex which implies both $x \in A(x)$ and $x \in \mu(x)$.

More generally we note that

$$x \in A(y) \Leftrightarrow y \in \mu(x).$$

If there is a circuit of the graph passing through x, then x is its own proper ancestor (and descendant). On the other hand if x does not lie on any circuit then x is not its own proper ancestor or descendant. Nevertheless we will still say that x is its own ancestor or descendant (in a broad sense).

Definition 18. The *length* of a path $[ab]$ is defined to be the number of arcs in the path.

Note that a path of length 1 is a set containing just one arc and as such it is to be distinguished from the arc itself.

A circuit of length 1 contains just one loop.

Let $\mu^i x$ denote the set of all vertices which are reachable from x by a path of length at most i.

Then $\mu^0 x = \emptyset$ and $\mu^1 x = \Gamma x$.

We will denote by

$$\mu x = \{y \mid y \in \lim_{r \to \infty} \mu^r x\} \tag{13}$$

the set of all proper descendants of x - and call it the μ-*transitive closure* of x or the *strict transitive closure* of x.

Note that $y \in \mu^i x \Rightarrow y \in \mu^j x \quad (j \geq i)$

and $y \in \mu^i x \Rightarrow y \in \mu x$.

A new relation μ can be defined on the underlying set X of the graph $G = (X,\Gamma)$ by setting

$$\mu = \{(x,y) \mid x,y \in X \text{ and } y \in \mu x\} \qquad (14)$$

This leads to a unique graph $T = (X,\mu)$ obtained from G.

Here the path [xy] in G corresponds to the arc (x,y) in T. Conversely $x\mu y$ is equivalent to "there is at least one path [xy] in G".

In view of Property 2, $(x,x) \in \mu$ if and only if there is a circuit passing through x. The graph T has loops only at those vertices which are their own proper ancestors (in G).

Another related and very useful concept, that of broad transitive closure, requires in addition *reflexivity*.

We put $\Gamma^i x = \mu^i x \cup \{x\}$

and $\hat{\Gamma} x = \mu x \cup \{x\}$

so that $\hat{\Gamma} x = \{y \mid y = x \quad \text{or} \quad y \in \mu x\}$ (15)

is the set of descendants of x in the broad sense mentioned earlier. $\hat{\Gamma} x$ is called the *transitive closure* of x or the *broad transitive closure* of x.

The relation of broad transitive closure on the underlying set X of the graph $G = (X,\Gamma)$ is given by

$$\hat{\Gamma} = \{(x,y) \mid x,y \in X \quad \text{and} \quad (x=y \text{ or } y \in \mu x)\} .$$

This leads to a unique graph $F = (X,\hat{\Gamma})$ obtained from the graph G.

Note that $x \in X \Rightarrow x \in \hat{\Gamma} x$, so that $(x,x) \in \hat{\Gamma}$ does not imply the existence of a circuit passing through x in this case.

Definition 19. A graph (X,γ) is said to be *strictly transitive* if for any path which is not a circuit there is an arc joining its origin to its terminal vertex, that is, if $a \neq c$, $a\gamma b$ and $b\gamma c \Rightarrow a\gamma c$. We say that the graph is *reflexive* if it has a loop at each vertex. We say that the graph is *broadly transitive* if it is both reflexive and strictly transitive.

The graph $T = (X,\mu)$ is strictly transitive and the graph $F = (X,\hat{\Gamma})$ is broadly transitive.

Definition 20. The graphs $F = (X,\hat{\Gamma})$ and $T = (X,\mu)$ are called the *transitive closure* and *μ-transitive closure* of G respectively.

For any $x \in X$, $\hat{\Gamma} x$ consists of all the descendants of x and the closure of $\hat{\Gamma} x$ is identical with $\hat{\Gamma} x$. The same is true of μx and the proper descendants of x. Thus we have

Property 3. The transitive closure of F and the μ-transitive closure of T are equal to F and T respectively.

If G_s is the symmetric graph based on $G = (X, \Gamma)$ we can define not only the relations $\hat{\Gamma}$ and μ but also $(\hat{\Gamma}_s)$ which we will denote by $\hat{\Gamma}_s$. Here $x\hat{\Gamma}_s y$ if $y = x$ or $]xy[$ is a chain in G.

This new relation represents the existence of chains in G (or paths in G_s) while $\hat{\Gamma}$ and μ represent the existence of paths in G.

$\hat{\Gamma}_s$ is reflexive, transitive and symmetric and is therefore an *equivalence relation*.

1.4 Connectivity, Equivalence and Preorder

The concept of connectivity will be introduced first for chains of a graph or paths of a symmetric graph and will then be extended to the case of paths in any graph. This will then give way to stronger and stronger concepts. Corresponding to these concepts there are equivalence relations between the vertices of the same path or of the same set of paths (partial subgraphs).

Definition 21. A graph is *unconnected* if it has two vertices which are not linked by any chain (Fig. 1.2) that is, $\exists i \exists j$ such that $i\hat{\Gamma}_s j$ is not valid.

Definition 22. A graph is said to be *connected*, if any two vertices can be linked by a chain in the graph. (Fig. 1.3)

$\forall i \forall j, \; i \hat{\Gamma}_s j$

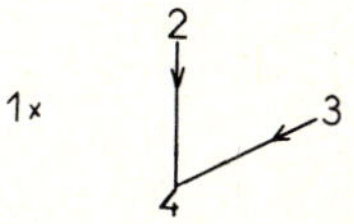

Fig. 1.2

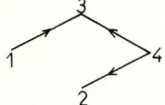

Fig. 1.3

Definition 23. A graph has *lower quasi-strong connectivity* (l.q.s connectivity) and is said to be *lower quasi-strongly connected* (l.q.s.c.) if there is a vertex which is an ancestor of every other vertex (Fig. 1.4).

$\exists i_o \forall j, \; i_o \mu j \quad (j \neq i_o)$

Definition 24. Similarly, a graph is *upper quasi-strongly connected* (u.q.s.c.) if $\exists k_o \forall j, \; j\mu k_o \; (j \neq k_o)$ (Fig. 1.5).

Fig. 1.4

Definition 25. A graph is *quasi-strongly connected* if it is both l.q.s.c. and u.q.s.c. (Fig. 1.6)

$\exists i_o \ \exists k_o \ \forall j, \ i_o \mu j \text{ and } j \mu k_o$

Fig. 1.5

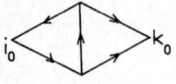

Fig. 1.6

Property 4. An l.q.s.c. graph cannot have more than one vertex without ancestors.

Because the only possibility is the vertex which is ancestor to all the others as described in Definition 23.

Definition 26. A graph is said to be *semi-strongly connected* (or *total*) if there is a path linking any two distinct vertices, that is,

$$\forall i \ \forall j (j \neq i), \ i\mu j \ \text{ or } \ j\mu i.$$

Definition 27. A graph is said to be *strongly connected* if there is a path from any given vertex to any other given vertex (Fig. 1.7).

It follows from this definition that $\forall i, \ i\mu i$ and $\forall i \ \forall j, \ [i...j...i]$ is a circuit. Hence there is a circuit passing through any two vertices of a strongly connected graph.

$\forall i \ \forall j, \ i\mu j \text{ and } j\mu i$

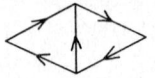

Fig. 1.7

Property 5. Strong connectivity implies semi-strong connectivity. If X is finite, semi-strong connectivity implies quasi-strong connectivity. Quasi-strong connectivity implies l.q.s. and u.q.s. connectivities. Lower quasi-strong connectivity and upper quasi-strong connectivity each imply weak connectivity. Furthermore, for a symmetric graph, weak connectivity implies strong connectivity.

We will prove only the three less obvious statements.

1. Suppose we have a finite graph in which

$$\forall i \ \forall j(j \neq i), \ i\mu j \ \text{ or } \ j\mu i \ .$$

If $\exists i_o, \ \forall j, \ i_o \mu j$, then the graph is l.q.s.c. Otherwise, since X is finite, there must be an i_o and a k for which we have neither $i_o \mu k$ nor $k \mu i_o$. But then the graph would not be semi-strongly connected.

Similarly semi-strong connectivity in a finite graph may be shown to imply u.q.s. connectivity.

However it is possible to construct an infinite graph which is connected and semi-strongly connected but which is not quasi-strongly connected. (Fig. 1.8)

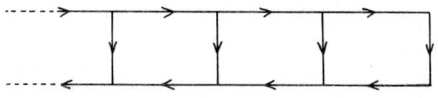

Figure 1.8

2. Suppose $\exists i_o; \forall j, i_o \mu j$.

Let k and ℓ be vertices such that $i_o \mu k$ and $i_o \mu \ell$. Then $k \hat{\Gamma}_s i_o$ and $i_o \hat{\Gamma}_s \ell$, whence $k \hat{\Gamma}_s \ell$.

3. If G is symmetric and weakly connected then

$$\forall i \; \forall j, \; i \hat{\Gamma}_s j \;\Rightarrow\; i \hat{\Gamma} j \text{ and } i \hat{\Gamma} j \;\Rightarrow\; j \hat{\Gamma} i$$

and, for $i \neq j$, $i \mu j$ and $j \mu i$ and so $i \mu i$. Thus G is also strongly connected.

The different types of connectivity listed above all correspond to relations defined on the graph $G = (X, \Gamma)$ and deduced from Γ. Two of them correspond to equivalence relations – strong and weak connectivity. The equivalence class containing the vertex a in each of these two cases will be referred to as the *strongly (weakly) connected class* of a. The graph (Y, Γ') generated by the quotient set Y of X by the equivalence relation of strong (or weak) connectivity is at most semi-strongly connected (or it is unconnected).

Definition 28. Two vertices a and b are in the same *weak class* E_a (under weak connectivity) whenever they coincide or are linked by a chain.

Property 6. If $x \in E_a$, then $a \in E_x$.

For $]xa[$ is a chain, whenever $]ax[$ is a chain.

It follows that

$$x \in E_a \;\Rightarrow\; E_x = E_a .$$

Thus

$$E_a = \{x \mid a \hat{\Gamma}_s x\} \tag{16}$$

is characterized by an equivalence relation.

Definition 29. The *connected component* C_a containing the vertex a in the graph G is the subgraph $C_a=(E_a,\Gamma_a)$ which is given by

$$x \in E_a \text{ if }]ax[\text{ is a chain in G or } x = a$$

and $\quad (x,y) \in \Gamma_a$, if $\{x \in E_a \text{ and } (x,y) \in \Gamma\}$. $\hfill (17)$

That is, C_a is the subgraph[1] generated by

$$E_a = \{y \mid a\hat{\Gamma}_s y\}.$$

In the literature C_a (resp. E_a) is also called the "weakly connected component" (resp. class) or "component".

Property 7. The connected components of a graph $G = (X,\Gamma)$ constitute a partition of G, that is,

1. $C_a \neq \emptyset$,   2. $C_a \neq C_b \Rightarrow C_a \cap C_b = \emptyset$,   3. $\bigcup_{a \in X} C_a = G$.

We will omit the straightforward proof of this result.

Definition 30. Two vertices a and b are in the same *strong class* E_a (under the equivalence of strong connectivity) whenever a and b coincide or there is a circuit passing through a and b.

Then $\quad a \in \mu b$ and $b \in \mu a \quad$ and therefore $\quad a \in \mu a$ and $b \in \mu b$.

Hence if $a, b \in E_a$, then $a\mu b$ and $b\mu a$.

Consequently every vertex in the strongly connected class (or *strong class*) of a is both an antecedent and a successor of a in T. However even when μa is empty, a must not lie in an empty class. In order to define the equivalence class of a under strong connectivity it is therefore necessary to use $\hat{\Gamma}a$ instead of μa.

Consequently $E_a = \{x \mid a\hat{\Gamma}x \text{ and } x\hat{\Gamma}a\}$ is characterized by an equivalence relation.

From the definition it is clear that if $x \in E_a$, then $a \in E_x$.

Further if $E_a \neq E_b$ but $E_a \cap E_b \neq \emptyset$, then there would be a vertex x for which $x \in E_a$ and $b \in E_b$. But then E_a would be equal to E_b, because $[axb]$ and $[bxa]$ are paths in G. This gives a contradiction and hence

$$E_a \neq E_b \Rightarrow E_a \cap E_b = \emptyset$$

---

[1] Notice that we are using the word "class" to mean a set of vertices and "component" to mean the subgraph on those vertices. This is in contrast with the usual convention which is to use "component" to mean the set.

Finally the union of all the E_a is equal to X just as in the case of the weakly connected classes.

Thus we have established the following result:

Property 8. The strong classes constitute a partition of the set of vertices of a graph.

In the case of a graph without circuits, each vertex constitutes a disjoint strongly connected class on its own.

We define the *strongly connected component* (or *strong component*) containing the vertex a in G to be the subgraph $C_a = (E_a, \Gamma'_a)$ (cf. Definition 29), where $x \in E_a$ if x and a lie on a circuit in G or x = a and

$$\Gamma'_a = \{(x,y) \mid x,y \in E_a \text{ and } x\Gamma y\} .$$

We observe that the strongly connected components do not always constitute a partition of the graph into subgraphs. For example if G has no circuits, then Γ'_a is empty. The concept of a co-circuit will later enable us to construct the set of arcs having their extremities in two distinct strongly connected classes.

A *strongly connected graph* (or *strong graph*) is a graph which has only one strong class. In such a graph every vertex is both a proper ancestor and a proper descendant of every other vertex and there is a circuit passing through each arc. It can also be characterized by the property of being weakly connected and cocircuitless.

Does the concept of equivalence with respect to quasi-strong connectivity allow the inclusion of vertices having no ancestors? We would have x and y in the same set Q_a (lower quasi-strong connectivity) if $a\hat{\Gamma}x$ and $a\hat{\Gamma}y$. But here $x \in Q_a$ does not imply $a \in Q_x$ and $Q_a \cap Q_b \neq \emptyset$ does not imply $Q_a = Q_b$; that is, these sets Q_a do not constitute a partition of the set of vertices. Consequently we shall not speak of equivalence in the sense of quasi- or semi-strong connectivity.

Besides *equivalence* relations (reflexive, transitive and symmetric) we can also investigate the conditions under which *preorder* relations (reflexive and transitive) become *order* relations (antisymmetric).

$\hat{\Gamma}$ is a preorder defined on the graph (X,Γ).

In a graph without circuits, $x\mu y \Rightarrow x \not{\mu} y$ and μ is antisymmetric. Conversely if μ is anti-symmetric, then the graph does not have any circuits.

μ is reflexive only for graphs in which every vertex lies on a circuit. Thus μ cannot be both reflexive and antisymmetric and is therefore never an order. However we know that μ is an equivalence relation on the vertices of a strongly connected graph. In such a graph the quotient of X over μ reduces to a single vertex.

The preorder Γ is total, that is such that

$$\forall x \; \forall y, \; x\hat{\Gamma}y \quad \text{or} \quad y\hat{\Gamma}x$$

if and only if the graph is semi-strongly connected.

$\hat{\Gamma}$ is a preorder with a unique first element if and only if

$$\exists i_o; \quad i_o\hat{\Gamma}x \quad \text{and} \quad \forall x(x \neq i_o), \; i_o \not\in \hat{\Gamma}x$$

That is, if the graph is l.q.s.c. and has a vertex without ancestors.

Definition 31. A graph G is said to be *finite* if its set of vertices is finite. The number $|X|$ will be denoted by n and referred to as the *order* of G when G is finite. The number of arcs is then at most n^2. It will be denoted by m.

The number of connected components will be denoted by p, (provided it is finite).

These symbols m, n and p will still to be used to designate the number of arcs, vertices and connected components in multigraphs and symmetric graphs. Further letting

            a   be the number of proper arcs
and   b   the number of loops,

we have the equation $a + b = m$.   (18)

In the case of a symmetric graph, the number $\bar{a}$ of distinct edges which are not loops is related to the number of arcs by the equation

(symmetric graphs) $2\bar{a} + b = m$.   (19)

Definition 32. A graph $G = (X,\Gamma)$ is said to be *complete* if there is at least one arc joining any two given vertices. That is

$$\forall x \; \forall y (x \neq y), \; (x,y) \not\in \Gamma \Rightarrow (y,x) \in \Gamma.$$

A complete graph is always semi-strongly connected.

The *degree* of a vertex x is the number of arcs having x as an extremity (initial or terminal, a loop being counted twice). The *indegree* and *outdegree* are the numbers of arcs entering into and leaving from x respectively. The degree, indegree and outdegree of the vertex x are denoted by d_x, $\partial\bar{x}$ and $\partial\overset{+}{x}$.

A graph is said to be *regular* if all its vertices have the same degree.

We shall prove that in a finite graph the number of vertices of odd degree is even.

A graph is said to be *homogeneous* if all its non-zero outdegrees are equal.

A vertex x of a graph is a *transmitter* if $\partial_x^- = \partial_x^+ = 1$ and there is no loop at x.

1.5 Graph Representations

1.5.1 Various definitions of graphs

It is possible to define graphs in many equivalent ways and we do not pretend that Definition 5 is the best for all possible research work and applications. As a rule there is a certain representation associated with each choice of definition.

For example, the *incidence matrix* will correspond to the definition of a graph as a *quadruple*

$$(X, A, d, a) , \qquad (20)$$

where X is a set (vertices),
 A is a set (arcs),
 d is a mapping (departure) from A into X,
and a is a mapping (arrival) from A into X.

If $u \overset{d}{\mapsto} x_1$ and $u \overset{a}{\mapsto} x_2$, then u may be identified with the arc (x_1, x_2). In the case when $x_1 = x_2$, then u is identified with a loop at x_1. The mappings d and a enable us to specify A as a subset of X^2.

- A graph may also be defined as a *triple*

$$(X, A, E) , \qquad (21)$$

where X is a set (vertices),
 A is a set (arcs),
and E is an injection from A into X^2.

If $u \overset{E}{\mapsto} (x_1, x_2)$, then u may be identified with the ordered pair (x_1, x_2).

This definition puts the stress once more on the individuality of the mathematical entity "arc" without direct reference to the vertices.

An s-graph is a triple (X, A, E') in which E' is a mapping and this becomes a graph if and only if E' is injective.

- The pair (X, Γ) can also be interpreted as

 X is a set
and Γ is a mapping from X into $\mathfrak{P}(X)$, the set of all subsets of X

$$X \xrightarrow{\Gamma} \mathfrak{P}(X) \qquad (22)$$

Because $\Gamma x \in \mathfrak{P}(X)$.

Given a graph G, we will sometimes denote the sets of vertices and arcs in G by $X(G)$ and $\Gamma(G)$ respectively.

• We can also define a graph to be a triple

$$(X, Q, \Psi), \qquad (23)$$

where X is a set (vertices),
Q is a set containing at least two elements (if Q has only two elements they will be for example $\{0,1\}$ or $\{FALSE, TRUE\}$),
and Ψ is a mapping from X^2 into Q.

If the ordered pair $(e_1, e_2) \in X^2$, then its image under Ψ, i.e. $\Psi(e_1, e_2)$, will be an arc of the graph whenever $\Psi(e_1, e_2) \neq z$, where z is a particular element of Q. When $|Q| = 2$, the set of ordered pairs (e_1, e_2) for which $\Psi(e_1, e_2) \neq z$ is the set of arcs Γ. Giving the mapping Ψ is equivalent to defining a relation on X. This is why Definitions 1.4 and 1.5 use pairs and not triples - the other element Q = (FALSE, TRUE) being implicitly understood.

1.5.2   Graph Isomorphisms

Graphs G and G* are said to be *isomorphic* if there is a bijection between the sets of elements of G and G* which preserves incidence.

Using Definition 1.5, $G = (X, \Gamma)$ and $G^* = (X^*, \Gamma^*)$ will be isomorphic if there are bijections b_1 and b_2 such that $X \xleftrightarrow{b_1} X^*$ and $\Gamma \xleftrightarrow{b_2} \Gamma^*$, that is such that $\forall x, y \in X$, $\forall x^*, y^* \in \Gamma^*$ if $(x,y) \in \Gamma$, $(x^*, y^*) \in \Gamma^*$ and $(x,y) \xrightarrow{b_2} (x^*, y^*)$, then $x \xrightarrow{b_1} x^*$ and $y \xrightarrow{b_1} y^*$.

Using Definition 4 of §1.5.1, we would then need bijections

$$X \xleftrightarrow{b_1} X^* \quad , \quad Q \xleftrightarrow{b_3} Q^* \quad \text{and} \quad \Psi \xleftrightarrow{b_2} \Psi^*$$

and we say that the graphs have the same valuation if $Q = Q^*$.

Deciding whether two given graphs are isomorphic is an important algorithmic problem. From the theoretical point of view, we will sometimes give a graph only up to isomorphism it being a representative of its equivalence class under isomorphism. It is also often of interest to discuss isomorphisms between subgraphs of a given graph as well as automorphisms of the graph itself.

To have a graph given in the form (X, Γ) amounts to knowing some integers such as m, n, p etc. between which we are able to establish certain relationships in the form of *formulae*. However most processes which lead to the generation of graphs will be based on algorithmic properties which will be deterministic and will enable us to analyse or synthesize the structure up to isomorphism.

The most usual representation, known as a topological graph, uses points (in a plane) for the vertices and curves or arrowed lines for the arcs and arrowless lines for the edges in the case of a symmetric graph. The lines may be straight or curved and, since the way in which the points and lines are disposed in the plane is quite arbitrary, we can only be certain that two diagrams represent distinct graphs if and only if there is no bijection giving a correspondence between the points and lines of one with those of the other. Of particular use in this connection are various observable parameters such as the numbers of vertices and arcs, the number of arcs entering into or leaving from a given vertex, etc.

1.5.3   Adjacency matrices

Let $G = (X,\Gamma)$ be a finite graph of order n and let

$$A = \{a_{ij} \mid (i,j) \in I^2, a_{ij} \in \{0,1\}\}$$

be a matrix based on the index set I of order n. Further suppose there is a bijection between X and I under which the image of $x_i \in X$ is $i \in I$ and a bijection between G and A under which the image of the ordered pair (x_i, x_j) is a_{ij}.

Then if

$$\forall i,j \in I, (x_i, x_j) \in \Gamma \Leftrightarrow a_{ij} = 1$$

and

$$\forall i,j \in I, (x_i, x_j) \notin \Gamma \Leftrightarrow a_{ij} = 0,$$

A is said to be the *adjacency matrix* of G.

The matrix A has n rows and n columns and so is a square array in which the *diagonal* entries a_{ii} are such that $a_{ii}=1$ if there is a loop at x_i and $a_{ii}=0$ otherwise. A is a *symmetric* (or *antisymmetric*) matrix if and only if G is a symmetric (or antisymmetric) graph. If G has m arcs, then A has m non-zero entries.

The adjacency matrix $S = \{s_{ij}\}$ of the symmetric graph G_s based on G is given by the rule

$$s_{ij} = 1 \text{ if and only if } a_{ij} = 1 \text{ or } a_{ji} = 1$$

A given graph can have many different adjacency matrices associated with it - different ways of indexing the vertices will in general lead to different matrices.

For, let $G = (X,\Gamma)$ and let

$$A = \{a_{ij} \mid (i,j) \in I^2\}, \quad B = \{b_{kl} \mid (k,l) \in L^2\}$$

be matrices such that $B = P^{-1}AP$, where P is a permutation matrix (in the sense of matrix theory).

Then there is a bijection between the ordered pairs (i,j) of I^2 and the ordered pairs (k,l) of L^2 such that one and only one entry b_{kl} corresponds to a given entry a_{ij} and therefore to one and only one ordered pair (x_i,x_j) of X^2. There is therefore a bijection between those entries b_{kl} of B which have value 1 and the arcs of G.

Thus if the rows and columns of the adjaceny matrix of a graph are permuted, the transformed matrix may still represent the same graph.

Except when stated otherwise, a graph and any of its adjacency matrices will usually be denoted by the same symbol.

1.5.4 The incidence matrix of a graph

The incidence matrix of a graph is an n x m matrix C which is given by

$c_{ij} = 1$, if x_i is the initial vertex of the arc u_j,

$c_{ij} = -1$, if x_i is the terminal vertex of the arc u_j,

$c_{ij} = \lambda$, if u_j is a loop at x_i, where λ is a non-zero integer,

$c_{ij} = 0$ otherwise.

When G is symmetric, the arcs (x_i,x_k) and (x_k,x_i) correspond to different columns, j and l say, such that $c_{ij} = c_{kl} = 1$ and $c_{il} = c_{kj} = -1$.

If G is an s-graph we sometimes write

$c_{ij} = s$, if x_i is the initial vertex of the arc $j \in \Gamma_s$

and $c_{ij} = -s$, if x_i is the terminal vertex of the arc $j \in \Gamma_s$.

1.5.5 Computer representation of graphs

Corresponding to the preceding representations there are different types of methods for processing graphs on a computer.

The adjacency matrix lends itself most easily to the programming of problems which involve the finding of certain paths (e.g. the shortest path, the longest path, paths originating at a given vertex, etc.). Regrettably, the use of the data is often uneconomical.

If each available bit in a memory word could be used to specify the presence or absence of an arc, the processing of adjacency matrices would be very economical, but in practice the use of high level languages only rarely allows the storing of more than one entry a_{ij} per word. However some programs do give rise to a quicker algorithm when the elements of the graph obtained after processing are to be arranged in an *incidence matrix*. The so-called *associated lists* allow some economy of storage space; it is sufficient to order the indices j of the columns followed by the (different) indices i of the rows for which $a_{ij} = 1$ and then those for which $a_{ij} = -1$. In this way 3m memory units suffice for storage. Orderings or changes of the arcs correspond to address modifications. If every vertex has at most one predecessor, then n memory units will suffice (predecessor lists).

The technique of *push-down lists* is also used for graph representations.

1.5.6 Valuations

Earlier (Definition 1.18) we defined the length of a path to be the number of arcs in the path. This concept may be generalised by introducing a mapping $\ell = \Gamma \to R$ called a *valuation* of the arcs.

Definition 33. A *valuation* V on a graph (X,Γ) is a mapping either from the set of arcs or from the set of vertices into some set Q. For example

$$V_A : \Gamma \to R \quad \text{or} \quad V_S : X \to R$$

Definition 34. A *valuated graph* is a triple (X,Γ,V) in which (X,Γ) is a graph and V is a valuation or a family of valuations.

The valuations involved here will usually take their values in some restriction of R, either R^+, Z^+, Q^+, $\{0,1\}$ or the interval $[0,1]$.

Definition 35. The *length* of an arc $u \in \Gamma$ is defined to be the real number $\ell(u)$ given by the valuation ℓ on the arcs. The *length* of the path $P = u_1, u_2, \ldots, u_p$ is the real number

$$L = \sum_{i=1}^{p} \ell(u_i) .$$

L is also called the *weighted length* of the path.

Usually the length will be a positive number which can be interpreted as a distance or a cost or a time. The absence of a path between the vertices x_1 and x_2 can then be thought of as the presence of a path of infinite length linking x_1 to x_2 (and x_2 to x_1).

The original definition of the length of a path is included in this general context, by means of the constant mapping

$$\forall u \in \Gamma, \; 1 : u \longmapsto 1(u) = 1 .$$

1.5.7 Coding

Let d be the maximum outdegree in a graph (X,Γ) and let x be a vertex with outdegree equal to d. The d arcs originating at x, $(x,x^1), (x,x^2), \ldots, (x,x^d)$ say, have the same first

projection and are distinguished from one another only by their second projections which are such that $x\Gamma x^i$ (i = 1,...,d). However it can also be of interest to *code* these arcs directly starting from x. We are thereby led to form a partition of the arcs
$$\Gamma = \bigcup_{i=1}^{d} \Gamma_i$$
in such a way that for all i, (X, Γ_i) is a partial graph in which the outdegree of every vertex is 0 or 1. We then establish an order between the successors of x and say that the arcs are *labelled*. We have thus realized a very particular valuation.

1.6 Paths, Circuits and Cocircuits

1.6.1 Chains and concatenation

We shall investigate certain properties which characterize paths amongst chains or which provide us with methods for constructing chains out of other given chains.

Property 9. An elementary chain is not a path if and only if there exists a vertex which is the initial (or the terminal) vertex of two arcs in the chain.

Any chain which has only one arc is clearly a path. Suppose therefore that there is a chain of length ℓ (≥ 2) which is not a path. It is clear that the condition is sufficient.

Let $\mu = [x_i \ldots x_j]$ be a path of maximal length $k(<\ell)$ in which the arcs are consecutive arcs of the chain $]x_1 \ldots x_{\ell+1}[$ and suppose that j>i.

If i = 1., then $j \leq \ell$: x_j cannot be the initial vertex of an arc of the chain, otherwise μ could be extended. Hence x_j the terminal vertex of two arcs of the chain.

If i>1, then (for j>i as well as for j<i) x_i cannot be the terminal vertex of any arc of the chain, otherwise μ could again be extended. Hence x_i is the initial vertex of two arcs of the chain.

Since $|\mu|<\ell$, the result can be extended to cycles, noting that $x_i \neq x_j$.

Property 10. An elementary cycle is not a circuit if and only if there exists a vertex which is an initial (or a terminal) vertex of two arcs of the cycle. In that case there is also a vertex which is a terminal (or an initial) vertex of two arcs of the cycle.

Let $[x_0 \ldots x_p \ldots x_{p+q}]$ be a path formed by the union of two sets of arcs:

$$\{(x_0,x_1),\ldots,(x_{p-1},x_p)\} \cup \{(x_p,x_{p+1}),\ldots,(x_{p+q-1},x_{p+q})\}.$$

Then we will write

$$[x_o x_p] \cdot [x_p x_{p+q}] = [x_o \ldots x_{p-1} x_p x_{p+1} \ldots x_{p+q}],$$

without repeating the vertex x_p and we say that $[x_o x_{p+q}]$ is the result of the *concatenation* of the other two paths.

Similarly if $u_j \in \Gamma$ and if

$$\forall i \ (i=1,2,\ldots,p+q-1), \ pr_1 u_{i+1} = pr_2 u_i,$$

then $[u_1 \ldots u_p] \ [u_{p+1} \ldots u_{p+q}] = [u_1 \ldots u_{p+q}]$ is a path of p+q arcs obtained by concatenation.

Two chains having a common extremity may also be concatenated in the same way.

Note however that the concatenation of two paths through a common extremity x_p will only produce a path if x_p is the terminal extremity of one and the initial extremity of the other. Otherwise, when x_p is the initial (or terminal) extremity of both paths, the result is a chain.

The operation of concatenation can be repeated and it transpires that, if μ_1, μ_2 and μ_3 are three paths (chains), then $(\mu_1 \cdot \mu_2) \cdot \mu_3 = \mu_1 \cdot (\mu_2 \cdot \mu_3)$. Thus we have

Property 11. Concatenation is associative.

Let G be a finite graph of order n having a path of length p.

If G has an elementary circuit γ of length $k(1<k<n)$, then the non-simple path obtained by the concatenation of $(n\%k)+1$ circuits[1] identical to γ is a path of length greater than n.

If G has an elementary path μ of length $p \geq n$ which is not a circuit, then the set of arcs of μ must have p+1 distinct vertices. Since p+1>n, μ cannot be an elementary path. We can extract from μ a set of consecutive arcs (distinct or not) in which the initial vertex of the first coincides with the terminal vertex of the last, that is, μ contains a subset of arcs forming a circuit. Thus we have the following result:

Property 12. A finite graph G of order n contains a path of length p>n if and only if G contains a circuit.

Finding the connected classes of a vertex a in a finite graph (X,Γ) requires the determination of the broad transitive closures $\hat{\Gamma}a$, for the strong class, and $\hat{\Gamma}_s a$, for the weak class.

---

[1] % denotes the operation of Euclidean division, that is, n%k is the integer for which $0 \leq n-(n\%k)k<k$.

By definition

$$\mu a = \lim_{r \to \infty} \mu^r a \quad \text{and} \quad \mu^r a \subset \mu^{r+1} a.$$

We show that if $x \notin \mu^n a$, then $x \notin \mu a$, (n is the order of the graph).

If $x \notin \mu^p a$ for $p \geq n$, then there is no path of length p from a to x. Since any path of length p+1 (>n) contains a circuit, so will any path of length greater than p and no vertex will be in $\mu^{p+1} a$ without already being in $\mu^p a$. Hence if $x \notin \mu^p a$, then $x \notin \mu a$.

It follows that $\mu^n a = \mu a$ and in the same way that $\Gamma^n a = \Gamma a$. Furthermore $\Gamma^{n-1} = \mu^{n-1} a \cup \{a\}$. For if $x \neq a$ and $x \in \Gamma^{n-1} a$, then a path $[ax]$ of length n would pass through n+1 vertices and so could not be elementary. Moreover if $x \notin \Gamma_s^n a$, then there is no elementary chain of length n linking a to x and hence $x \notin \hat{\Gamma}_s a$.

Thus we have established

Property 13. In a finite graph of order n, the transitive closure μ (or $\hat{\Gamma}$ or Γ_s) is given by the sets of paths and elementary chains of length at most n (n-1, n-1).

Consequently algorithms for finding paths, elementary chains or connected classes can be obtained by means of the graphs

$$(X, \mu^n), \quad (X, \Gamma^{n-1}) \quad \text{or} \quad (X, \Gamma_s^{n-1}).$$

We say that a path is a *maximal path of the graph* if it cannot be increased in length by concatenation with any other path.

1.6.2   Cocircuits, cocycles and cycles

Definition 36. Let A be a subset of the set of vertices of the graph $(X;\Gamma)$ such that there are no arcs entering into A. Then the *cocircuit* $\omega^+(A)$ is the set of all those arcs which leave from A and enter into X-A.

We will write $\omega^+(a)$ if A happens to consist of only one vertex a. The arcs of $\omega^+(A)$ enter into X-A and from this point of view they constitute the cocircuit $\omega^-(X-A)$. If B is a weakly connected class, then $\omega^+(B) = \omega^-(B) = \emptyset$.

If suppression of the arc (a,b) causes the subgraph g generated by the sets of vertices A and B linked to a and b to lose its connectivity, then $\omega^+(A) = \{(a,b)\}$ in g.

Definition 37. A *cocycle* $\omega(A)$ of a symmetric graph is the set of all edges having only one extremity in A. A *cocycle* $\omega(A)$ of a graph is the set of all arcs having only one extremity in A.

FUNDAMENTAL PROPERTIES OF GRAPHS

These definitions refer to what we could call simple cocircuits and cocycles. In general $\omega^+(A)$ and $\omega^-(A)$ will denote the sets of arcs which leave from and enter into A. Then $\omega(A) = \omega^+(A) \cup \omega^-(A)$.

Definition 38. A cocircuit is said to be *elementary* if it consists of arcs which link two connected subgraphs.

Suppose that the vertices of a connected graph G are partitioned into $X = A \cup B \cup D$ in such a way that the subgraphs generated by the three sets of vertices S_A, S_B, S_D are mutually disjoint.

Suppose further that there is no arc in G linking B to D and that $\omega^-(A) = \emptyset$. Then the set $\omega(A)$ of arcs leaving from A and entering into B or D is a cocircuit. But $\omega^+(A)$ can be decomposed into a sum $\omega^-(B) + \omega^-(D)$ of disjoint cocircuits consisting of the arcs linking the connected classes B and X-B on the one hand and D and X-D on the other, i.e. $\omega^-(B)$ and $\omega^-(D)$ are elementary cocircuits.

Definition 39. A cocycle is said to be *elementary* if it consists of the arcs linking two connected subgraphs.

An elementary cocycle is an elementary cocircuit when every arc enters into the same given set of vertices.

The suppression of all the arcs of an elementary cocycle then creates one and only one supplementary weak component. In Figure 1.9, $\omega(B)$ is an elementary cocycle whereas $\omega(A)$ is not elementary.

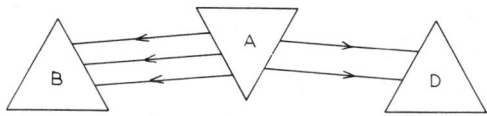

Figure 1.9

Property 14. Every arc belongs either to an elementary circuit or to an elementary cocircuit but not to both.

    a. Let $(x,y) \in \mu = [xy...tx]$, $x \in A$, $y \in X-A$ where μ is an elementary circuit. Then, whatever the number of vertices of μ in A, $\omega(A)$ contains at least one arc entering into A and at least one leaving from A. Hence (x,y) does not belong to a cocircuit.

    b. Suppose the arc (x,y) does not belong to any circuits. Then x and y belong to disjoint strongly connected classes A and B such that $x \in A$, $y \in B$ and $A \cap B = \emptyset$. Let C be the set of all those vertices to which there is no path from B. C is not empty because $A \subset C$ and $x \in A$. Hence $\omega^+(C)$ is a cocircuit containing (x,y).

Definition 35. The *cycle rank* $k(G)$ and the *cocycle rank* $\ell(G)$ of the graph G are the integers

$$k(G) = m-n+p \quad \text{and} \quad \ell(G) = n-p$$

related by the equation $k(G) + \ell(G) = m$.

Property 15. Let G be an s-graph and G' an s-graph obtained from G by adjunction of an arc (a,b) between two of the vertices a and b of G.

If the two vertices belong to the same connected component C_a in G, then

$$k(G') = k(G)+1 \quad \text{and} \quad \ell(G') = \ell(G) \;.$$

If not (i.e. $C_a \neq C_b$), then

$$k(G') = k(G) \quad \text{and} \quad \ell(G') = \ell(G)+1$$

Further $k(G) \geq 0$ and $\ell(G) \geq 0$.

The graph $G_o = (X, \emptyset)$ consisting of n isolated vertices has $m=0$, $p=n$ and $k=\ell=0$.

The adjunction to an s-graph of an arc (a,b) (or a loop (a,a), if $a=b$) transforms m into $m'=m+1$, preserves n and reduces p by one or not according as $C_a \cap C_b \neq 0$ or not. The property follows immediately.

Property 16. $k(G)$ is equal to the number of elementary independent cycles. $\ell(G)$ is equal to the number of elementary independent cocycles.

Hence G is aclyclic if and only if $k(G) = 0$ and G has no cocycles if and only if $p=n$.

In particular any arc which is not a loop belongs to an elementary cocycle.

EXERCISES

1. Let $C_1 = [u_1 \; u_2 \; u_3 \; u_4 \; u_5]$
and $\quad C_2 = [u_1 \; u_6 \; u_7 \; u_8 \; u_9 \; u_{10}]$
be circuits defined by the arcs and let

$$C_3 = [a \; b \; c \; d \; e \; f]$$

be a path defined by the vertices such that $a = pr_1 u_9$ and $f = pr_2 u_3$.

(i) Construct the elementary cycles of the graph G generated by $\{C_1, C_2, C_3\}$. Is there another circuit possibly elementary?

(ii) What sort of connectivity does G have?

(iii) What happens if we replace the path C_3 by the path $C_4 = $]f e d c b a[?

(iv) Let G' be the graph generated by $\{C_1, C_2, C_3\}$ in the case when $d = pr_2u_1$. Does the connectivity change? How many arcs must be reversed (that is, the order of the projections is reversed) in order to suppress all the circuits?

(v) Construct the transitive closures of G' in both the strict and broad senses.

2. Prove the following statements.

(i) Every vertex in a strongly connected graph is the initial vertex of at least one arc and also the terminal vertex of at least one arc.

(ii) A cocircuit has no subset of arcs which is a path of length greater than 1.

3. Let]$ax_1...x_\ell b$[be an elementary chain which is not a path. Show that each of the following properties P_i implies the existence of certain classes of vertices on the chain.

P_1: $a\Gamma x_1$ and $x_\ell \Gamma b$ $\Rightarrow$ classes r and t.

P_2: $a\Gamma x_1$ and $b\Gamma x_\ell$ $\Rightarrow$ class t.

P_3: $x_1 \Gamma a$ and $x_\ell \Gamma b$ $\Rightarrow$ class r.

P_4: $x_1 \Gamma a$ and $b\Gamma x_\ell$ $\Rightarrow$ classes r and t.

Where $x \in r$, if x is the initial vertex of two arcs on the chain and $x \in t$, if x is the terminal vertex of two arcs on the chain.

4. Characterize as far as possible those graphs which have the following three properties:

(a) weakly connected, non-symmetric,

(b) there exist exactly two vertices which are the origins of paths leading to all the other vertices of the graph,

(c) there exist exactly three vertices which are terminal vertices of paths originating from all the other vertices of the graph.

5. Construct an algorithm for finding the maximal number of independent cocircuits in a circuitless graph. Is it possible

to obtain such an algorithm for a graph with circuits by using the equivalence relation of strong connectivity?

Chapter Two

LATTICOIDS AND ARBORESCENCES

2.1 Circuitless Graphs

Theorem 1. The following seven properties are equivalent in a finite graph $G = (X, \Gamma)$:

S1. G is circuitless.

S2. The main diagonal in the adjacency matrix of the strict transitive closure T consists entirely of zeros.

S3. The relations Γ and μ on X are antisymmetric.

S4. For any $A \subseteq X$ ($A \neq \emptyset$), $A - \Gamma(A)$ is not empty.

S5. There exists an ordering of the vertices in which each vertex appears before all its proper descendants and after all its proper ancestors.

S6. G has a triangular adjacency matrix.

S7. There is an elementary cocircuit passing through any given arc of G.

Proof. S1 $\Rightarrow$ S2: If G has no circuits, then every vertex a has $(a,a) \notin \mu$. Hence the adjacency matrix of T cannot have any 1's on the main diagonal.

S2 $\Rightarrow$ S3: If the main diagonal of T consists entirely of zeros, then the relation μ is not reflexive because $(x,x) \notin \mu$. Hence if G has a path $[xy]$, then $[yx]$ is not a path in the graph, that is

$$(x,y) \in \mu \;\Rightarrow\; (y,x) \notin \mu$$

and in particular

$$(x,y) \in \Gamma \;\Rightarrow\; (y,x) \notin \Gamma.$$

S3 $\Rightarrow$ S1: If $[xy]$ is a path in G and if G had a circuit passing through x and y, then $[yx]$ would be a path in G in contradiction of S3. It follows that G cannot have a circuit.

S1 $\Rightarrow$ S4: If $A \subset X$ were such that $A - \Gamma(A) = \emptyset$, then we would have

$$\forall a_o \notin A, \; \Gamma a_o - A \neq \emptyset \qquad (\alpha)$$

In this case let the vertices $a_1, a_2, \ldots$ be chosen in such a way that

$$a_1 \in \Gamma a_0 - A,\ a_2 \in \Gamma a_1 - A,\ldots, a_{i+1} \in \Gamma a_i - A,\ldots$$

In view of (α) this sequence can be extended indefinitely. The (i+2)nd vertex considered in A is a_{i+1} and, if $i+2 > |A|$, then $a_{i+1} \in \{a_0, a_1, \ldots, a_i\}$ and there would be a circuit in G. But G is supposed to be circuitless and we deduce that $A - \Gamma A$ cannot be empty.

S4 $\Rightarrow$ S5: Since $A - \Gamma A$ is not empty for all $A \subseteq X$, there must be a vertex $a \in X$ for which $\Gamma^{-1}a = \emptyset$. This a may be placed first in a descending sequence of the vertices. If there are many vertices a_i for which $\Gamma^{-1}a_i = \emptyset$, then they may be placed in arbitrary order in the first places of the sequence. We then form $G - \{a_i\}$ and, since the subgraph thereby obtained still has property S4, we can choose those vertices which in $G - \{a_i\}$ have no ancestors to put them as immediate successors of the a_i in the sequence. This process may be repeated until the remaining set is empty, at which stage all the vertices will have been included in the sequence. The ordering thus produced is not unique in that at any stage there may be many vertices of G or $G - \{a_i\}$ which have no ancestors. There is however an equivalence between the many different orderings given by property S5.

S5 $\Rightarrow$ S6: Let $1, 2, \ldots, n$ be the ranks of the vertices as given by an ordering from property S5. Then

$$i > j \Rightarrow (a_i, a_j) \notin \Gamma.$$

Hence the adjacency matrix of G in which the rows and columns are indexed by the vertices in order of increasing rank has all its non-zero entries above the main diagonal and is therefore upper triangular.

S6 $\Rightarrow$ S7: Since the matrix is triangular, upper triangular say, a path $[x_1\ x_2 \ldots x_p]$ in G is such that the ranks of its vertices increase with their positions along the path. Consequently $x_p \neq x_1$ and we cannot have x_p an ancestor of x_q for any $q < p$. Hence there is no circuit in G.

S1 $\Rightarrow$ S7: This is an immediate consequence of Property 1.14.

Definition 1. Any vertex which has no proper ancestors will be called a <u>root</u> and any vertex which has no proper descendants will be called a *terminal*.

Corollary 2. Every finite circuitless graph has both a root and a terminal.

The first part of the corollary has already been established while proving S4 $\Rightarrow$ S5. The second can be deduced by inverting Γ which process does not create any circuits. Note that if any weakly connected component has only one vertex, then that vertex has neither proper ancestors nor proper descendants.

Let F be a set of n elements partially ordered by an order

relation ϕ. The graph $\omega = (F, \Gamma)$, where (x_1, x_2) is an arc if and only if $x_1 \phi x_2$ is called an *order graph*. Following the usual notation, we will denote this graph by $\omega = (F, \phi)$.

The properties of the partial order can be extended to the order graph.

Property 1. A graph is an order graph if and only if it is broadly transitive and circuitless except for loops.

We can specify a partial order ϕ on F by ω.

Let $\omega_1 = (F, \phi)$ and $\omega_2 = (F, \phi_2)$ be two order graphs. We say that the partial orders ϕ_1 and ϕ_2 are *compatible* if the transitive closure graph of $\omega_1 \cup \omega_2$ is broadly transitive and has no circuits except loops.

We say that the order ϕ_1 is *included* in the order ϕ_2 if ω_1 is a partial graph of ω_2. We say that the order graph ω_2 is *deduced* from ω_1 by the *adjunction* of the arc $u = (x,y)$, where $x, y \in F$, if ϕ_1 is included in ϕ_2 and if the only arc belonging to ω_2 and not to ω_1 is the arc u.

If ω_2 is deduced from ω_1, then there exists another graph ω_3 which we can deduce from ω_1 by adjoining the arc (y,x) to ϕ_1; ω_2 and ω_3 are not compatible.

Similarly two distinct total orders ω and ω' are not compatible because the transitive closure of $\omega \cup \omega'$ has a circuit which is not a loop.

Theorem 3. The following four properties L1 to L4 are equivalent and characterize a type of finite graph Λ to be called a *latticoid*:

    L1 Λ is an l.q.s.c. circuitless graph.

    L2   The matrix T has only zeros in its main diagonal and has only one row consisting entirely of 1's in its non-diagonal entries.

    L3 Λ is circuitless and any two vertices are linked by at least one chain consisting of either a path or two paths having these vertices as terminal vertices and a fixed vertex as origin.

    L4 Λ is weakly connected, circuitless and has only one root.

This concept was introduced without the finiteness condition by Berge who called it a latticoid graph.

Proof. L1 $\Rightarrow$ L2: Since Λ has no circuits, T has only zeros in its main diagonal. Since Λ is l.q.s.c., there is a vertex which is an ancestor of all the others and this corresponds in T to a row of 1's except for the diagonal place. If there were two rows of this form, then T would have two non-zero symmetrically placed entries and the graph would have a circuit.

L2 ⇒ L3: Since for all i, i∤i, Λ has no circuits. Only one row of T has all its non-diagonal entries equal to 1. Thus if vertices x and y are not linked by a path, then there is a unique vertex i such that $x \in \mu i$ and $y \in \mu i$.

L3 ⇒ L4: Since there are no circuits, there is at least one root. If there were two distinct roots, they would not be linked by a path nor by a chain consisting of two paths ending in them. Hence there can be only one root.

L4 ⇒ L1: Let α be a root. If $x \in X-\{\alpha\}$ and $x \notin \mu\alpha$, then either x has no ancestors or it is the successor of a vertex without ancestors and distinct from α and x because $x \notin \mu x$. But this is a contradiction because |X| is finite. Hence $x \notin \mu\alpha$ and Λ has a vertex which is the origin of paths linking it to each of the other vertices.

Property 2. A latticoid is antisymmetric and has at least one terminal.

This follows from property S3 of Theorem 1 and Corollary 2.

The subgraph of the latticoid $\Lambda = (X, \Gamma)$ with root α which is generated by $X-\{\alpha\}$ is a circuitless graph in which all the connected components are circuitless.

Conversely, any connected circuitless graph $G = (X', \Gamma')$ which has r(>1) vertices without ancestors may be realised as the subgraph of a latticoid $\Lambda = (X, \Gamma)$ obtained by the adjunction of an extra vertex α to G and r arcs linking α to each of the vertices without ancestors in G.

Thus the discussion of connected circuitless graphs can be reduced to that of latticoids.

Remark. A *lattice* is a set X partially ordered by a relation R in such a way that any two elements $x, y \in X$ have both (a) a greatest lower bound and (b) a least upper bound. Thus the graph (X, Γ) shown in Figure 2.1 generates a lattice (X, R), where the relation R (aRx if $(a, x) \in R$) is none other than the local transitive closure $\hat{\Gamma}$. This is related to the fact that (X, Γ) is (a) lower and (b) upper quasi-strongly connected and has no circuits. However these three conditions are not in themselves sufficient. The graph (X, Δ) shown in Figure 2.2 has broad transitive closure $\hat{\Delta}$ but this is not a lattice because x and y have lower bounds α, a, b, where α is the smallest while a and b are not comparable. Thus there is no greatest lower bound of x and y. What we can say however is that a q.s.c. graph always contains a partial graph whose broad transitive closure is a lattice. Similarly an l.q.s.c. circuitless graph contains a partially ordered set such that any two elements have a greatest lower bound which may be the root (semi-lattice structure). Such a graph ought to be called a *lower semi-latticoid* but we have agreed to call it more simply *latticoid*.

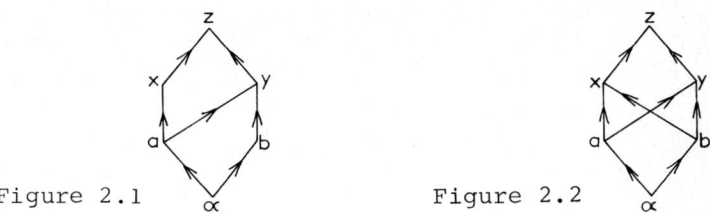

Figure 2.1   Figure 2.2

Definition 2. A latticoid is said to be *Dedekindian* if it satisfies the Jordan-Dedekind condition, that is, if all paths linking any two given vertices in the graph have the same length.

2.2 Arborescences and Trees

Theorem 4. Let $A = (X, \Gamma)$ be a finite graph of order n. The following six properties A1,...,A6 are equivalent and characterize what is known as an *arborescence* .

A1. A is lower quasi-strongly connected (l.q.s.c.) and is acyclic.

A2. A is acyclic, n-1 of its vertices are the terminal extremities of only one arc and the nth vertex is not the terminal extremity of any arc.

A3. A is l.q.s.c. and has n-1 arcs.

A4. A is l.q.s.c., n-1 of its vertices are the terminal extremities of only one arc and the nth vertex is not the terminal extremity of any arc.

A5. A is l.q.s.c. and the suppression of any arc destroys this property.

A6. A has a unique vertex which is linked to any other vertex by a unique path of which it is the origin (Figure 2.3).

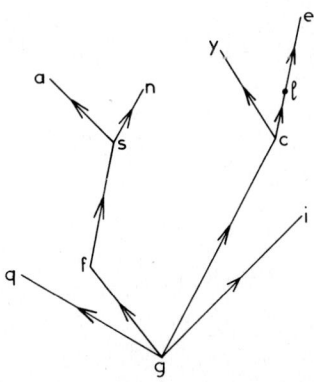

Figure 2.3

Proof. A1 ⇒ A2: The condition "A is l.q.s.c." can be expressed in the form

$$\forall x \; \forall y, \; \exists \alpha \; ; \; x \in \hat{\Gamma}\alpha \quad \text{and} \quad y \in \hat{\Gamma}\alpha \; .$$

Suppose the vertex t was the terminal extremity of two arcs (x,t) and (y,t). In view of the l.q.s. connectivity, there are paths [α...x] and [α...y], and thus there would be two paths [α...xt] and [α...yt] linking α to t and hence there would be a cycle.

If α was the terminal extremity of an arc (β,α), then the path [α...β] and the arc (β,α) would form a cycle.

If there was another vertex γ which was not the terminal extremity of any arc, then the graph would not be l.q.s.c. Thus we have shown that A1 ⇒ A2.

A2 ⇒ A3: The number of arcs in Γ is equal to the number of arcs entering into X. Thus $m = n - 1$. Let α be the vertex which has no arcs entering into it. Since A is acyclic, the cycle rank is $k = 0$. In particular there are no loops or circuits of length 2, that is, the graph is antisymmetric and the number of edges is equal to the number of arcs. Hence $p = k + n - m$ is equal to 1, and A is therefore connected. For all t, there exists a chain]α...t[. If $t \notin \hat{\Gamma}\alpha$, some vertex on the chain]α....t[would be the terminal extremity of two arcs in contradiction of A2. Hence $t \in \hat{\Gamma}\alpha$ for all t and this proves A3.

A3 ⇒ A4: A is l.q.s.c. so every vertex except perhaps one is the terminal extremity of only one arc and the nth vertex is not the terminal extremity of any arc.

A4 ⇒ A5: The suppression of an arc (β,γ) of A causes γ to have no ancestors. The resulting graph therefore has two vertices α and γ without ancestors, that is, $\alpha \notin \hat{\Gamma}\gamma$ and $\gamma \notin \hat{\Gamma}\alpha$. Thus this graph is not l.q.s.c.

A5 ⇒ A6: Since A is l.q.s.c., every vertex t distinct from a certain vertex α is the terminal extremity of at least one path [α...t] from α. If there were two different paths [α...γt] and [α...∂t] then the suppression of one of the arcs (γ,t) or (∂,t) would not destroy l.q.s. connectivity. Therefore any vertex distinct from α is the terminal extremity of a unique path from α.

A6 ⇒ A1: Let α be the vertex which is linked to each of the other vertices of X by unique paths; thus

$$\forall x \; \forall y, \; (x,y \in X) \Rightarrow x \in \hat{\Gamma}\alpha \quad \text{and} \quad y \in \hat{\Gamma}\alpha \; .$$

The graph A is therefore l.q.s.c.

Suppose there exists a cycle]u...vwu[in A which does not pass through α. Then since the paths which end at u, v and w are unique, u would be an ancestor of v and w;

[α...u...vw] would be a path and)w,u(would be an edge.

If the last arc is (w,u), then we have two paths ending at u; [α...u] and [α...vwu].

If the last arc is (u,w), then we have two paths ending at w; [α...uw] and [α...vw].

Thus if there is a cycle in A, then, by hypothesis A6, it must pass through α. It can be formed from a path [a...v] and the edge)α,v(.

If the arc is (α,v), then v is linked to α by two paths from α.

If the arc is (v,α), then [α...v] and [α...vα...v] are two distinct paths from α to v.

In particular if both (α,v) and (v,α) ∈ Γ, then v would be the terminal extremity of a path of length 3 with origin α, viz. [αvαv]. Thus A is acyclic.

This completes the proof of the theorem.

Property 3. A graph G has an arborescence as a partial subgraph if and only if it is l.q.s.c.

For, if G is not l.q.s.c., then *a fortiori* no partial graph can be l.q.s.c. If G is l.q.s.c., then the suppression of any arc of G results in a new graph which may or may not be l.q.s.c. If it is l.q.s.c., then we can proceed to suppress another arc and so on. When it is no longer possible to suppress an arc without losing l.q.s. connectivity, then the resulting graph (possibly G itself) is an arboresence (because of A5).

Property 4. Any arborescence is a Dedekindian.

For, a finite acyclic l.q.s.c. graph has no circuits and the Jordan-Dedekind condition is obviously satisfied in an arborescence.

Consequently we have: a graph G has a latticoid as a partial graph if and only if it is l.q.s.c.

Definition 3. A *degenerate arborescence* or *trivial graph of order 1* is a graph ({a}, ∅) having only one vertex and no loops.

Definition 4. A *forest* is a graph in which each connected component is an arborescence (possibly degenerate).

If all the components are degenerate arborescences, then we say that the forest itself is degenerate.

Property 5. Any arborescence is acyclic and the adjunction of an arc between any two vertices creates one and only one cycle.

We use A3, i.e. A is l.q.s.c. and has n-1 arcs. Since A is connected
$$p = 1 \quad \text{and} \quad k = (n-1) - n + 1 = 0,$$
that is, A has no cycles. The adjunction of an arc would result in a graph with cycle rank $k'=k+1=1$ and therefore with one and only one cycle.

This property implies the following Property 6 which is weaker than A5.

Property 6. An arborescence is connected and this property is destroyed by the suppression of any one arc.

For, since A is acyclic, k is zero. If A was not connected, the adjunction of an arc would not necessarily create a cycle in contradiction of Property 5. Hence $p=1$. The suppression of an arc which did not reduce the cycle rank would reduce the cocycle rank and the partial graph deduced from an aborescence in this way has cocycle rank $\ell=n-p' < n-1$. Hence $p'>1$ and the partial graph has two connected components.

Trees. The concept of a *tree* may be deduced from that of an arborescence by replacing terms as shown by the following table (S)

the concept: in place of	the stronger concept:
arc or edge	arc
chain	path
weak connectivity	l.q.s. connectivity

The substitution of Property 5 in place of A4 for aborescences leads to the following results:

Theorem 5. Let $A = (X, \Gamma)$ be a finite graph of order n. Then the following six properties T1,...,T6 are equivalent and characterize a *tree*.

T1. A is connected and acyclic.

T2. A is acyclic and has n-1 arcs.

T3. A is connected and has n-1 arcs.

T4. A is acyclic and the adjunction of an arc between any two vertices creates one and only one cycle.

T5. A is connected and this property is destroyed by the suppression of any arc.

T6. Any pair of vertices is linked by one and only one chain.

Weak connectivity is equivalent to the statement

$$\forall x\ \forall y\ ,\quad]x\ldots y[\quad \text{is a chain}$$

or to

$$\forall x\ \forall y \neq \alpha,\quad]\alpha\ldots x[\quad \text{and}\quad]\alpha\ldots y[\quad \text{are chains.}$$

T1 ⇒ T2 and T2 ⇒ T3 are established from the defining equation for the cycle rank

$$k = m - n + p$$

In the first case, p=1 and k=0 implies m=n-1. In the second case, m=n-1 and k=0 implies p=1.

The other implications (T5 ⇒ T6, T6 ⇒ T1) can be deduced from the analogous proofs (A5 ⇒ A6 and A6 ⇒ A1) by use of table (S) but without giving a privileged place to the vertex α. Property 5 has already been established starting from weak connectivity.

Since weak connectivity is implied by l.q.s. connectivity, T1 is a property of arborescences. Moreover T1 implies that a tree is antisymmetric. Thus we have

Property 7. Every tree is antisymmetric and every arborescence is a tree.

There is no equivalence between the two types of graph, but it is possible to characterize an arborescence as a tree which is l.q.s. connected.

Property 8. A graph G has a tree as a partial graph if and only if it is connected. Furthermore:

Property 9. Every tree has at least two vertices of degree 1, at least one vertex without ancestors and at least one vertex without descendants.

Definition 5. An *endpoint* of a tree is any vertex of degree 1.

If a tree did not have two distinct endpoints, it would be possible to extend every simple chain indefinitely from at least one of its extremities. For all the simple chains in a tree are elementary because there are no cycles. Since a tree is finite, it follows that there must be at least two endpoints. The trivial graph of order 1 may also be thought of as a *degenerate tree*.

Let (X,Γ) be a latticoid with root α and N terminals and suppose that the numbers q_k of vertices having a_k outgoing arcs (where k runs over a finite set of indices K) are known but the value of $|X|$ is not known. Further suppose that the indegrees $\partial \bar{x}(x \in X)$ are given and let $g = \sum_{x \in X-\{\alpha\}} (\partial\bar{x}-1)$. It is clear that g=0 if and only if (X,Γ) is an arborescence. The number of arcs leaving the vertices of (X,Γ) is $\sum_{k \in K} a_k q_k$.

Of these, $g + \sum_{k \in K} q_k - 1$ enter into non-terminal vertices while N enter into the terminals.

Hence

$$N + g = \sum_{k \in K} a_k q_k - \sum_{k \in K} q_k + 1$$

that is

$$N \leq \sum_{k \in K} a_k q_k - \sum_{k \in K} q_k + 1$$

With equality if and only if (X, Γ) is an arborescence.

In particular when $|K| = 1$, then all non-terminal vertices have the same outdegree q_1 and

$$N \leq (a_1 - 1) q_1 + 1 .$$

In the case of an arborescence, the number of vertices is not arbitrary when $|K| = 1$ but N must be congruent to 1 modulo $(a_1 - 1)$. N can take any integer value if and only if $a_1 = 2$. An arborescence is *homogeneous* if all the non-terminal vertices have the same outdegree a.

Definition 6. If $a = 2$, we say that the arborescence is *dichotomous* and if $a > 2$ we say that it is *polychotomic*.

Property 10. A latticoid having q_k vertices of outdegree a_k ($k \in K$, where K is a finite family of indices) has N terminals, where

$$N \leq \sum_{k \in K} (a_k - 1) q_k + 1 .$$

If it is homogeneous ($a_k = a$ for all k), then

$$N \leq (a - 1) q + 1 ,$$

and if $a = 2$ then $n \leq q + 1$.

Equality holds in each of these if and only if the latticoid is an arborescence.

Furthermore if the outdegrees of the root and the other non-terminal vertices are given, then it is possible to determine the order of the latticoid; $n \leq N + \sum_{k \in K} q_k$.

Property 11. A latticoid having q_k vertices of outdegrees a_k ($a_k > 0$, $k \in K$) is a graph of order n, where

$$n \leq \sum_{k \in K} a_k q_k + 1 ,$$

with equality if and only if the latticoid is an arborescence.

Theorem 6. A tree is not an arborescence if and only if one of its vertices is the terminal extremity of at least two arcs.

Proof. The condition is obviously sufficient.

A weakly connected graph is not l.q.s.c. if it has no vertex which is the origin of paths leading to each of the other vertices. Moreover, if the graph is acyclic, every vertex is linked to each of the others by a unique chain and at least one of these chains is not a path. By Property 1.9, in such a chain, there is a vertex which is the initial (or the terminal) extremity of two arcs.

Let a be a vertex from which there are chains leading to each of the other vertices and let x be a vertex which is the initial extremity of two arcs along one of the chains from a. Then either x is the origin of paths leading to all other vertices (in which case the tree is an arborescence) or x is the initial extremity of the first arc of a chain passing through other vertices. Thus on this chain there will be a vertex which is the extremity of two arcs.

The condition is therefore also necessary.

Theorem 7. A latticoid is not an arborescence if and only if one of its vertices is the terminal extremity of at least two arcs.

Proof. The condition is sufficient.

The root of the latticoid is the origin of paths leading to each of the other vertices. If these paths are unique, then we have an arborescence. Otherwise we have two paths [acb] and [adb] ($c \neq d$) linking the root a to some vertex b, and then b is the terminal extremity of at least two arcs.

Theorem 8. A latticoid is not an arborescence if and only if there are two vertices linked by more than one chain.

Proof. If the vertices a and b are linked by two chains, then there would be a cycle. Thus the condition is sufficient.

If the latticoid Λ is not an arborescence, then there is a cycle in Λ. Hence there are vertices which are linked by more than one chain. Thus the condition is also necessary.

We can also prove two further conditions each of which is equivalent to the preceding ones.

Property 12. A tree is not an arborescence if and only if there are vertices a and b linked by a chain which is neither a path nor a combination of two paths ending at a and b.

Property 13. A latticoid is not an arborescence if and only if it has more than n-1 arcs.

Thus we observe several extremal aspects of arborescences;

latticoids of given order having the minimal number of arcs, latticoids in which the set of paths has no redundancy, latticoids and trees in which no vertex is the terminal extremity of more than one arc.

Definition 7. A *proper subgraph* of a graph G is a subgraph of G having at least two vertices and distinct from G.

Definition 8. A *sub-latticoid* is a proper l.q.s.c. subgraph of a latticoid. A *sub-arborescence* is a proper connected subgraph of an arborescence. A *sub-tree* is a proper connected subgraph of a tree.

Theorem 9. A sub-latticoid, a sub-arborescence and a sub-tree are respectively a latticoid, an arborescence and a tree.

Proof. A sub-latticoid is l.q.s.c. and circuitless and is therefore a latticoid. A sub-arborescence and a sub-tree are connected and acyclic. Consequently they are trees. If some vertex of a sub-arborescence was the terminal extremity of two arcs, it would not be a subgraph of an arborescence A. If each vertex of a sub-arborescence S was the terminal extremity of just one arc, then S would not contain the root of A but there would be a path linking the root of A to a vertex x of S. Then, in A, x would be the terminal extremity of two arcs which is a contradiction. If S had two vertices a and b without ancestors, then there would be a chain]ab[in which the extremities would be the extremities of arcs in the chain and which would therefore have a vertex being the terminal extremity of two arcs. It follows that a sub-arborescence is an arborescence.

From this we deduce the following result.

Corollary 10. A proper subgraph of an arborescence is a forest and, if it is connected, then it is an arborescence.

This theorem and its corollary explain why the definitions were made with the connectivity as weak as possible. Thus for example, a weakly connected subgraph of a latticoid is not necessarily a latticoid. The following theorem elucidates the link between connectivity, proper subgraphs and graphs.

Theorem 11. If in an l.q.s.c. (respectively connected) graph every proper subgraph of the same connectivity is a latticoid (respectively an arborescence or a tree), then either the same is true for the graph or else the graph contains no arcs except those which constitute a Hamiltonian circuit (respectively a Hamiltonian circuit or a Hamiltonian cycle).

Proof. 1. Suppose that $G = (X, \Gamma)$ is l.q.s.c. and that every proper l.q.s.c. subgraph of G is a latticoid.

If γ is a circuit in G, then it must be broken by the suppression of any vertex and hence it is a Hamiltonian circuit. If G has a Hamiltonian circuit γ and an arc (a,b), then either there must be a second Hamiltonian circuit (in which case a and b are joined by two arcs so that G is an s-graph and not a graph) or there is a non-Hamiltonian circuit. Hence G cannot have any arcs except those of γ unless it is a latticoid.

2. Suppose $G = (X,\Gamma)$ is connected and every proper subgraph of G is an arborescence.

If γ is a cycle in G, then it must be broken by the suppression of any vertex and hence it is Hamiltonian. If G has a Hamiltonian cycle γ and an arc (a,b), then it must also have a non-Hamiltonian cycle which is impossible. G is therefore connected and acyclic (it is a tree) or else it has a Hamiltonian cycle γ. If γ is not a circuit or if G is not an arborescence, then there exists at least one vertex a which has two arcs entering into it and some connected subgraph generated by S, where $a \in S$, is not an arborescence.

3. In the case of trees, we show in the same way that, if G is not a tree, then it has a Hamiltonian cycle and no other arcs.

2.3 Arborescences and Data Processing

Arborescences and the associated concepts of root and rank (to be defined) come into various types of problem from enumeration techniques to questions of formal grammars, programming and coding and we shall briefly indicate some typical examples.

Definition 9. The *rank* of a vertex in a Dedekindian is the length of the path linking the root to the vertex. The *maximal rank* or *height* of a Dedekindian is the length of the longest path.

2.3.1 Simplexes and arborescences

An enumeration of the subsets of a finite set E of order N can be facilitated by the construction of an auxiliary arborescence from which they may be arranged in order without omission or repetition.

A simplex is the graph of the set $\mathfrak{P}(E)$ of all subsets of E, that is, $S = (\mathfrak{P}(E),\Gamma)$, where $\mathfrak{P}(E) = \{\xi_i\}$ is the set of subsets having p elements for $p=0,1,\ldots,N$. The number of subsets with p elements is equal to the binomial coefficient $\binom{N}{p}$.

Here $\xi_i \Gamma \xi_j$ if and only if the subset ξ_j consists of ξ_i and just one more element of E not in ξ_i.

It is well known that S is a hypercube of R^N having a vertex without ancestors - the empty set $\emptyset$ - and a vertex without descendants - the whole set E. Moreover $\emptyset$ is an ancestor of all other vertices and E is a descendant of all other vertices. Thus S is a quasi-strongly connected (and circuitless) graph. In particular, S is l.q.s.c. and has a partial graph A which is an arborescence. We shall construct this arborescence directly without recourse to the hypercube.

The required relation Δ of $A = (\mathfrak{P}(E), \Delta)$ will be more restrictive than Γ. In order to define it, we use the natural order

$$x_h \quad (h = 1, 2, \ldots, N)$$

of the elements of E and we suppose that the subset ξ_i is ordered accordingly, that is,

$$\xi_i = \{x_{i,1}, x_{i,2}, \ldots, x_{i,p}\} \text{ and } x_{i,k} < x_{i,k+1},$$

the relation < being a strict total order on the x_h.

Now we put $\xi_i \Delta \xi_j$ if and only if the subset ξ_j consists of the subset ξ_i together with just one more element $x_{j,p+1}$ of E not in ξ_i for which

$$x_{j,p+1} > x_{i,p}$$

Thus

$$\xi_j = \{x_{j,1}, x_{j,2}, \ldots, x_{j,p}, x_{j,p+1}\}$$

with $x_{j,k} = x_{i,k}$ for all $k \leq p$.

Then the graph $A = (\mathfrak{P}(E), \Delta)$ is an arborescence. For $\emptyset$ has no ancestors. If ξ_1 is a subset of order 1, then $\emptyset \Delta \xi_1$. If ξ_i does not contain x_N, then there is an arc $\xi_i \Delta \xi_j$, where $\xi_j = \xi_i \cup \{x_k\}$ for any k with $x_p < x_k \leq x_N$, x_p being the greatest vertex (in the given order) of ξ_i. The vertex ξ_i has a unique predecessor: the vertex ξ_{i-1} obtained from ξ_i by leaving out its greatest element. Hence every ξ_i, including E but excluding $\emptyset$, has only one arc entering it. This means that A is an arborescence.

The endpoints of A are those subsets which contain the greatest element x_N of E.

Both S and A have $1 + \sum_{p=1}^{N} \binom{N}{p} = 2^N$ vertices which form $\mathfrak{P}(E)$. There are 2^{N-1} vertices containing any given element, in particular x_N of E, and A therefore has 2^{N-1} endpoints.

The direct construction of A may be obtained through the following algorithm:

- the root is the subset $\emptyset$.

- the vertices of rank 1 are the N single-element subsets $\{x_h\}$.

- any subset which contains x_N has no successor.

- any subset with maximal element x_p ($x_p < x_N$) has N-p successors according to the rule for forming arcs $\xi_i \Delta \xi_j$.

Let $E = \{a, b, c, d\}$.

Then on the one hand we obtain

- the simplex S, shown in Figure 2.4, where the arcs belonging to both Γ and Δ are indicated by solid lines and those belonging only to Γ by dotted lines.[1]

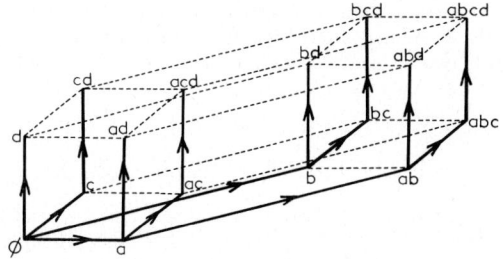

Figure 2.4

and on the other hand

- A, directly as shown in Figure 2.5

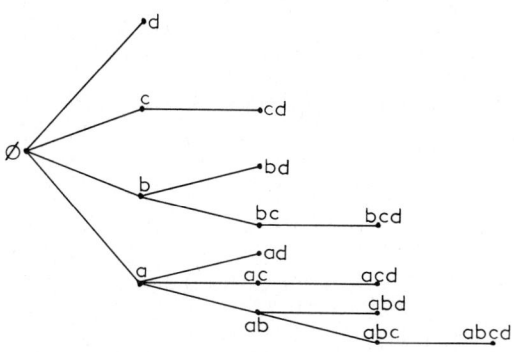

Figure 2.5

There are $\binom{N}{p}$ elements of order p $(0 < p \leqslant N)$ in A each corresponding to a subset of E with p elements.

2.3.2 Monoids and Arborescences

Let X be an alphabet of a letters. The words of the monoid X* may be formed by concatenation of one or more of the letters.

---

[1] Vertices such as $\xi_i = \{a, d\}$ are indicated simply by ad etc.

Supposing that there is an "alphabetical" order on the letters of X and that the repetition of letters is forbidden, the enumeration of the irreducible words of at most a letters may be effected by construction of the arborescence A, the partial graph of the simplex on X.

More generally the words of X* may be thought of as the vertices of an infinite graph in which a partial subgraph is an arborescence. The most economical way of forming the words of S is to generate them in a unique manner, that is by using an arborescence.

2.3.3 Chains, Paths and Arborescent Procedures

An exhaustive enumeration of the distinct chains of a finite graph may be effected by means of the following procedure:

Let $u_1, u_2 \ldots, u_m$ be the arcs of G.

- The chains of length 2 containing u_1 are formed by putting together with u_1 those arcs which have an extremity in common with u_1. This condition corresponds to two rows in the incidence matrix S of the arcs and this makes it possible to effect a choice which may be formalized by the set of arcs of an auxiliary arborescence.

- Similarly the chains of length p will be formed by putting together with the chains of length p-1 those arcs having an extremity in common with the "free" extremity of the last arc. If no chain of length p can be formed in this way, then the enumeration is complete. If a non-simple chain of length p+1 appears, this detects the presence of a cycle of length at most p and the chain of length p+1 must be rejected. All the chains of G can be generated starting from the empty chain ∅ with any arc being taken at the first.

In order to generate the set of subsets $\mathcal{P}(E)$ or the set of words of at most a letters (without repetition) in X*, the choice of a new subset or word succeeding a vertex already constructed was based on the set of elements E or X among the greatest ones. In general the arcs of a graph G do not constitute a set admitting a strict total order and the construction of a chain of length p+1 from a chain γ of length p will be possible if and only if there is an arc not in γ which is adjacent to the last arc of γ.

We say that the *procedure used* is *arborescent*.

For we can construct an auxiliary arborescence A having the empty chain ∅ as its root and the distinct chains of length p in G as vertices of rank p.

However the equality of two chains in the sense of graphs, that is, the equality of the two sets consisting of the same arcs, does not guarantee that a chain of G with first arc u_i is a sub-chain of a chain starting with an arc adjacent to u_i.

The procedure described above enables us to construct all the chains. However it is a little more interesting to give a procedure which produces the chains of longest length.

If the vertex a of A represents the chain $]u_1 u_2 u_p[$ of G and if b represents a chain of length 2 in G being a sub-chain $]u_{p-1} u_p[$ of the preceding one, then it is clear that in A there will be isomorphic sub-arborescences coming out from a and b. In the new procedure we will construct an arborescence A' in which b will be an endpoint and only the longest distinct chains will be produced.

These algorithms will be equally applicable to the paths of G, a path of length 1 then being sufficient to be a sub-chain of a path of length p.

Such *arborescent procedures* - that is, *exhaustive* ones - are lengthy to process but they do have the advantage of finding *all* the *required solutions* and of giving a sure answer if there is no solution (no circuit, no Hamiltonian or Eulerian path, no cycle etc.). There are many programming problems for which arborescent procedures can easily be developed. Against their inconvenience of having numerous processing steps we may set the certainty of obtaining a solution in a finite time, which, it is true, can increase very rapidly with the order of the graph under consideration.

Example: Let G be a graph of order 4 having 5 arcs as shown in Figure 2.6

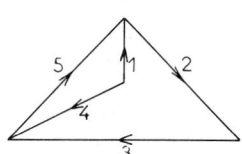

Figure 2.6

We represent the arborescence A' of distinct longest paths by the solid lines and the arborescence A of distinct paths by dotted lines in Figure 2.7.

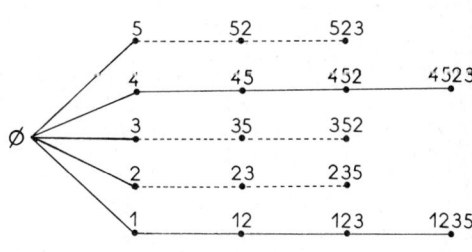

Figure 2.7

2.3.4 Coding

In an arborescence every vertex except the root is the terminal extremity of only one arc. But the vertices which are not endpoints can be the initial extremity of one or more arcs. Let $A = (X,\Gamma)$ be a polychotomic arborescence in which every vertex which is not an endpoint is the initial extremity of a arcs. Then we can think of A as the union of a partial graphs $A_i = (X,\Gamma_i)$; $i = 1,\ldots,a$, where

$$x \Gamma_i y \Rightarrow x \not\Gamma_j y \text{ for } j \neq i,$$

by following the procedure described at the end of §1.5.6.

Each of these partial graphs is acyclic and their connected components have no vertices which are terminal extremities of two arcs. Thus the A_i are forests.

Let $[x_0 x_p]$ be a path of A linking the root x_0 to a vertex of rank p. This path consists of a sequence of p arcs belonging to one or other of the forests A_i. It is possible to code this path by a p-letter word in which the letters come from the set $\{0,1,\ldots,a-1\}$. The first letter (the left one for example) will indicate the forest containing the first arc (x_0,x_1) and similarly for the kth letter and the kth arc. Any other path $[x_0 x_p]$ of length p will be coded in the same way by a p-letter word with letters from $\{0,1,\ldots,a-1\}$ and with at least one different from the corresponding letter of $[x_0 x_p]$. (That of an arc having the same initial extremity as an arc of $[x_0 x_p]$.)

Conversely consider an arbitrary p-letter word (assuming A has maximal rank at least p). If the kth letter is an i, then the arcs linking the vertices of rank k-1 and k in the forest A_i may be selected for the decoding. Proceeding in this way for all k from 1 to p we will produce a unique path of A, the root being coded by the empty word.

Given x_p every vertex on the path $[x_0 x_p]$ will be coded by an h-letter word (h<p) consisting of the first h letters of the code of x_p. Any two vertices in which neither is an ancestor of the other will always have at least one different letter.

If we code all the vertices of A, the code words relative to endpoints will begin with the code words relative to their ancestors. But if we code only the endpoints then no prefix (or word formed from the first letters) will be another code word. An arbitrarily long sequence of code words can be decoded in a unique way. If a sequence of letters represents an endpoint of A, then the following letters will represent the start of a new path leading to the endpoint coded later (instantaneous code).

Moreover it is possible to define two strict total orders on the vertices of an arborescence - $\prec$ and γ say:

a. if x and y have a predecessor z in common such that $z\Gamma_i x$ and $z\Gamma_j y$ with i<j, then $x \prec y$ and $x \gamma y$,

b. if $x\hat{\Gamma}y$, then $x \prec y$ and $x \gamma y$,

c. Let t be the common ancestor of x and y such that $t\Gamma_i \bar{x}$,

$\overline{x}\hat{\Gamma}x$, $t\Gamma_j y$, $\overline{y}\hat{\Gamma}y$ and $i<j$, then $x \prec y$,

   d. let $r(x)$ and $r(y)$ be the ranks of x and y and let x', y' be the predecessors of x and y. Then, if $r(x)<r(y)$ or if $x'\gamma y'$, then $x\gamma y$.

We note that condition a. is a special case of c. and is unnecessary in defining $\prec$.

In the order $\prec$, three vertices on the same path can be ordered consecutively whereas under γ three vertices of the same rank can be ordered consecutively.

For example, the vertices of the arborescence A' in §2.3.3 (Figure 2.7) can be coded and ordered according to $\prec$:

$$\emptyset, 0, 00, 000, 0000, 1, 2, 3, 30, 300, 3000, 4$$

and according to γ:

$$\emptyset, 0, 1, 2, 3, 4, 00, 30, 000, 300, 0000, 3000.$$

If the arborescence is not homogeneous, then we adopt an analogous coding in which a is the maximal outdegree of the vertices in the arborescence.

The order $\prec$ will be frequently encountered when discussing arborescent procedures. $\prec$ is sometimes called Tarry's order and γ scriptural order.

2.3.5 Finite and Infinite Graphs

By suppressing the condition of finiteness in the definitions of an arborescence, we introduce the concept of an infinite graph A having Property 5, properties A1, A5, A6 and property A'2 (respectively A'4).

A is acyclic (respectively l.q.s.c.), has a vertex without ancestors and all other vertices have indegrees equal to 1.

Any such graph obtained by indefinitely lengthening at least one of the paths coming out of the root of an arborescence will be called an *infinite arborescence*.

It is possible to generate other infinite graphs.

Let $A_i = (X_i, \Gamma_i)$; $i=1,2,\ldots,p$, be a family of l.q.s.c. acyclic graphs with roots x_i such that $x_{i+1}\Gamma_i x_i$ and that A_i is a subgraph of A_{i+1}. As long as p is finite, A_p remains l.q.s.c., that is, A_p is an arborescence, but if p tends to infinity, then in the limit A_∞ no longer has a vertex without ancestors. It is a weakly connected acyclic graph and therefore an infinite graph of tree type but it is no longer of arborescence type. However no vertex is the terminal extremity of two arcs.

The first infinite graph is encountered when discussing the

paths of a flow diagram (order graph for programming) for which a required condition can never be satisfied.

An *infinite latticoid* is an l.q.s.c. infinite graph without circuits. It is possible to show that an infinite latticoid satisfies condition L3.

Any infinite arborescence is an infinite latticoid.

It is possible to generate infinite latticoids with or without cycles by starting from a latticoid or an arborescence (possibly infinite) in which at least one vertex has a denumerable infinity of successors while still preserving property L3 or A1, A'2, A'4, A5 and A6.

2.4 Transportation Networks

Definition 10. A *transportation network* is a quasi-strongly connected finite graph (X,Γ) in which there are no loops, there is a unique root and a unique terminal and for which there is a valuation by non-negative integers on the arcs.

The valuation $\gamma: \Gamma \to N$ is known as the *capacity* of the arcs, and we note that $c_{ij} \geqslant 0$.

The vertex α without ancestors is known as the *source* of the network and the vertex z without descendants as the *sink* of the network.

Definition 11. A *flow* is a valuation on Γ taking integral values for which:

1. $\forall u \in \Gamma,\ \phi_u \geqslant 0$

2. $\sum_{u \in \omega^-(x)} \phi_u = \sum_{u \in \omega^+(x)} \phi_u$, for $x \in X$ and $x \neq \alpha, z$.

3. $\forall u \in \Gamma,\ \phi_u \leqslant c_u$.

From this we deduce that the sum

$$\phi_z = \sum_{u \in \omega^+(\alpha)} \phi_u = \sum_{u \in \omega^-(z)} \phi_u$$

is preserved (or transported if we think of it physically as a quantity of matter) from the source to the sink and ϕ_z is called the *value* of the flow. The ϕ_u are called flows.

The resolution of a transportation problem consists essentially in determining the maximal value of the flow when (X,Γ) and the capacities c_u are given.

Condition 2. requires the *conservation of flow* at every vertex except the source and sink.

Let A be a subset of X such that

$$z \in A \quad \text{and} \quad \alpha \in X-A.$$

Then a *cut-set* of the transportation network G is the set $\omega^-(A)$ consisting of those arcs which enter into A. A cut-set is therefore a cocircuit if and only if there is no arc leaving A.

The *capacity of the cut-set* will be the sum of the capacities of its arcs:

$$C(\omega^-(A)) = \sum_{u \in \omega^-(A)} C_u$$

and, since we have chosen $z \in A$,

$$\phi_z \leq C(\omega^-(A)).$$

Consequently when there is a cut-set V such that $\phi_z = C_V$, then simultaneously we have that the flow ϕ_z takes the maximal value and the cut-set V is of minimal capacity.

Theorem 12. In a transportation network the maximal value of flow is equal to the minimal capacity of a cut-set:

$$\max_{\phi} \phi_z = \min_{\substack{A \subset X-\{\alpha\} \\ z \in A}} C(\omega^-(A)).$$

Ford and Fulkerson have used this theorem to produce an algorithm for determining the maximal flow from a given initial flow.

EXERCISES

Prove the following statements:

1. A finite graph without circuits has no path of infinite length.

2. If the strict transitive closure T of a graph G is irreflexive, then G is circuitless.

3. If a graph is circuitless, then any weakly connected subgraph has a vertex without ancestors and a vertex (possibly the same one) without descendants.

4. If every finite proper subgraph has a vertex without ancestors and a vertex without descendants, then either the graph is circuitless or it has a Hamiltonian circuit and no arcs off this circuit. Further if there is an isolated vertex, then the graph is circuitless.

5. The strict transitive closure of an arborescence of height $h(\geqslant 2)$ is a latticoid but not a Dedekindian.

6. Show that in a finite l.q.s.c. circuitless s-graph there exists a vertex a whose only predecessor is the root and a vertex b all of whose successors are terminals (distinct or not).

Chapter Three

OPERATIONS ON GRAPHS

The purpose of this chapter is to make it possible to discuss the creation of new graphs and the modification of a given graph in order to meet the requirements of specialization or generalization.

3.1 General Definitions

Definition 1. An *operation* on a pair of graphs is any composition law of the two graphs the product (or result) of which is again a graph.

Definition 2. A *vertex-preserving operation* is any operation on a pair of graphs having the same set of vertices in which the product also has the same set of vertices.

Definition 3. A *unary operation* on a graph $G = (X, \Gamma)$ is a rule whereby a second graph may be constructed from G as a partial subgraph of the complete symmetric and reflexive graph generated by X. The specification of such an operation and of G makes possible a new graph uniquely related to G.

Definition 4. A *graph transformation* is an operation on a pair of graphs in which the product is a partial subgraph of their union.

If the second graph can be characterized by an arc (or arcs) and the extremities of this arc (these arcs) we will say improperly that the transformation is a law composing the graph with one (several) of its arcs.

In order not to increase an already very rich terminology we will use the same name for an operation on a pair of graphs and the product graph of the operation.

We have already introduced the concept of *elementary operations* in which the composition law is the same for the sets of vertices and of arcs (cf. §1.3) and these can also be vertex-preserving operations when applied to graphs having the same set of vertices. As we have seen, union and intersection are associative and commutative operations.

When the *union* and *intersection* are applied to graphs $G_1 = (X, \Gamma_1)$ and $G_2 = (X, \Gamma_2)$ defined on the same set X, they are vertex-preserving operations. We note that the distributive laws also hold:

$$(G \cup H) \cap K = (G \cap K) \cup (H \cap K)$$

$$(G \cap H) \cup K = (G \cup K) \cap (H \cup K) \qquad (1)$$

3.2 Unary Operations

Certain operations define starting from a graph G a partial graph of the complete symmetric and reflexive graph generated by G. We start by studying several types of graph defined in general on the same set of vertices as $G = (X,\Gamma)$.

• The *empty graph* is the graph $\emptyset = (\emptyset,\emptyset)$ having no vertices and no arcs.

• A *trivial graph* is any graph in which the set of arcs is empty.

Thus every vertex is isolated and, if X is finite, then $p=n$ and $m=k=\ell=0$. We will denote this graph by

$$O_X = (X,\emptyset) \qquad (2)$$

If $|X|=1$, then we will write O_e for $(\{e\},\emptyset)$.

• A *diagonal graph* is any graph which has a loop at every vertex and no other arcs.

If X is finite, then $m=n=p=k$ and $\ell=0$.

A diagonal graph will be denoted by

$$I_X = (X,I) \qquad (3)$$

where $I \subseteq X^2$ is the set of all the loops, i.e.

$$I = \{(x_i,x_i) \mid x_i \in X\}.$$

If $|X|=1$, we will write I_e for $(\{e\},\{(e,e)\})$ or $I_e=(\{e\},I)$.

The adjacency matrix of O_X is a square matrix of order $|X|$ in which all entries are zero (zero-matrix). The adjacency matrix of I_X is the identity matrix of order $|X|$.

Let Γ_a be the set of proper arcs in Γ and Γ_b the set of loops in Γ. Then $\Gamma = \Gamma_a \cup \Gamma_b$ and $\Gamma_a \cap \Gamma_b = \emptyset$.

• The *irreflexive graph associated with* G is the graph $G_a = (X,\Gamma_a)$.

• The *reflexive graph associated with* G is the graph $G_b = (X,\Gamma_b)$ – a partial graph of I_X.

• The *inverse* G^{-1} is the graph corresponding to the inverse of Γ (considered as a relation). $G^{-1} = (X,\Gamma^{-1})$, where

$$x\Gamma y \Leftrightarrow y\Gamma^{-1}x$$

so that $G \cap G^{-1}$ is a symmetric graph which may have loops. $G = G^{-1}$ if and only if G is symmetric.

• The *symmetric graph* G_s based on G has already been

defined in a unique manner (§1.2).

The following unary operations enable us to produce graphs of the above types. We can also think of these as functions of the graph. They are:

- *Trivialization*: $G \longmapsto 0_X$
- *Diagonalization*: $G \longmapsto I_X$
- *Suppression of loops*: $G \longmapsto G_a$
- *Restriction to loops*: $G \longmapsto G_b$
- *Inversion*: $G \longmapsto G^{-1}$
- *Symmetrization*: $G \longmapsto G_s$
- The *graph*: $G^p = (X, \Gamma^p)$ in which Γ^p is defined by means of the set e_p of all paths of length at most p in G; that is

$$x_i \Gamma^p x_k \quad \Leftrightarrow \quad [x_i x_k] \in e_p,$$

is the result of a unary operation. There is a unique arc of G^p corresponding to the set of paths of length at most p linking x_i to x_k.

- The *strict transitive closure* $T = (X, \mu)$ in which μ is defined by means of the routing, or set of all elementary paths e of G, that is

$$x_i \mu x_k \quad \Leftrightarrow \quad [x_i x_k] \in e,$$

also results from a unary operation. Each arc in T corresponds to the sets of paths linking x_i and x_k.

T may also be obtained by taking the union of the graphs G^p for $p = 1, 2, \ldots, |X|$.

The broad transitive closure F can be defined by a unary operation. It is also $F = T \cup I_X$.

- The results of certain unary operations can often be defined by means of vertex-preserving operations, remarking that

$$G \cap I_X = G_b,$$
$$G_a \cup G_b = G, \quad G_a \cap G_b = 0_X. \quad (4)$$

3.3 Transformations

According to Definition 4 (§3.1) a transformation of a graph has as its result a partial subgraph of the union of a pair of graphs. It can involve an operation of *local character* applying to some vertices and certain arcs or an operation of *global character* affecting the whole set of vertices of the graph and producing a new graph which represents the first in the sense of a certain equivalence.

Since the first chapter, we have met with graph transformations and unary operations which also make it possible to define graphs which we may consider as transforms of the first. Defining a subgraph (X',Γ') or a partial graph (X,Γ'') of a graph $G = (X,\Gamma)$ are transformations of G.

1. *Suppression of an arc* is the transformation which substitutes $(X,\Gamma-\{u\})$ for (X,Γ), where $u \in \Gamma$ (Figure 3.1).

A partial graph is the result of a sequence of suppressions of arcs.

2. *Suppression of a vertex* is the transformation which produces the subgraph $(X-\{a\},\Gamma')$, where $a \in X$ and $x \neq a \Rightarrow \Gamma'x = \Gamma x-\{a\}$.

3. *Adjunction of an arc* preserves the vertices and adds a new arc: $(X,\Gamma') = (X,\Gamma \cup \{(x,y)\})$, where $x,y \in X$ and $(x,y) \notin \Gamma$.

4. *Adjunction of a vertex* preserves the arcs and adds an isolated vertex: $(X',\Gamma) = (X \cup \{\xi\},\Gamma)$, where $\xi \notin X$.

5. *Posing a vertex* adds a vertex of degree 2 by modifying the adjacency: $(X',\Gamma') = (X \cup \{\xi\},\Gamma')$, where $\xi \notin X$, $(a,b) \in \Gamma$ and $\Gamma' = (\Gamma-\{(a,b)\}) \cup \{(a,\xi),(\xi,b)\}$.

The cycle and cocycle ranks are both preserved by adjunction of a vertex whereas at least one of them is altered by suppression or adjunction of an arc or by posing a vertex.

It is left as an exercise for the reader to determine the variations in the numbers m, n, p, k and ℓ under each of these five transformations.

In order to preserve the reflexivity or symmetry of a graph G and its transform we adjoin a loop at the same time as we add a vertex and we suppress or adjoin two arcs inverse to each other. For *symmetric posing* of a vertex we define the new graph by
$$(X',\Gamma') = (X \cup \{\xi\},\Gamma'),$$
where
$$\xi \notin X, \quad (a,b) \in \Gamma, \quad (b,a) \in \Gamma$$
and
$$\Gamma' = (\Gamma-\{(a,b),(b,a)\}) \cup \{(a,\xi),(b,\xi),(\xi,a),(\xi,b)\}.$$

6. We define the *doubling of a vertex* η as the transformation which separates η into two adjacent vertices in such a way that $\partial^-\xi = \partial^{-1}\eta$ and $\partial^+\zeta = \partial^+\eta$: $X' = (X-\{\eta\}) \cup \{\xi,\zeta\}$, where
$$\xi,\zeta \notin X, \quad \eta \in X, \quad (\xi,\zeta) \in \Gamma'$$
and
$$\forall x,y \in X - \{\eta\}, \text{ if } (x,y) \in \Gamma, \text{ then } (x,y) \in \Gamma',$$

if $(x,\eta) \in \Gamma$, then $(x,\xi) \in \Gamma'$ and if $(\eta,y) \in \Gamma$, then $(\zeta,y) \in \Gamma'$. In particular if $(\eta,\eta) \in \Gamma$, then $(\zeta,\xi) \in \Gamma'$, that is a loop at η (thought of as a circuit of length 1) is transformed into a circuit of length 2.

7. The *contraction at a subset* $A (\subsetneq X)$ of the graph

$$G = (X,\Gamma)$$

is the transformation which produces the graph $G_A = (X_A, \Gamma_A)$ obtained from G by replacing the set A by a single vertex ξ and giving ξ the adjacency relationships between A and X-A.

Accordingly G_A is such that $X_A = (X-A) \cup \{\xi\}$, where $\xi \notin X$, $\Gamma_A \xi = \Gamma A - A$, if G has no circuits in the subgraph generated by A and $\Gamma_A \xi = (\Gamma A - A) \cup \{\xi\}$ otherwise.

For $x \in X-A$, if $\Gamma x \cap A = \emptyset$, then $\Gamma_A x = \Gamma x$ and, if $\Gamma x \cap A \neq \emptyset$, then $\Gamma_A x = (\Gamma x - A) \cup \{\xi\}$.

8. The *reduction* of a graph consists in making all the possible contractions at every strong class but without forming any loops. The result is called a *reduced graph*.

9. An *exchange of arcs* is the transformation which substitutes the arcs (x,b) and (a,y) for the arcs (x,y) and (a,b) while preserving the vertices and all other arcs.

This transformation preserves m and n but may alter p and thus also k and ℓ. It will be used in the theory of questionnaires when G is a latticoid and y and b are roots of disjoint sub-latticoids. Paths such as

$$[x_0 \ldots xy \ldots z] \quad \text{and} \quad [x_0 \ldots ab \ldots c]$$

will be transformed into the paths

$$[x_0 \ldots xb \ldots c] \quad \text{and} \quad [x_0 \ldots ay \ldots z]$$

and for example the longest path in the new latticoid may be shorter than that of G. Thus two sub-arborescences with roots b and y will be eventually permuted. It is for this reason that this transformation is called *arborescence transfer* in the case when b and y are roots of l.q.s.c. acyclic subgraphs.

10. *Exterior substitution of an arc* is the transformation which replaces an arc (a,b) by an arc (x,b) while preserving the vertices and all the other arcs (Figure 3.1).

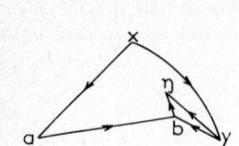
Suppression of an arc

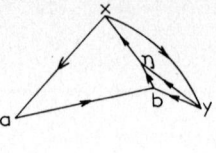
Suppression of a vertex

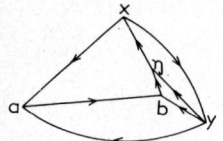
Adjunction of an arc

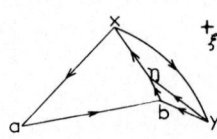
Adjunction of a vertex

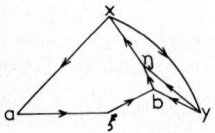
Posing a vertex

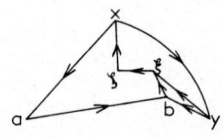
Doubling a vertex

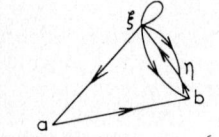
Contraction at {x,y}

Reduction (strong equivalence)

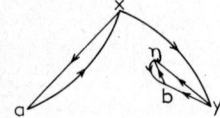
Exchange of arcs:
(η,x) and (a,b)

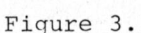
Figure 3.1

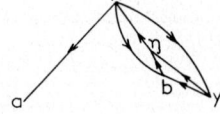
Substitution of an arc:
(a,b) replaced by (x,b)

This transformation enables us to increase the outdegree of one vertex by 1 and reduce the outdegree of another by 1.

Interior substitution of an arc is the transformation which replaces an arc (a,y) by an arc (a,b) while preserving the vertices and all the other arcs.

This affects the indegrees in the same way as the previous one affects the outdegrees.

Note that an exchange of arcs may be effected by two exterior substitutions of arcs or two interior substitutions of arcs. In practice we will use the first type of substitution without explicitly saying that it is "exterior".

11. Contractions of arborescences and Prüfer Codes

The machine representation of an arborescence A can be realized by a triangular adjacency matrix. Since every vertex except one is the terminal extremity of only one arc, every column except one has only one non-zero entry and the wastage of storage space is very great. The wastage can be reduced to one memory position by representing an arborescence of order n by using the n positions in the memory bank forming the so-called predecessors list. In this if 1, 2,...,n are the vertices of A, then position i will contain the name (or the address) of the predecessor of i. It is possible to reduce the wastage to zero by adopting the following coding due to Prüfer.

In order to form the *Prüfer code* with the aid of n-1 "characters" which will be the labels of the predecessors, we start with the list of predecessors and suppose that the labelling of the vertices has been fixed in some arbitrary manner using the integers 1,2,...,n. The first character of the code is the label of the predecessor of the terminal vertex having the smallest label. We suppress this terminal and then the second character is the label of the predecessor of the terminal with the smallest label in the resulting arborescence of order n-1.

Proceeding by repitition of this process, we stop when taking the origin of the unique remaining arc, that is the root of A. A vertex of outdegree d_i^+ appears in d_i^+ characters of the Prüfer code and there are as many characters as arcs, that is n-1.

Decoding the Prüfer code leads to the construction of n-1 arcs, whose origins have as labels the n-1 characters of the code. The terminal extremity of the first arc is the vertex whose label is the smallest integer not in the code. The terminal extremity of the ith arc, whose origin is the ith character, is the vertex whose label is the smallest integer which (1) has not already been attached to the terminal extremity of one of the (i-1) earlier arcs and (2) which is not present among the n-1 characters not yet decoded.

The predecessor list is constructed as the decoding proceeds character by character (the first encoded character is also the first to be decoded, for example, the one furthest to the left).

The economy in storage space is of course an illusion when taken together with the encoding and decoding work involved. But this does illustrate the utilisation of successive contractions (or of simple suppressions of endpoints) when coding and the construction of an arborescence arc by arc with the possible formation of a forest contained in A are carried out.

In particular we will effect transfers of sub-arborescences reduced to a unique vertex. The transformed graph will be isomorphic to the original graph but we will cause at the same time a modification of the coding of these terminal vertices or of their valuation.

Example:

The arborescence with predecessor list 8, 5, 1, 1, ∅, 8, 1, 5 (where ∅ indicates that the vertex labelled 5 is the root) can be represented by the Prüfer code 5 1 1 8 1 8 5 and presented diagrametically as shown in Figure 3.2.

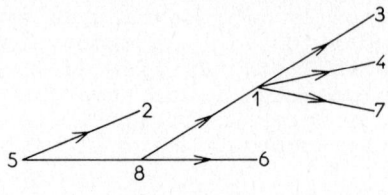

Figure 3.2

3.4 Cartesian Operations

Definition 5. A *Cartesian operation* on a pair of graphs $G = (X,\Gamma)$ and $H = (Y,\Delta)$ is any operation which produces a third graph $P = (XY,\Psi)$ in which the vertices are generated by the Cartesian product of the sets X and Y.

Cartesian operations will be given by defining the law of composition which produces Ψ from Γ and Δ.

Notation:

The characteristic parameters of G will be denoted by lower-case letters indexed by a "1".

The characteristic parameters of H will be denoted by lower-case letters indexed by a "2".

The characteristic parameters of P will be denoted by capital letters.

We will also have occasion to use another graph $K = (Z,\Lambda)$ for which we will use the index 3.

For example the numbers of vertices are related by the equation
$$N = n_1 n_2 \tag{5}$$
for all Cartesian operations on a pair of finite graphs.

Vertices in the graphs will be denoted by

$$x_i, x_k \in X; \qquad y_j, y_\ell \in Y; \qquad x_i, y_\ell \in XY,\ldots$$

3.4.1 Product and sum

Definition 6. P is the (*direct*) *product* of G and H and is denoted by GH when

$$x_i y_j \Psi x_k y_l \Leftrightarrow x_i \Gamma x_k \quad \text{and} \quad y_j \Delta y_l \qquad (6)$$

The arcs of Ψ are quadruples - elements of the product set $(XY)(XY)$ - in which the 1st and 3rd components form an ordered pair in Γ and the 2nd and 4th components form an ordered pair in Δ. All the quadruples formed in this way from Γ and Δ belong to Ψ. We will think of Ψ as the Cartesian product of the relations Γ and Δ and we will write $GH = (XY, \Gamma\Delta)$. Thus the direct product can also be defined as an elementary operation. However we note that $(x_i y_j, x_k y_\ell) \in (XY)(XY)$, whereas, if $u \in \Gamma$ and $v \in \Delta$, then $uv \in (X^2)(Y^2)$ and strictly speaking Ψ is only in 1-1 correspondence with $\Gamma\Delta$.

The direct product is known under various names: product (Berge), cardinal product (Čulik), tensorial product (Capobianco), Kronecker product (Weichsel) and conjunction (Harary). We will sometimes say "product" when there is no risk of ambiguity (Figure 3.7).

Property 1 GI_e is isomorphic to G.

For corresponding to any $x \in X$, there is a unique vertex $xe \in X\{e\}$ and to any arc $(x_i, x_k) \in \Gamma$, there is a unique $(x_i e, x_k e) \in \Psi$. In this case Ψ will be denoted by ΓI or sometimes more simply by Γ.

We also have

$$GH \cong G \quad \Rightarrow \quad H \cong I_e, \qquad (7)$$

where $\cong$ denotes isomorphism. For there is a one-to-one mapping between the elements of GH and those of G and so H has only one vertex and one arc (actually a loop) and is therefore of the form $(\{e\}, I)$.

Similarly we see that $I_e H \cong H$.

Thus we have the following result: Up to isomorphism, the diagonal graph of order 1 is the neutral element with respect to direct products.

If $G = I_x$ and $H = I_y$, then

$$GH = I_x I_y = (XY, I) = I_{xy}$$

Similarly $\qquad O_x O_y = O_{xy}$.

Further the relation $\emptyset$ is absorbent so that

$$O_x H = G O_y = O_{xy}$$

and the graph $\emptyset$ is absorbent relative to direct product

$$\emptyset H = G \emptyset = \emptyset$$

If $H = I_y = (Y, I)$, then GI_y is the union of $|Y|$ disjoint graphs all isomorphic to G. Similarly $I_X H$ is the union of $|X|$ disjoint graphs all isomorphic to H.

Definition 7 The direct product $P = GI_Y$ (or $I_X H$) is known as the *extension* of G by H (or of H by G).

The extension[1] of a graph G of order n_1 by a graph H of order n_2 is thus a graph of order $n_1 n_2$ containing n_2 disjoint subgraphs isomorphic to G.

The neutral element ($n_2 = 1$) is a special case of this.

Property 2 The direct product is associative.

For, let G, H, K be three graphs, then it is clear that $(XY)Z = X(YZ)$ and $(\Gamma\Delta)\Lambda = \Gamma(\Delta\Lambda)$ (up to a one-to-one correspondence).

The product of sets is not commutative. XY consists of the ordered pairs $z = (x, y)$ with first projections in X, whereas YX consists of the ordered pairs $z' = (y, x)$ with first projections in Y. Similarly the elements of $\Gamma\Delta$ are ordered pairs $(x_i y_j, x_k y_\ell)$ while those of $\Delta\Gamma$ are ordered pairs $(y_j x_i, y_\ell x_k)$. However there is a bijection from XY onto YX and from $\Gamma\Delta$ onto $\Delta\Gamma$ and we therefore have.

Property 3 The direct products GH and HG are isomorphic but not equal.

We know that the Cartesian product of sets is distributive both on the left and the right over both the union and the intersection of sets, that is

$$(X \cup Y)Z = XZ \cup YZ \quad \text{and} \quad (X \cap Y)Z = XZ \cap YZ .$$

We find in the same way that

$$(\Gamma \cup \Delta)\Lambda = \{ac; a \in \Gamma \cup \Delta \quad \text{and} \quad c \in \Lambda\}$$

$$\Gamma\Lambda \cup \Delta\Lambda = \{ac; (a \in \Gamma \text{ and } c \in \Lambda) \text{ or } (a \in \Delta \text{ and } c \in \Lambda)\}$$

whence $((X \cup Y)Z, (\Gamma \cup \Delta)\Lambda) = (XZ \cup YZ, \Gamma\Lambda \cup \Delta\Lambda)$,

that is
$$(G \cup H)K = GK \cup HK .$$

Similarly we can show that

$$(G \cap H)K = GK \cap HK$$

as well as $G(H \cup K) = GH \cup GK$ and $G(H \cap K) = GH \cap GK$.

---

[1] We say that GI_Y is the extension to the *right* of G and $I_X H$ is the extension to the *left* of H.

Property 4 The direct product of graphs is distributive over the union and intersection of graphs.

Example Let $X=Y=\{a,b,c\}$, $Z=\{1,2,3\}$, $\Gamma=\{(a,b),(a,c)\}$, $\Delta=\{(a,b),(b,c)\}$ and $\Lambda=\{(1,2),(2,3)\}$. Then

$$(\Gamma \cup \Delta)\Lambda = \{(a1,b2),(a2,b3),(a1,c2),(a2,c3),(b1,c2),(b2,c3)\}$$

and $(\Gamma \cap \Delta)\Lambda = \{(a1,b2),(a2,b3)\}$.

The vertices a3 and c1 are isolated in $(G \cup H)K$, while the vertices a3, c1, b1, c2 and e3 are isolated in $(G \cap H)K$. This is all shown in Figure 3.3 where the solid lines indicate the arcs of $(G \cap H)K$ and the dotted lines indicate the arcs of $(G \cup H)K$ which are not in $(G \cap H)K$.

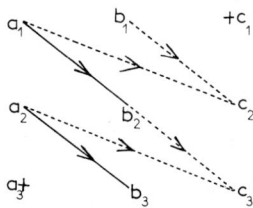

Figure 3.3

Definition 8 The *sum* G+H of the graphs G and H is the Cartesian operation given by the union of the extensions of G by H and of H by G, that is

$$G + H = GI_Y \cup I_X H \qquad (8)$$

The sum (Berge, Weichsel) is also called the Cartesian product (Sabidussi, Harary and Wilcox). (Figure 3.8)

Property 5 G+H and H+G are isomorphic.

This is an immediate consequence of Property 3 and commutativity of the union of graphs.

Since the direct product and union are associative, we also have

Property 6 The sum of graphs is associative.

We note that

$$G + H + K = GI_{YZ} \cup I_X HI_Z \cup I_{XY} K, \qquad (9)$$

and ask which arcs are common to GH and G+H. Now

$$GH \cap GI_Y = O_{XY} \quad \Leftrightarrow \quad H \subseteq \bar{I}_Y \text{ or } G=O_X$$

and

$$GH \cap I_X H = O_{XY} \quad \Leftrightarrow \quad G \subseteq \bar{I}_X \text{ or } H=O_Y$$

Property 7 The direct product GH and the sum G+H of non-trivial graphs G and H have no arcs in common if and only if G and H have no loops.

If $H_b \neq 0_Y$, then $GH_b \cap GI_Y$ has a_1b_2 proper arcs and b_1b_2 loops. In that case $GH \cap (G+H)$ can then have both loops and proper arcs. But

$$GI_Y \subseteq GH \quad \Leftrightarrow \quad I_Y \subseteq H$$

and

$$I_X H \subseteq GH \quad \Leftrightarrow \quad I_X \subseteq G .$$

Property 8 $G+H \subseteq GH$ if and only if G and H have a loop at every vertex.

Decomposing G and H into their respective reflexive and irreflexive graphs, we obtain

$$GH = (G_a \cup G_b)(H_a \cup H_b) .$$

Now $\quad G_a H_b \subseteq HI_Y \; , \; G_b H_a \cup G_b H_b \subseteq I_X H$

and therefore $\quad GH \subseteq G+H \quad \Leftrightarrow \quad G_a H_a \subseteq G+H$.

But $\quad G_a H_a \subseteq I_X H \quad \Leftrightarrow \quad G_a = 0_X \text{ or } H_a = 0_Y$

and $\quad G_a H_a \subseteq GI_Y \quad \Leftrightarrow \quad G_a = 0_X \text{ or } H_a = 0_Y$.

Property 9 $GH \subseteq G+H$ if and only if one of the graphs G or H has no proper arcs. $GH = G+H$ if and only if one of the graphs is diagonal and the other has a loop at every vertex.

Property 10 Let $G_s = (X_s, \Gamma_s)$, $s \in S$, be a finite family of graphs, then the sum over S of these graphs is the graph (X, Γ), where $X = \prod_{s \in S} X_s$ and

$$\Gamma = \bigcup_{s \in S} [(\prod_{j=1}^{s-1} I_j) \Gamma_s (\prod_{\substack{j \in S \\ j > s}} I_j)] \tag{10}$$

and I_j is the set of loops in I_{X_j}.

Here it is understood that the product of the I_j's on the left does not appear in the case when $s=1$ and that, similarly when $|S|$ is finite, the product of the right does not appear in the case $s=|S|$. Further in order to simplify the notation, it is assumed that the indices s are arranged in increasing order.

For $|S|=2$, (10) reduces to the definition of the sum of two graphs.

For $|S|=3$, (10) is true because of the associativity of the sum (9).

Suppose it is true for $|S|=p$. We show that it is then also true for $|S|=p+1$.

Now we may write

$$\Gamma = \{ \bigcup_{s \leqslant p} [(\prod_{j=1}^{s-1} I_j) \Gamma_s (\prod_{j=s+1}^{p} I_j)] \} I_{p+1} \cup [(\prod_{j=1}^{p} I_j) \Gamma_{p+1}] \quad ,$$

an expression which is essentially the same as (10) and is identical with the original definition given by Berge.

3.4.2 Classes of vertices

The status of the connectivity and routing in a graph resulting from a Cartesian operation can sometimes be decided directly from the corresponding properties of the component graphs. It is then possible to establish formulae which link certain characteristics of GH directly to those of G and H. In other cases we often need to have recourse to algorithmic methods. A detailed study of the vertices will enable us to obtain practical formulae.

We partition the set of vertices of $G = (X,\Gamma)$ into disjoint classes.

Let ∂_x^+ be the outdegree of $x \in X$, ∂_x^- the indegree of x and β_x the number of loops at x ($\beta_x=0$ or 1).

Definition 9 We classify the vertices into eight types as shown and named below:

$E_q = \{x;\ \partial_x^+ > 0,\ \partial_x^- > 0,\ \beta_x = 0\}$, *arbitrary*

$E_r = \{x;\ \partial_x^+ > 0,\ \partial_x^- = 0,\ \beta_x = 0\}$, *roots*

$E_s = \{x;\ \partial_x^+ = 0,\ \partial_x^- = 0,\ \beta_x = 0\}$, *isolated*

$E_t = \{x;\ \partial_x^+ = 0,\ \partial_x^- > 0,\ \beta_x = 0\}$, *terminals*

$E_{q'} = \{x;\ \partial_x^+ > 0,\ \partial_x^- > 0,\ \beta_x = 1\}$, *arbitrary looped*

$E_{r'} = \{x;\ \partial_x^+ > 0,\ \partial_x^- = 0,\ \beta_x = 1\}$, *looped roots*

$E_{s'} = \{x;\ \partial_x^+ = 0,\ \partial_x^- = 0,\ \beta_x = 1\}$, *isolated and looped*

$E_{t'} = \{x;\ \partial_x^+ = 0,\ \partial_x^- > 0,\ \beta_x = 1\}$, *looped terminals*

It is clear that for all $x \in X$, there is a k for which $x \in E_k$. We will say that x and y are equivalent if they are in the same class; that is

$$x \equiv y \quad \Leftrightarrow \quad \exists\, k;\ x,y \in E_k \ .$$

The quotient of X over this equivalence relation is a set which we will denote by

$$E = \{q,\ r,\ s,\ t,\ q',\ r',\ s',\ t'\}.$$

It is possible for some of the classes to be empty. For example, if G is diagonal, then $X = E_{s'}$.

If $x \in E_k$, we will say that x is of *type* k or is in class k.

If $x \in E_{q'} \cup E_{r'} \cup E_{s'} \cup E_{t'}$, then x is said to be looped. Otherwise x is not looped.

According to Definitions 6 and 8 there is an arc uv coming out of the vertex xy in GH if and only if u comes out of x <u>and</u> v comes out of y. Similarly there is an arc uv coming out of the vertex xy of G+H if and only if u comes out of x <u>or</u> v comes out of y. It is the same for entering arcs.

A vertex xy of GH is arbitrary if either both x and y are arbitrary or if one is arbitrary and the other is looped. xy is isolated if either one of them is isolated or if one is a root and the other is a terminal. A vertex xy of GH is a root if and only if one is a root and the other is neither a terminal nor isolated. Continuing in this way we can build up the table of classes of GH shown in figure 3.4.

T(GH)

G\H	q	r	s	t	q'	r'	s'	t'
q	q	r	s	t	q	q	q	q
r	r	r	s	s	r	r	r	r
s	s	s	s	s	s	s	s	s
t	t	s	s	t	t	t	t	t
q'	q	r	s	t	q'	q'	q'	q'
r'	q	r	s	t	q'	r'	r'	q'
s'	q	r	s	t	q'	r'	s'	i'
t'	q	r	s	t	q'	q'	i'	t'

Figure 3.4

We see that for any $c \in E = E'' \cup E'$

$$(c,c) \to c \quad \text{(idempotence)}$$
$$(c,s) \to s \quad \text{(absorption)}$$
$$(c,s') \to c \quad \text{(neutrality)}$$

If $H = I_Y$, then all vertices of H are of type s' and, since s' is neutral, $x_i y_j$ is of *the same class as* x_i. That is, the table reduces to one column (s').

We move now to consider the *union* of a pair of graphs $G_1 = (X, \Gamma_1)$ and $G_2 = (X, \Gamma_2)$ based on the *same* set of vertices X.

Let d_1 be the degree of $x_i \in X$ as a vertex in G_1 and d_2, d_{12} its degrees in G_2 and $G_1 \cup G_2$. It is clear that $d_{12} \geq \max(d_1, d_2)$. The absorbing class then corresponds to the vertices which are richer in types of arcs (entering, leaving, loops) - it will be q'.

The composition table can be built with the aid of this inequality. For example if x has no distinct ancestor in G_1 and no successor in G_2 but has at least one successor in G_1, a loop in G_1 and a predecessor in G_2, then $(r',t) \to q'$. (Figure 3.5)

OPERATIONS ON GRAPHS

$T(G+H)$

G\H	q	r	s	t	q'	r'	s'	t'
q	q	q	q	q	q'	q'	q'	q'
r	q	r	r	q	q'	r'	r'	q'
s	q	r	s	t	q'	r'	s'	t'
t	q	q	t	t	q'	q'	t'	t'
q'	q'	q'	q'	q'	q'	q'	q'	q'
r'	q'	r'	r'	q'	q'	r'	r'	q'
s'	q'	r'	s'	t'	q'	r'	s'	t'
t'	q'	q'	t'	t'	q'	q'	t'	t'

Figure 3.5

q is absorbing for non-looped vertices; s' is neutral for looped vertices; s is neutral for all 8 types.

Since I_X and I_Y are diagonal, $GI_X \cup I_YH$ has the same table (Figure 3.5) as $G_1 \cup G_2$ but the rows do not have the same significance, the vertices of $G_1 \cup G_2$ being elements of a set while those of G+H are ordered pairs.

The enumeration of the vertices by class can be deduced from the above tables.

We will use capital letters to denote the numbers of elements in each class - Q for q,..., T' for t'. Those for the product graph will carry no subscript, those for G and H will have subscripts 1 and 2 respectively.

The number of vertices in GH is

$$N = Q + R + S + T + Q' + R' + S' + T'.$$

If the law gives $(a,b) \to k$, that is, if the operation produces a vertex of type k whenever its projections are of types a and b, then there will be A_1B_2 vertices of type k in G_1G_2. By summing the numbers obtained for each position where a k occurs, we obtain an enumeration of the vertices in class k. From which we deduce the following results:

For GH:

$$Q = Q_1Q_2 + Q_1(Q_2' + R_2' + S_2' + T_2') + Q_2(Q_1' + R_1' + S_1' + T_1'),$$

$$R = R_1(n_2 - S_2 - T_2) + R_2(n_1 - S_1 - T_1) - R_1R_2,$$

$$S = S_1n_2 + S_2n_1 + R_1T_2 + R_2T_1 - S_1S_2, \qquad (11)$$

$$T = T_1(n_2 - R_2 - S_2) + T_2(n_1 - R_1 - S_1) - T_1T_2,$$

$$Q' = Q_1'Q_2' + Q_1'(R_2' + S_2' + T_2') + Q_2'(R_1' + S_1' + T_1') + R_1'T_2' + R_2'T_1',$$

$$R' = R_1'R_2' + R_1'S_2' + R_2'S_1',$$

$$S' = S_1'S_2',$$

$$T' = T_1'T_2' + T_1'S_2' + T_2'S_1',$$

For G+H:

$$Q = Q_1Q_2 + Q_1(R_2 + S_2 + T_2) + Q_2(R_1 + S_1 + T_1) + R_1T_2 + R_2T_1,$$

$$R = R_1R_2 + R_1S_2 + R_2S_1 \qquad (12)$$

$$S = S_1 S_2,$$
$$T = T_1 T_2 + S_1 T_2 + S_2 T_1,$$
$$Q' = Q'_1 n_2 + Q'_2 n_1 + Q_1(R'_2 + S'_2 + T'_2) + Q_2(R'_1 + S'_1 + T'_1) - Q'_1 Q'_2$$
$$+ R_1 T'_2 + T_1 R'_2 + R'_1 (T_2 + T'_2) + T'_1 (R_2 + R'_2),$$
$$R' = R'_1 R'_2 + R'_1 (R_2 + S_2 + S'_2) + R'_2 (R_1 + S_1 + S'_1) + R_1 S'_2 + R_2 S'_1,$$
$$S' = S'_1 S'_2 + S_1 S'_2 + S_2 S'_1,$$
$$T' = T'_1 T'_2 + T'_1 (S_2 + S'_2 + T_2) + T'_2 (S_1 + S'_1 + T_1) + T_1 S'_2 + T_2 S'_1,$$

3.4.3 Connectivities

Theorem 1 The arc uv of GH belongs to a circuit in GH if and only if the arcs u and v belong to circuits in G and H respectively.

Proof The condition is necessary: Suppose
$$[x_1 y_1, \ldots, x_\ell y_h, \ldots, x_1 y_1]$$
is a circuit in GH. Then there exists a circuit
$$[x_1 \ldots x_\ell \ldots x_1]$$
in G and a circuit of the same length
$$[y_1 \ldots y_h \ldots y_1]$$
in H which may be elementary, simple or non-simple.

If $u = (x_1, x_2)$ is an arc in the circuit
$$G_\gamma = [x_1 x_2 \ldots x_\gamma x_1]$$
in G and $v = (y_1, y_2)$ is an arc in the circuit

$$C_\partial = [y_1 y_2 \ldots y_\partial y_1]$$

in H, where $\gamma \leq \partial$, then the path $[x_1y_1, x_2y_2, \ldots, x_\gamma y_\gamma]$ [1] can be extended to the path

$$[x_1y_1, x_2y_2, \ldots, x_\gamma y_\gamma, x_1 y_{\gamma+1}, \ldots, x_{\partial-\gamma} y_\partial, x_{\partial-\gamma+1} y_1]$$

in GH. This new path can again be extended up to $x_1 y_{2\gamma-\partial+1}$ and so on up to the vertex $x_\varepsilon y_\varepsilon$, where ε is the least common multiple of γ and ∂, whose successor will be $x_1 y_1$. Hence the arc uv is in a circuit which we note will be elementary if C_γ and C_∂ are elementary.

Corollary 2 The arc uv belongs to a cocircuit of GH if and only if either u belongs to a cocircuit of G or v belongs to a cocircuit of H.

Corollary 3 The direct product of a pair of strongly connected graphs has no cocircuits, all the weak components are also strong components and every vertex lies on a circuit.

A detailed study of the vertices of such a graph will show that they are all of types q or q' and, since the arcs entering into each vertex cannot belong to a cocircuit, it is easy to deduce the second part of the corollary.

Let G and H be arborescences. Then G has $n_1 = 1 + T_1 + Q_1$ vertices and H has $n_2 = 1 + T_2 + Q_2$ vertices because $R_1 = R_2 = 1$ and $S_1 = S_2 = 0$. Therefore according to (11) GH has

$$R = 1 + Q_1 + Q_2$$

$$T = T_1 T_2 + T_1 Q_2 + T_2 Q_1$$

$$S = T_1 + T_2$$

$$Q = Q_1 Q_2 .$$

The graph GH has no circuits (Theorem 1) and it has $n_1 n_2 - (T_1 + T_2)$ non-isolated vertices. Each connected component not reduced to a single vertex has at least one root and one terminal. There are therefore at most $S + \min(R,T)$ components. But $T \geq R$, with equality if and only if $T_1 = T_2 = 1$, and therefore the number of connected components in GH is $P \leq S+R$.

But there are Q+T vertices having exactly one arc entering into each of them because $\partial\bar{1}j \leq 1$, so there cannot be a vertex having an arc of a path from one root entering it and another arc of a path from another root also entering it. There are therefore at least R connected subgraphs in which all vertices except one are terminal extremities of only one arc and which

---

[1] We assume that $x_i = x_j$ if $i \equiv j$ (mod γ) and $y_k = y_h$ if $k \equiv h$ (mod ∂).

have one root. That is, there are at least R arborescent subgraphs of GH. Hence $P \geq S+R$.

It follows that $P=S+R$ or, alternatively, $P=n_1+n_2-1$.

The longest path in GH is a set of h consecutive arcs obtained from the product of h arcs of G and h arcs of H which is possible for $h=\min(h_1,h_2)$, where h_1 and h_2 are the heights of G and H.

Theorem 4 The direct product of two arborescences is a forest having $1+Q_1+Q_2$ disjoint arborescences, T_1+T_2 isolated vertices and containing a path of maximal length $\min(h_1,h_2)$.

Definition 10 A *looped latticoid* (resp. *looped arborescence*) is an l.q.s.c. graph without circuits (resp. without cycles) except for a loop at the vertex without proper ancestors which is a looped root.

Let $\hat{G}$ and $\hat{H}$ be looped latticoids having x_o and y_o as looped roots.

For every x in $\hat{G}$ and y in $\hat{H}$, $[x_o \ldots x_n]$ and $[y_o \ldots y]$ are paths and $[x_o y_o, \ldots, xy]$ is a path in GH. Because there exists a simple or non-simple path from x_o to x which is of the same length as a path (simple or non-simple) from y_o to y. Thus $\hat{G}\hat{H}$ is l.q.s.c. and it has a loop at the root. Further by Theorem 4, GH has no circuits except the loop $(x_o y_o, x_o y_o)$.

Hence $\hat{G}\hat{H}$ is a looped latticoid.

Furthermore, if the only cycles of $\hat{G}$ and $\hat{H}$ are the loops at x_o and y_o, then $\forall x, \forall y, \bar{\partial}_{xy}=1$. Thus no vertex is the terminal extremity of two arcs and consequently there cannot be two different paths from $x_o y_o$ to a vertex xy. There cannot be a path from xy to $x_o y_o$ because there would then be a circuit. If $]xy\ x_o y_o[$ was a chain, some vertex $\xi\eta$ of this chain would either be the terminal extremity of two arcs which is not allowed or the origin of two arcs. In the latter case xy and $x_o y_o$ or two other vertices, one from the chain $]xy,\ldots,\xi\eta[$ and the second from the chain $]\xi\eta,\ldots,x_o y_o[$, would be terminal extremities of two arcs which is again not allowed. Hence GH has no cycles except the loop at the root.

Thus we have established the following result:

Theorem 5 The direct product of a pair of looped latticoids (resp. looped arborescences) is a looped latticoid (resp. a looped arborescence).

Definition 11 A *looped Dedekindian* is a looped latticoid in which all elementary paths from the looped root to any given vertex have the same length. The *height* of a looped Dedekindian and of a looped arborescence is the length of the longest elementary path in the graph.

OPERATIONS ON GRAPHS

Let $\hat{G}$, $\hat{H}$ and $\hat{G}\hat{H}$ be looped arborescences of heights h_1, h_2 and h and assume that $h_1 > h_2$. Then we can construct an elementary path in GH of length h_1 by associating the arcs of a path of length h_1 in $\hat{G}$ with the arcs of a path in $\hat{H}$ formed either from h_1-h_2 loops (y_0,y_0) and the h_2 arcs of a longest path in $\hat{H}$ or from h_1-h_2+k loops (y_0,y_0) and h_2-k arcs of a shorter path in $\hat{H}$. The elementary path so constructed is maximal in GH. Hence $h = \max(h_1, h_2)$.

If $\hat{G}$ and $\hat{H}$ are looped Dedekindians of heights h_1 and h_2, then the above formula also gives the height of $\hat{G}\hat{H}$. Moreover any two elementary paths of $\hat{G}\hat{H}$ from x_0y_0 to x_1y_1 are similarly obtained by composing the arcs of a path in $\hat{G}$ with the arcs of a path in $\hat{H}$ one of the paths being necessarily elementary.

Property 11 The direct product of a pair of looped Dedekindians of heights h_1 and h_2 is a looped Dedekindian of height $\max(h_1, h_2)$.

Similarly the length of the path $[x_0y_0, xy]$ is equal to $\max($length of $[x_0x]$, length of $[y_0y])$.

Remarks

1. The direct product of a pair of arborescences being a partial graph of the direct product of a pair of looped arborescences (obtained by adjunction of a loop) has no cycles. However the product of a pair of trees can have a cycle.

For example if G is an arborescence of order 3 with root A and terminals B and C and if H is an u.q.s.c. tree of order 3 having a terminal 1 and roots 2 and 3, then GH contains the cycle $]A2,B1,A3,C1,A2[$ and has no other arcs (Figure 3.6)

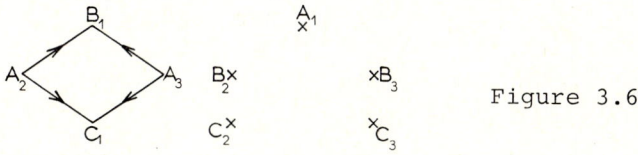

Figure 3.6

2. Let G and H be arborescences obtained from the looped arborescences $\hat{G}$ and $\hat{H}$ by suppression of loops. Now $\hat{G}$ has n_1 arcs, $\hat{H}$ has n_2 arcs and $\hat{G}\hat{H}$ therefore has $n_1 n_2$ arcs, that is, the $(n_1-1)(n_2-1)$ arcs of GH together with one further loop and n_1+n_2-2 proper arcs. These latter arcs make connections between the n_1+n_2-1 connected components of GH without creating any cycles; n_1-1 of them form a subgraph of $\hat{G}\hat{H}$ isomorphic to G and the other n_2-1 from a subgraph isomorphic to H. The endpoints of these subgraphs are the isolated vertices of GH and the common root of them is the looped root of $\hat{G}\hat{H}$. Figure 3.7 which illustrates this remark could be captioned "the tree hides the forest".

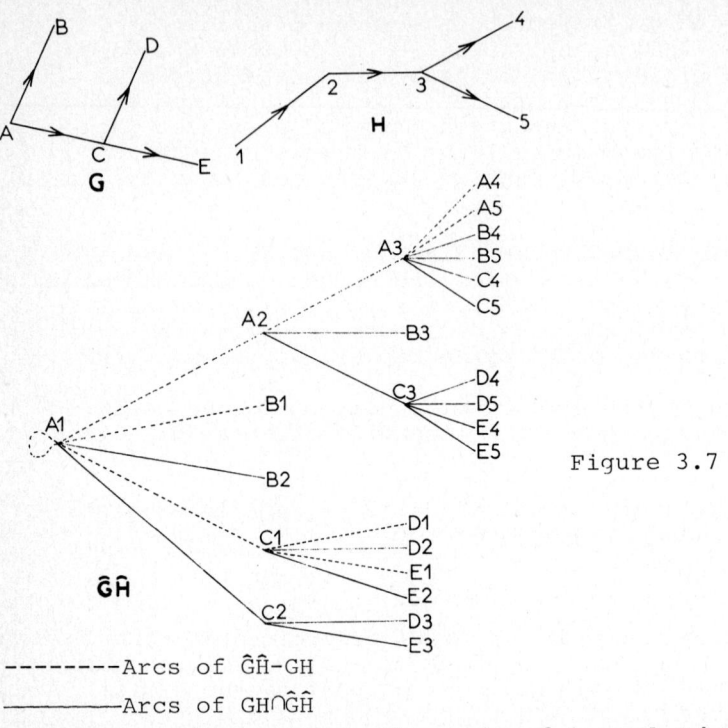

Figure 3.7

-------- Arcs of $\hat{GH} - GH$

———— Arcs of $GH \cap \hat{GH}$

Property 12 The product of a pair of looped arborescences is a looped arborescence which covers the product forest of the arborescences obtained by suppression of loops.

Let p_1, p_2, P be the numbers of weak components of G, H and G+H respectively. Then Aberth (1966) has proved the following result:

Theorem 6 The number of connected components of the sum of graphs G and H is equal to the product of the numbers of connected components of G and H; that is

$$P = p_1 p_2 .$$

Proof For $x \in X$ and $y \in Y$, GI_y has p_1 components, GI_y has $p_1 n_2$ components, $I_x H$ has $\ell_2 = (n_2 - p_2)$ cocycles linking two components of GI_y and therefore $GI_y \cup I_x H$ has $p_1 n_2 - \ell_2$ components.

It follows that $GI_y \cup I_x H$ has $p_1 n_2 - \ell_2 p_1$ components (and not $p_1 n_2 - \ell_2 n_1$, because the cocycles of $I_x H$ reduce the number of components of G+H by ℓ_2 per component of G).

Therefore $P = p_1 n_2 - (n_2 - p_2) p_1$, from which we deduce the theorem.

From this we deduce that the cocycle rank of G+H is

$$L = n_1 n_2 - p_1 p_2 = \ell_1 \ell_2 + p_1 \ell_2 + p_2 \ell_1$$

and that the cycle rank of G+H is

$$K = m_1n_2 + m_2n_1 - b_1b_2 - n_1n_2 + p_1p_2$$
that is
$$K = k_1n_2 + k_2n_1 - b_1b_2 + l_1l_2 \ .$$

Let $\widehat{G+H}$ be the transitive closure of G+H. Then we have:

Property 13 The graph G+H has a path from x_iy_j to x_ky_ℓ if and only if G has a path $[x_ix_k]$ and H has a path $[y_jy_\ell]$ one of them being possibly of length zero ($x_i=x_k$ or $y_j=y_\ell$ but not both).

Property 14 G+H has a circuit if and only if either G or H has a circuit.

Property 15 The transitive closure $\widehat{G+H}$ has the direct product GH as a partial graph.

For, if $[x_ix_k]$ is a path of length t_1 in G and $[y_jy_\ell]$ is a path of length t_2 in H, then there exists a path $[x_iy_j,\ldots,x_iy_\ell,\ldots,x_ky_\ell]$ of length t_1+t_2 in G+H. Conversely if $[x_iy_j,\ldots,x_ky_\ell]$ is a path of length t in G+H, then there exists an integer t_1 such that $[x_ix_k]$ is a path of length t_1 in G and $[y_jy_\ell]$ is a path of length $t-t_1$ in H and $0 \leq t_1 \leq t$. Property 13 follows from this. Note that if $t_1=0$, then $x_k=x_i$ and, if $t_1=t$, then $y_j=y_\ell$. (x_iy_j,x_ky_ℓ) is then an arc in $\widehat{G+H}$ if and only if (x_i,x_k) and (y_j,y_ℓ) are arcs in $\widehat{G}$ and $\widehat{H}$. The case of loops (x_i,x_i) in G corresponds to the case $t_1=0$. Property 14 is a corollary of Property 13 on taking loops into consideration.

If $x_i\Gamma x_k$ and $y_i\Delta y_\ell$, then G+H has paths $[x_iy_j,x_iy_\ell,x_ky_\ell]$ and $[x_iy_j,x_ky_j,x_ky_\ell]$ and hence, if (x_iy_j,x_ky_ℓ) is an arc in GH, it is also an arc of G+H. This completes the proof of Property 15.

Strong components of G+H

Let f_1 and f_2 be the numbers of strong components of G and H.

Let $C_1, C_2,\ldots,C_{f1}$ and $D_1, D_2,\ldots,D_{f2}$ be circuits of G and H each belonging to a different strong component.

Further let $C_i = [x_1x_\ell x_1]$ and $D_j = [y_1\ldots y_k\ldots y_1]$. Then for all $y \in D_j$, there exists a path $[x_1y,\ldots,x_\ell y,\ldots,x_1y]$ in GI_y and for all $x \in C_i$ there exists a path $[xy_1,\ldots,xy_k,\ldots,xy_1]$ in I_xH. Hence there exists a circuit passing through every xy with $x \in C_i$ and $y \in D_j$. There also exists a circuit passing through any xy in which x belongs to the same strong component as C_i and y belongs to the same strong component as D_j. But if $z \in D_k \neq D_j$, H has no circuit passing through y and z and hence $[xy,\ldots,xz,\ldots xy]$ is not a circuit of G+H.

Thus corresponding to any pair (C_i,D_j) there is exactly one strong class and finally $F=f_1f_2$. In particular we have

Property 16 The sum of a pair of strongly connected graphs is strongly connected.

The sum of a pair of latticoids is a graph having the following partition of the vertices:

$$R = 1, \quad T = T_1 T_2, \quad S = 0,$$

$$Q = Q_1 Q_2 + Q_1(1+T_2) + Q_2(1+T_1) + T_1 + T_2,$$

and $\quad P = p_1 p_2 = 1.$

Hence this sum is weakly connected, has no circuits and has only one root. It is therefore a latticoid (Theorem 2.3).

If G and H are Dedekindians containing respectively the paths

$$[x_0 x_1 x_\ell], \quad [x_0 x_1' x_\ell] \quad \text{with } \ell \text{ arcs}$$

and $\quad [y_0 y_1 y_m], \quad [y_0 y_1' y_m] \quad \text{with } m \text{ arcs},$

then
$$[x_0 y_0, x_0 y_1, \ldots, x_\ell y_m],$$
$$[x_0 y_0, x_0 y_1', \ldots, x_\ell y_m], \ldots$$

are paths with $m+\ell$ arcs and there are no paths of any other length from $x_0 y_0$ to $x_\ell y_m$.

The sum of two arborescences is therefore also a Dedekindian which has

$$n_1(n_2-1) + n_2(n_1-1) = 2n_1 n_2 - (n_1+n_2)$$

arcs among which are all the proper arcs of $\hat{\hat{GH}}-GH$.

However the sum of two arborescences is an arborescence if and only if
$$2n_1 n_2 - (n_1+n_2) = n_1 n_2 - 1,$$

that is, if $\quad n_1(n_2-1) = n_1 - 1$

and therefore if $\quad n_2 = 1 \text{ or } n_1 = 1.$

In these cases G+H is isomorphic to $G(n_2=1)$ or to $H(n_1=1)$. Otherwise G+H is a Dedekindian latticoid. Summarizing these results we have:

Property 17 The sum of two latticoids is a latticoid. The sum of two Dedekindians is a Dedekindian. The sum of two arborescences is an arborescence if and only if one of G or H is a degenerate arborescence.

3.4.4 <u>Valuations</u>

Let $S=\{s_\sigma\}$ and $A=\{a_\rho\}$ be families of valuations one on the vertices and the other on the arcs, that is families of mappings into the real numbers R such that

$$\forall \sigma, s_\sigma : X \to R \text{ and } \forall \rho, a_\rho : \Gamma \to R.$$

We will then write the graph G in the form of a quadruple (X, Γ, S, A). Usually S and A will contain only 1 or 2 elements. If $|S|=|A|=1$, we will denote the mappings from X into R and from Γ into R by s and a respectively and write (X, Γ, s, a) for the quadruple. Sometimes the mappings will take their values in some restriction of R possibly N (positive integers) or the interval [0,1] in which latter case we speak of probabilities. In some problems S or A may be empty leaving only one valuation. Sometimes also there may be a natural relationship between the value of a vertex and those of its incident arcs as for example in the case of the flow out of a vertex in a transportation network (cf. §2.4).

Let G = (X,Γ,S,A) and H = (Y,Δ,T,B) be graphs in which

$$i, k \in X, \quad u \in \Gamma \quad ; \quad j, \ell \in Y, \quad v \in \Delta$$

with valuations s(i), s(k), t(j), t(ℓ); a(u), b(v).

Further let K = (Z,Λ,M,N) be a graph produced from the preceding ones by one of the operations defined in this chapter. We will write $\mu \in M$ and $\tau \in N$.

When $K = I_X$, we put $\mu(i) = s(i)$

and

$$\tau(i,i) = \begin{cases} a((i,i)) & , \text{ if } (i,i) = u \in \Gamma \\ 1 & , \text{ if } (i,i) \notin \Gamma \end{cases}$$

When $K = G \cup H$, we put $\mu(i) = \max(s(i),t(i))$ with the convention that, for $i \in X-Y$, $t(i)=-\infty$ when working over R and $t(i)=0$ when working over N or [0,1].

We then put $\tau(u) = \max(a(u),b(u))$ with the same convention if u is an arc of G or H, that is,

$$b(u) = -\infty \text{ (resp. 0), if } u \in \Gamma-\Delta.$$

When $K = G \cap H$, we put $\mu(i) = \min(s(i),t(i))$ and μ, which is to be a mapping from $X \cap Y$ into some set (R, N or [0,1]), is then defined for every vertex i of $X \cap Y$. Similarly we put $\tau(u) = \min(a(u),b(u))$.

When K = GH, we put

$$\mu(i,j) = s(i)t(j) \quad \text{and} \quad \tau(u,v) = a(u)b(v).$$

We then find that for G+H = $GI_Y \cup I_XH$

$$\mu(i,j) = s(i)t(j).$$

If $(i,k) \in \Gamma$ is a proper arc or a loop and H has no loop at j, then $\tau(ij,kj) = a((i,k))$. If $(j,\ell) \in \Delta$ is a proper arc or a loop and G has no loop at i, then $\tau(ij,i\ell) = b((j,\ell))$.

However, if $(h,h) \in \Gamma$ and $(\ell,\ell) \in \Delta$ are loops, then the valuation for the loop of the sum G+H common to GI_Y and I_XH is given by
$$\tau(h\ell,h\ell) = \max(a((h,h)), b((\ell,\ell))).$$

Example Let G and H be the graphs shown in Figure 3.8 with the valuations on the arcs as indicated by the encircled numbers. We then obtain the valuations for GH and G+H.

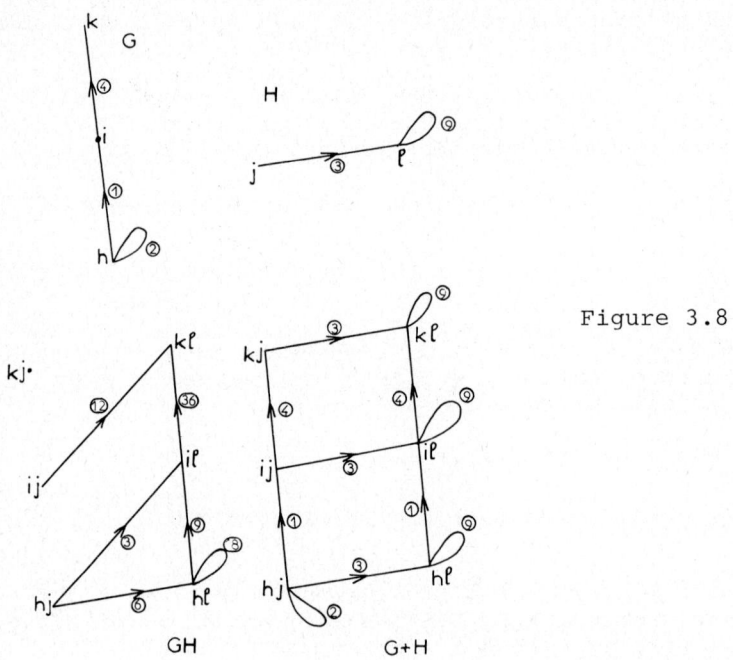

Figure 3.8

There is another kind of mapping which can be associated with the arcs of a graph with respect to the adjacency matrix. If we denote the entries of the adjacency matrix of G by 0 if $i\not\Gamma k$ and by γ if $i\Gamma k$ and similarly for H, 0 and ∂, then we are led to write the entries of the adjacency matrix as follows:

For I_X;
$$a_{ii} = \begin{cases} \gamma, & \text{if } i\Gamma i, \\ 1, & \text{if } i\not\Gamma i, \end{cases}$$

$$a_{ik} = 0, \text{ if } i \neq k.$$

For $G \cup H$;
$$a_{ik} = \begin{cases} 0, & \text{if } i\not\Gamma k \text{ and } i\not\Delta k, \\ \gamma, & \text{if } i\Gamma k \text{ and } i\not\Delta k, \\ \partial, & \text{if } i\not\Gamma k \text{ and } i\Delta k. \end{cases}$$

If we thought of $G \cup H$ as an s-graph, then one of the arcs (i,k) could have been given the valuation γ and the other ∂. On the other hand, if we admit that G is a graph and there is no ordering between γ and ∂, then we cannot choose between γ and ∂ and we will represent the entry a_{ik} by the ordered pair

$$a_{ik} = (\gamma, \partial), \text{ if } i\Gamma k \text{ and } i\Delta k.$$

Consequently in the sum G+H the loop $(h\ell,h\ell)$ obtained by composition of the loops (h,h) of G and (ℓ,ℓ) of H must be valuated by the ordered pair (γ,∂) (Figure 3.8). More generally we will take the ordered pairs $(\gamma,0)$, $(0,\partial)$, (γ,∂), where the projection γ or ∂ indicates the origin of the arc and $(0,0)$, when there is no arc.

Yet another type of mapping consists in the distribution of the vertices into eight classes $\{q,r,s,t,q',r',s',t'\}$ and deducing the classes of the vertices in the product graph. This was the concern of §3.4.2.

We may then try analogously to *characterise the arcs* by ordered pairs in which the first element will be one of q, r, q', r' and the second one of q, t, q', t' giving 16 classes of proper arcs. If the first and second element are the same, we obtain 4 classes of loops (q',q'), (r',r'), (s',s') and (t',t'). Note that the ordered pair (q',q') indicates that the origin and terminal vertex of the arc are arbitrary looped vertices. If they coincide, the arc is a loop otherwise it is a proper arc. We shall set up the composition tables of the arcs and find the algebraic properties analogous to those of the composition tables of the classes of vertices. In particular the composition table of the arcs of the product of graphs without loops has 4 rows and 4 columns (rq, rt, qq, qt) and has a neutral element (qq), an absorbing element (rt), idempotence, and the product (rq,qt) gives the absorbing element (rt). This table is isomorphic to the composition table of the vertices of the product of graphs.

3.5 Latticoid Operations

A *latticoid operation* is an operation on a pair of latticoids the product of which is also a latticoid.

Definition 12 Let G and H be latticoids with roots a and b respectively and in which the sets of terminal vertices are $E = \{\xi_1,\ldots,\xi_T\} \subseteq X$ and $D = \{\eta_1,\ldots,\eta_{T'}\} \subseteq Y$. Then the graph $G_\triangle H = GI_b \cup I_E H$ is called the *latticoid product* of G by H.

We note that GI_b has the root ab which is also a root of G+H, GI_b is isomorphic to G, and $I_E H$ is a graph consisting of $|E|=T$ subgraphs isomorphic to H having roots $\xi_1 b, \xi_2 b,\ldots,\xi_T b$. It follows that the union of GI_b and $I_E H$ is a latticoid graph having a subgraph isomorphic to G, the terminal vertices of which are roots of the latticoid subgraphs. $G_\triangle H$ is therefore l.q.s.c. and circuitless.

Property 18 The latticoid product $G_\triangle H$ is a latticoid.

Note that $G_\triangle H$ has $|ED|$ terminal vertices and is of height $h(G)+h(H)$.

The graph $H_\triangle G$ has the same number of terminal vertices and the same height as $G_\triangle H$ but is not isomorphic to it. However it is possible to set up a bijection between the paths $[ab\ldots\xi_i\eta_j]$ of $G_\triangle H$ and $[ba\ldots\eta_j\xi_i]$ of $H_\triangle G$.

It is straightforward to establish associativity.

Property 19 $(G_\Delta H)_\Delta K = G_\Delta(H_\Delta K)$.

Let $G = (X,\Gamma)$ and $H = (Y,\Delta)$ be latticoids and let $x_i y_j$ and $x_k y_\ell$ be vertices of $G_\Delta H$ and therefore also of $G+H$.

Then we shall show that the subgraph of $G+H$ generated by the vertices of $G_\Delta H$ coincides with $G_\Delta H$. It is sufficient to show that either $(x_i y_j, x_k y_\ell)$ is an arc in $G_\Delta H$ *and* in $G+H$ or it is an arc in neither of them.

There are four possible cases:

a. $x_i y_j \in X\{b\}$ and $x_k y_\ell \in X\{b\}$:

Then $y_j = y_\ell = b$ and $(x_i b, x_k b)$ is an arc of $G+H$ and of $G_\Delta H$ if and only if $x_i \Gamma x_k$.

b. $x_i y_j \in X\{b\}$ and $x_k y_\ell \in EY$:

Then $y_j = b$ and $x_k = \xi_k \in E$. Hence if $(x_i b, \xi_k y_\ell)$ is an arc of $G+H$, then either

(i) $x_i = \xi_k$, in which case $b \Delta y_\ell$ and $(\xi_k b, \xi_k y_\ell)$ is an arc of $G+H$ and $G_\Delta H$,

or (ii) $y_\ell = b$ and $x_i \Gamma \xi_k$ and this case reduces to case a.

c. $x_i y_i \in EY$ and $x_k y_\ell \in X\{b\}$:

Then $x_i = \xi_i \in E$, but ξ_i has no descendants so $(\xi_i y_j, x_k y_\ell)$ will be an arc of $G+H$ only if $x_k = \xi_i$. Further $y_\ell = b$. But if $y_j = b$, then $x_i = x_k$ and $y_j = y_\ell$ and there is no arc. Otherwise y_j is a descendant of b and there is no arc again.

Thus case c does not give rise to an arc in $G+H$ and *a fortiori* nor to one in $G_\Delta H$.

d. $x_i y_j \in EY$ and $x_k y_\ell \in EY$:

Then either (i) $x_i = \xi_i$ and $x_k = \xi_k \neq \xi_i$, in which case $x_i \neq x_k$ and $x_i \not\Gamma x_k$ and therefore $(x_i y_j, x_k y_\ell)$ is not an arc in $G+H$, or (ii) $x_i = \xi_i = x_k$ in which case, if $y_i \not\Delta y_\ell$, $(x_i y_j, x_k y_\ell)$ is not an arc in $G+H$ and if $y_i \Delta y_\ell$, it is an arc in both $G_\Delta H$ and $G+H$.

Thus we have shown that any arc of $G+H$ with both its extremities in $G_\Delta H$ is an arc in $G_\Delta H$.

Property 20   The latticoid product $G_\Delta H$ is a subgraph of the sum $G+H$.

Definition 13   The *restricted latticoid product* is the restriction of $G_\Delta H$ obtained by using a subset of the terminals of G, that is, if $E' \subseteq E$ and $E' \neq \emptyset$, then $G_\Diamond H = GI_b \cup I_{E'} H$.

In particular when $|E'|=1$, the restricted product $G_\Diamond H$

consists of the union of two subgraphs ismorphic respectively to G and H having one vertex in common, terminal in GI_b, root in $I_E \cdot H$ and arbitrary in $G_\Diamond H$.

If G is generated by a unique path $[a, \xi]$ and H is generated by a unique path $[b\eta]$, then $G_\Delta H$ is a path $[ab...\xi b...\xi \eta]$ ismorphic to the path obtained by concatenation of G and H. It follows that:

If G and H are Dedekindians (resp. arborescences), then $G_\Delta H$ and $G_\Diamond H$ are Dedekindians (resp. arborescences).

Definition 14 The *latticoid sum* $G \boxplus H$ of disjoint latticoids $G = (X, \Gamma)$ and $H = (Y, \Delta)$ with roots a and b respectively is the latticoid (Z, Λ) in which

$$Z = X \cup Y \cup \{\xi\}, \text{ where } \xi \notin X \cup Y$$

and

$$\Lambda = \Gamma \cup \Delta \cup \{(\xi, a), (\xi, b)\}.$$

The latticoid $G \boxplus H$ therefore has one vertex and two arcs not in G or H. If G and H are arborescences of orders n_1 and n_2, then $G \boxplus H$ is an arborescence with $n_1 + n_2$ arcs.

Note that, if G has T_1 terminals and H has T_2 terminals, then G+H and $G_\Delta H$ have $T_1 T_2$ terminals whereas $G \boxplus H$ has $T_1 + T_2$ terminals.

The graphs $G \boxplus H$ and $H \boxplus G$ will be considered to be distinct if and only if the arcs coming out of a given vertex are distinguished by some coding system.

If G, H and K are Dedekindians, then the height of $G \boxplus H$ is $1 + \max(h_1, h_2)$, the height of $(G \boxplus H) \boxplus K$ is $1 + \max(1 + \max(h_1, h_2), h_3)$, and the height of $G \boxplus (H \boxplus K)$ is $1 + \max(h_1, 1 + \max(h_2, h_3))$. It follows that the latticoid sum is not associative.

EXERCISES

1. Let G_1 be a graph and G_3 its transform by one of the transformations listed in §3.3. Show that in accordance with Definition 3.4 there exists a graph G_2 such that G_3 is a partial subgraph of $G_1 \cup G_2$. Determine G_2 for all the transformations shown in Figure 3.1.

2. Form the Prufer code of the arborescence shown in Figure 3.9

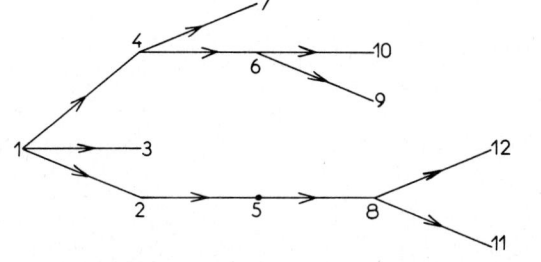

Figure 3.9

3. Decode the Prüfer codes of ten or eleven characters formed by the sequence 4 4 3 2 1 1 9 9 8 8 1. (The syllable 11 may be though of as being either of one or two characters.)

Interpret in one of the cases the order of enumeration of the vertices adopted while labelling the arborescence.

4. Determine the numbers Q, Q' for the product and the sum of two strongly connected graphs. Deduce Corollary 6, and assuming Theorem 6, establish Property 14 by a new method.

5. Let $A = (X,\Gamma)$ be an arborescence and $\Lambda = (Y,\Delta)$ a latticoid. Show that $A\Lambda$ is not a latticoid in general.

6. By definition all the vertices of a *bipartite graph* belong to the two classes r and t. Show that the composition laws of the classes of vertices of bipartite graphs for the product and the sum are not internal to $\{r,t\}$.

Let $G = (D_1, A_1, \Gamma)$ and $H = (D_2, A_2, \Delta)$ be bipartite graphs. Show that GH is the union of the bipartite graph $(D_1D_2, A_1A_2, \Gamma\Delta)$ and a trivial graph which is to be explicitly stated. Show that $G+H$ is not bipartite and has paths of length 2 at most. Partitioning the vertices of $G+H$ into

$$D_1D_2 \cup D_1A_2 \cup A_1D_2 \cup A_1A_2$$

we may construct an adjacency matrix of $G+H$ having 16 blocks $(|D_1D_2|X|D_1D_2|, |D_1D_2|X|D_1A_2|$ etc.). Show that 12 of these blocks contain only zeros and characterize the types of vertices in the other four blocks.

7. Let $G = (X,\Gamma)$, $H = (Y,\Delta)$, $K = (Z,\Omega)$ be isomorphic graphs such that

$$X = \{\bar{a},a\}, \quad \Gamma = \{(\bar{a},\bar{a}), (\bar{a},a), (a,a)\}$$

$$Y = \{\bar{b},b\}, \quad Z = \{\bar{r},R\}.$$

Form the double direct product GHK. This produces the "blood donors" graph; the vertices ($\overline{abr}$, $a\overline{br}$, $\overline{a}b\overline{r}$, $ab\overline{r}$, $\overline{ab}R$, $a\overline{b}R$, $\bar{a}bR$, abR) usually being known in medecine by the names (O^-, A^-, B^-, AB^-, O^+, A^+, B^+, AB^+). Generalize the results of Exercise 6 to graphs having vertices in the classes r' and t' for the product and double product and verify the formulas for enumerating the proper arcs and the loops in the "blood donors" graph.

8. Identify the classes of vertices in the adjacency matrix of a graph by considering the diagonal entries and the positions of the non-zero entries of the rows and of the columns. Use this to find an algorithm for determining the classes of vertices in the direct product and the sum of two graphs.

9. Let G and H be graphs with incidence matrices $\{a_{ik}\}$ and

$\{b_{i\ell}\}$. Show that the incidence matrix of the direct product GH is an $n_1 n_2 \times m_1 m_2$ matrix in which the rows and columns are those of the Kronecker product of the incidence matrices of G and H and in which the entries are

$$c_{ij,\,k\ell} = \begin{cases} 1, & \text{if } a_{ik} = b_{j\ell} = 1 \\ -1, & \text{if } a_{ik} = b_{j\ell} = -1 \end{cases}$$

or ($a_{ik}=1$ and $a_{hk}=0\,(h\neq i)$ and $b_{j\ell}=-1$)

or ($a_{ik}=-1$ and $b_{j\ell}=1$ and $b_{h\ell}=0\,(h\neq j)$)

and $c_{ij,k\ell}=0$, otherwise.

All this with the convention that the entry d_{ik} of the incidence matrix D takes the value -1 if the arc k enters into vertex i and +1 if the arc k comes out of vertex i or is a loop at vertex i.

10. Identify the classes of vertices in the incidence matrix of a graph. Use this to find an algorithm for determining the classes of vertices of the direct product of two graphs.

Chapter Four

GENERAL PROPERTIES OF QUESTIONNAIRES

4.1 _Preliminaries_

The theory of questionnaires is concerned with the elaboration of choice or decision models for the design of experiments or processes. Many different situations are covered by the theory. For example: making a series of clinical tests with a view to giving a medical diagnosis based on the combination of the various symptoms; asking a set of questions in order to elicit some specific information needed when taking an administrative decision; running through a technical check-list prior to take-off; choosing calls at the game of bridge; making provision for the various possible situations which may occur in a computer program. All these are related to the same concept, a questionnaire, formalized in the theory.

A complete system of N answers $e_1, e_2, \ldots, e_N$ is a set E of events of which one and only one can be the outcome of an experiment. The probabilities $p(e_j)$ of these events occurring satisfy the equation

$$\sum_{j=1}^{N} p(e_j) = 1 ,$$

and no event is excluded _a priori_ nor is any event certain, otherwise the system would have less than N answers.

If, when designing an experiment, we find a means of identifying the N elementary events which uses only one test, we say that we have asked a single question to distinguish the N events. If the experimental means do not allow us to identify more than a_M answers with any one test but do allow us to operate in several successive phases, then we say that we have a set of questions F at our disposal with bases $a \leq a_M$. Thus in order to identify N answers by means of questions with bases less than N, it is necessary to have recourse to a complex experiment which makes use of a sequence of questions. The answers to a question asked at a given phase of the experiment are called the outcomes of the question and may lead either to the finding of an elementary event or to the asking of a new question. Thus some answers may be identified with the help of a small number k_0 of questions and others with a larger number $h > k_0$. The questions asked are dependent on the results of previous questions and it is no longer necessary to pose new questions when all the events have been distinguished. The complex experiment is then complete and it is just such a plan of experiments which is called a questionnaire. The theory of questionnaires makes it possible to determine the identification criteria for each of the N answers of a complete system by asking on average the least number of questions possible or by taking certain compatibility relations into account.

For a given set of N answers and a set of questions with given bases, the theory tells us how to form the list of realizable experiments in the quickest or the most economical manner. We will also see how the theory suggests the very nature of the questions to be asked. The methods of giving practical specifications to the different outcomes of a question will be introduced through various applications. The outcomes will be the various possible answers to a general question of the type "what is the weight of this coin?" or the answers "yes" and "no" to a question of the type "is A greater than B?". The applications will illustrate some aspects of the scope of the theory which may be thought of as a branch of information theory in the case of discrete sources using graph theory and in particular latticoids and arborescences.

4.2 The Concept of a Questionnaire

4.2.1 Axioms and definitions

The definitions we give make it possible to undertake a general discussion of questionnaires. However deletion or modification of one of the axioms appearing in Definition 1 will later lead to extensions or restrictions of the theory.

Definition 1 A *questionnaire* is a valuated lower quasi-strongly connected graph,

$$Q = (X, \Gamma, P_\Gamma)$$

such that the set of vertices has the partition

$$X = E \cup F,$$

where E consists of terminal vertices, called *answers*
F consists of non-terminal vertices, called *questions*,
satisfying the axioms (q_1) to (q_5).

(q_1) (X,Γ) is a circuitless graph,

(q_2) (X,Γ) is a finite graph,

(q_3) no vertex of X is the origin of only one arc,

(q_4) there exists a mapping P_Γ from the set of arcs $\Gamma = \{(i,j); i \in F\}$

into the interval $]0,1[$,

$$P_\Gamma : (i,j) \mapsto p(i,j)$$

such that
$$\sum_{j \in \Gamma i} p(i,j) = \sum_{h \in \Gamma^{-1} i} p(h,i)$$

for all $i \in F$ with Γi and $\Gamma^{-1} i$ non-empty,

(q_5) this mapping is such that

$$\sum_{i \in E} \sum_{h \in \Gamma^{-1}i} p(h,i) = 1$$

We say that (X,Γ) is the *graph of the questionnaire* or that the triple (X,Γ,p_Γ) has for *support* the graph (X,Γ) satisfying the axioms $(q_1),(q_2)$ and (q_3). More simply the non-valuated graph (X,Γ) may also be called a questionnaire.

The partition of X requires both E and F to be non-empty (cf. Property 1.7). The lower quasi-strong connectivity and the axioms (q_1) and (q_2) may be interpreted as:

Property 1 The support of a questionnaire is a latticoid.

Definition 2 A *Dedekindian questionnaire* (resp. an *arborescent questionnaire*) is a questionnaire in which the support is a Dedekindian (resp. an arborescence).

A questionnaire with support subject to no particular condition will be known simply as a latticoid.

The general properties of latticoids and arborescences in particular, the combinatorial ones, will be applicable to questionnaires.

Definition 3 The root of the support of a questionnaire will be called *the root of the questionnaire* or *the first question* and the other questions will be called *internal vertices* or *internal questions*.

We will use α to denote the root and $\overline{F}=F-\{\alpha\}$ to denote the set of internal questions.

If $|\overline{F}|=0$, then, according to (q_3), the root has at least two successors being distinct terminals. Hence $|E| \geq 2$.

If $|\overline{F}|>0$ and if E contains only one element e_1, then either some question would have this terminal as its unique successor contrary to (q_3) or else every question would have a question as successor. But then the graph would either have a circuit or it would be infinite in contradiction of (q_1) and (q_2). Hence $|E|>1$.

Property 2 Any questionnaire has at least two answers. We will adopt the convention of writing $|E|=N$ and $|F|=M$.

Definition 4 The *basis* of a question is its out-degree: the successors of a question are called its *outcomes*.

Definition 5 A *homogeneous questionnaire* is a questionnaire in which every question has the same basis; a questionnaire which is not homogeneous is said to be *heterogeneous*.

The basis of a homogeneous questionnaire will be denoted by a, the basis of a question i of a heterogeneous questionnaire will be denoted by a_i. In a heterogeneous questionnaire we will agree to make a partition of F into (non-empty) classes F_k where k belongs to a family of indices K; a_k will then designate the basis of vertices of F_k and we will write $q_k=|F_k|$.

In a heterogeneous arborescent questionnaire there exists (cf. Property 2.10) a compatibility relation between the number of terminal vertices N and the bases of each class F_k:

$$N = \sum_{k \in K} a_k q_k - \sum_{k \in K} q_k + 1 \qquad (1)$$

A homogeneous arborescent questionnaire satisfies the strict compatibility relation:

$$N = (a-1) M + 1 \qquad (2)$$

The order of the support of an arborescent questionnaire is:

$$n = N + \sum_{k \in K} q_k ,$$

that is

$$n = \sum_{k \in K} a_k q_k + 1 .$$

If the questionnaire is latticoid, then $n \leq \sum_{k \in K} a_k q_k$, and and by Property 2.11, if it is homogeneous,

$$n \leq aM. \qquad (2')$$

The strict compatibility relation (2) implies that it is impossible to form a homogeneous arborescent questionnaire if a-1 is not a divisor of N-1. Thus when a-1 is a divisor of N-1, we will say that there exists a homogeneous arborescent questionnaire in the strict sense having N answers.

A questionnaire in which all questions except one have the basis a and in which the remaining question has a basis less than a will be said to be homogeneous in the *broad sense*.

Definition 6 An arborescent questionnaire in which all questions, except possibly one, have the same basis is called a *polychotomic questionnaire* or, if the basis is 2, 3 or 4, *dichotomic*, *trichotomic* or *tetrachotomic*.

If the strict compatability relation (2) holds, then an arborescent questionnaire has no (resp. at least two) questions with basis less than a and it is then polychotomic (resp. heterogeneous). It is only in the case when (2) does not hold that one and only one question of a polychotomic questionnaire is of basis less than that of the other questions.

Examples Let K_1, K_2, K_3 be the questionnaires shown in Figures 4.1, 4.2, 4.3 where the p(i,j) have been indicated except for a factor 10 (for clarity).

K_1 is Dedekindian and has a question which the terminal extremity of two arcs.

GENERAL PROPERTIES OF QUESTIONNAIRES

K_2 is homogeneous with basis 2, it has two questions which are the terminal extremities of two arcs and there are five distinct paths from the root α_2 to e_3, two of length 3 and three of length 4. (The dashed and dotted lines will be explained in §4.2.2.)

K_3 is trichotomic of height 2: it has an odd number of answers conforming with the strict compatability relation (2).

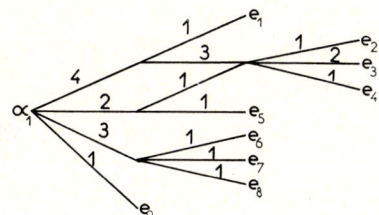

Figure 4.1

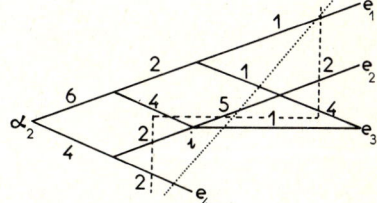

Figure 4.2

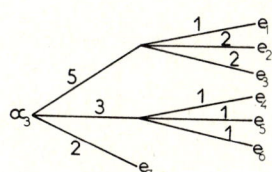

Figure 4.3

4.2.2   Cutsets of a Questionnaire

The valuation P_Γ of the arcs induces a valuation of the vertices P: X → $]0,1[$ defined by:

$$p(\alpha) = \sum_{j \in \Gamma\alpha} p(\alpha,j) \qquad \text{, if } i=\alpha$$

$$p(i) = \sum_{j \in \Gamma i} p(i,j) = \sum_{h \in \Gamma^{-1}i} p(h,i) \qquad \text{, if } i \in \bar{F} \qquad (3)$$

$$p(e) = \sum_{h \in \Gamma^{-1}e} p(h,e) \qquad \text{, if } e \in E.$$

Axiom q_5 then implies that

$$\sum_{e \in E} p(e) = 1 \qquad (4)$$

and we will write

$$p(E) = \sum_{e \in E} p(e) \;.$$

Axiom q_4, which expresses the conservation of the sum of valuations between α and e, leads to

$$p(\alpha) = 1 \qquad (5)$$

It is possible to associate a transportation network (X', Γ', Q') with a questionnaire in which the valuations have rational values by:

- the adjunction of a vertex z which will be the sink of the transportation network: $X' = X \cup \{z\}$ and $z \notin X$

- the adjunction of N arcs joining each of the answers of X to z.

$$\Gamma' = \Gamma \cup \bigcup_{e \in E} \{(e,z)\} \;.$$

The new graph (X', Γ') is quasi-strongly connected and loopless, it will become a circuitless transportation network if we define a valuation by non-negative integers (cf. Definition 2.10). Let k be the GCD of the $p(i,j)$ for $(i,j) \in \Gamma$.

We take as capacities ϕ'

$$\phi'(i,j) = kp(i,j) \quad \text{for} \quad i \in F, \; j \in E \cup \overline{F}$$

and
$$\phi'(e,z) = kp(e) \;.$$

Equation (2) of Definition 2.11 which expresses the conservation of flow then follows from (q_4). The transportation network has a flow in which each component ϕ'_u is equal to the capacity of the arc u.

The value of the flow is

$$\phi'_z = \sum_{e \in E} \phi'(e,z) = k$$

Ford-Fulkerson's theorem (Theorem 2.12) then shows that the value of the flow is equal to the minimum capacity of a cut-set because the flows are equal to the capacities.

Let us therefore define a cutset for Q, whether the valuations $p(i,j)$ are rational or not.

Definition 7 A *cutset of the questionnaire* Q is any set $\overline{\omega}_A$ consisting of the arcs entering into a subset $A \subset X$ such that $\alpha \in X - A$ and $E \subset A$.

If the valuations are rational, ω_A^- is also a cutset of the associated transportation network but we will employ the concept of a cutset of a questionnaire without any restriction on the valuation. By analogy with transportation networks the valuations of the arcs will now be called flows.

Let $B \subset \overline{F}$ be a subset of the vertices.

We will write $p(\omega_B^-)$, resp. $p(\omega_B^+)$, for the sum of the flows of the arcs entering into B, resp. coming out of B, that is

$$p(\omega_B^-) = \sum_{i \in B} \sum_{h \in \Gamma^{-1}i-B} p(h,i)$$

and

$$p(\omega_B^+) = \sum_{i \in B} \sum_{j \in \Gamma i-B} p(i,j) .$$

Since

$$\sum_{i \in B} \sum_{h \in \Gamma^{-1}B} p(h,i) = \sum_{i \in B} \sum_{j \in \Gamma i \cap B} p(i,j)$$

is the sum of the flows of the arcs having both their extremities in B, we obtain

$$p(\omega_B^+) - p(\omega_B^-) = \sum_{i \in B} (\sum_{j \in \Gamma i} p(i,j) - \sum_{h \in \Gamma^{-1}i} p(h,i)) ,$$

that is

$$p(\omega_B^+) = p(\omega_B^-) ,$$

which generalizes (q_4) to any set of internal questions.

We therefore define the *capacity of a cutset* $C = \omega_A^-$ of a questionnaire by

$$p(\omega_A^-) = \sum_{u \in \omega_A^-} p(u) . \tag{7}$$

Property 3 A cutset C of a questionnaire is minimal if and only if C is a cocircuit; its capacity is then $p(C) = 1$.

Proof Suppose there exists a minimal cutset C with capacity c which is not a cocircuit. Then there exists an arc $(a,b) \in \Gamma$ having origin $a \in A$ and extremity $b \in X-A$ with flow $p(a,b)$. Axiom (q_4) implies that a flow equal to $p(a,b)$ will be distributed on one or many arcs entering into A. There then exists a cutset C' having capacity $c-p(a,b)$ and C is not minimal. Any minimal cutset is therefore a cocircuit.

Now consider the cocircuit

$$\omega_{A_o}^- , \text{ where } A_o = E \cup \overline{F} .$$

According to (5) the sum of the valuations of the arcs of

$\omega_{\bar{A}_0}$, or the capacity of the cutset $\omega_{\bar{A}_0}$, is

$$p(\alpha) = 1.$$

We construct an arbitrary sequence of sets A_i such that

$$A_0 \supset A_1 \supset \ldots \supset A_j \supset A_{j+i} \ldots \supset E$$

by suppressing at least one vertex a_j between A_i and A_{i+1} ($\forall_j$) in such a way that $\omega_{\bar{A}_{j+1}}$ is a cocircuit ($\forall$ j).

We deduce from (6) and (7) that

$$p(\omega_{\bar{A}_{j+1}}) = p(\omega_{\bar{A}_j}) .$$

Any set A containing E and such that $\omega_{\bar{A}}$ is a cocircuit, can be generated in this way and is such that

$$p(\omega_{\bar{A}}) = p(\alpha) = 1 .$$

In particular

$$p(E) = 1 ;$$

Thus a cutset which is a cocircuit is minimal, and this completes the proof.

Example In Figure 4.2 we have represented the "trace" of two cutsets on the arcs of the questionnaire: by dotted lines a minimal cutset; by dashed lines a cutset which is not a cocircuit and has capacity 1.5.

With any minimal cutset $\omega_{\bar{A}}$ it is useful to associate the smallest set

$$A' \subset A \quad \text{for which} \quad \omega_{\bar{A}'} = \omega_{\bar{A}}$$

This set is the set of terminal extremities of the arcs $u \in \omega_{\bar{A}}$. For this set, such that

$$\sum_{i \in A'} p(i) = 1 ,$$

we will reserve the name *section* and we note that if Q is an arborescent questionnaire, then any answer is linked to α by a path passing through one and only one vertex of A', whereas any $q \in A'-E$ is the origin of paths leading to at least two answers. If the questionnaire is not arborescent, there exists at least one answer which is the terminal extremity of two paths issuing from α and consequently there is no mapping from E to any section.

Property 4 There exists a surjection from the set of answers to any section if and only if the questionnaire is arborescent.

4.2.3 Partitions of the Answers

Let Q be a questionnaire, E the set of its terminals and

$\mathfrak{P}(E)$ the set of all subsets of E.

With any vertex $i \in X$, we will associate the set

$$E(i) = \hat{\Gamma}i \cap E \qquad (8)$$

consisting of the terminal descendants of i; $E(i) \in \mathfrak{P}(E)$.

If $i = \alpha$, then $E(\alpha) = E$.

If $i = e \in E$, then $E(e) = \{e\}$.

Otherwise if $i \in F$, then it is clear that $|E(i)| \geq 2$.

Suppose Q is an arborescent questionnaire. Then, for any pair of questions i,k such that neither is an ancestor of the other, we have $E(i) \cap E(k) = \emptyset$.

If $\Gamma i = \{j; j = j_1, j_2, \ldots, j_{a_i}\}$

then $E(i) = \underset{j \in \Gamma i}{\cup} E(j)$, $i \neq k$ $E(i) \cap E(k) = \emptyset$

and $\forall i, E(i) \neq \emptyset$, that is,

the outcomes of question i induce a partition of the set $E(i)$ associated with i ($\forall i \in F$).

We may then consider that an arborescent questionnaire is a graph allowing a series of increasingly fine partitions of E for each path linking α to E, whatever the flow on the arcs of the graph.

Now suppose that Q is a Dedekindian questionnaire. Then corresponding to any "latticoid question" or to any answer z with two entering arcs there are at least two paths [xtz] and [xvz] from some vertex x such that

$$(x,t) \in \Gamma, (x,v) \in \Gamma, (t,v) \notin \hat{\Gamma}, (v,t) \notin \hat{\Gamma}.$$

It follows that $E(z)$ is contained in each of $E(x)$, $E(t)$ and $E(v)$ and hence $E(t) \cap E(v) \neq \emptyset$.

If Q is a latticoid but not a Dedekindian, then there exists a path [x...z] and a longer path [x...tz] such that

$$E(z) \subseteq E(t) \subseteq E(x) \quad \text{and} \quad E(t) \cap E(z) \neq \emptyset.$$

These results are summarized in Property 5 and illustrated in Figures 4.4, 4.5 and 4.6.

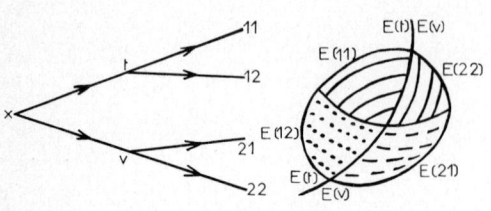

Figure 4.4 arborescent case

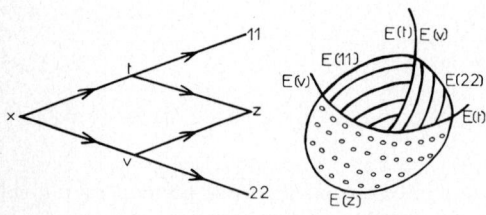

Figure 4.5 Dedekindian case

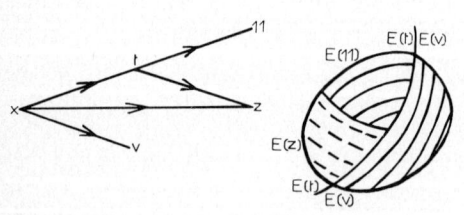

Figure 4.6 latticoid case
(non-Dedekindian)

Property 5 A questionnaire is arborescent if and only if the outcomes of any question i induce a partition of $E(i) = \hat{\Gamma}i \cap E$.

Any vertex $z \in X$ which is the terminal extremity of more than one arc then acts as a re-examination of the partition of answers by the question which is the origin of the two shortest paths entering into z.

If Q is latticoid, the subsets $E(j)$ defined for $j \in \Gamma_i$ belong to $\mathfrak{P}(E(i))$ but are not all disjoint.

If Q is arborescent, any question i performs a *disjunction*, that is a partition of $E(i)$, such that

$$\forall j,k \in \Gamma i, j \neq k \Rightarrow E(j) \cap E(k) \neq \emptyset.$$

In this case we can associate the pair $(i, \Gamma i)$ formed by a question i and its outcomes with a subset $E(i) \in \mathfrak{P}(E)$ and a partition A of $E(i)$ into a_i subsets, that is, with a pair $(E(i), A)$, where $A = \bigcup_{j \in \Gamma i} E(j)$ and $E(j) \in \mathfrak{P}(E(i))$ having the following properties:

(a) There is only one arc entering into any $i \in X - \{\alpha\}$ and thus

$$i \in E \cup \overline{F} \Rightarrow p(h,i) = p(i) . \qquad (9)$$

(b) If j and k are the outcomes of a dichotomic question i, then

$$E(i) = E(j) \cup E(k) \Rightarrow p(i) = p(j) + p(k) . \qquad (10)$$

(c) For any set $\{E(j)\}_{j \in J}$ of mutually disjoint elements of $\mathfrak{P}(E)$ we can define

$$p(\cup_{j \in J} E(j)) = \sum_{j \in J} p(E(j)) \qquad (11)$$

and if in particular $\{E(j); j \in J\} = \Gamma i$, then

$$p(E(i)) = \sum_{j \in \Gamma i} p(E(j)) .$$

Properties (b) and (c) mean that $p(E(i))$ is a measure on $\mathfrak{P}(E)$.

Remark If E is a set having cardinality equal to a continuum, we can define a measurable space to be a pair $(E, \mathcal{T})$, where $\mathcal{T}$ is a σ-ring of subsets of E. However since we will essentially be using discrete sets in this book, there is no point in introducing the concepts of σ-ring and σ-algebra to define measure and probability when it is sufficient to use the pair $(E, \mathfrak{P}(E))$. If E is denumerable, it will be understood that J is an infinite family. The properties (b) and (c) are called the additive and σ-additive properties respectively.

4.2.4 Probabilities in an arborescent questionnaire

We know that the pair $(E, \mathfrak{P}(E))$ constitutes a probabilizable space when E is denumerable or finite which is the case for the set of answers in an arborescent questionnaire Ω. In the language of probability theory the elements of E are called *elementary events or eventualities*[1] whereas the elements of $\mathfrak{P}(E)$ are called *events*. Thus the set $E(i)$ defined for any question of an arborescent questionnaire is an event. But the valuation $p(e)$ of any answer is a measure (according to (b) and (c)) such that $p(E)=1$ so that the triple $(E, \mathfrak{P}(E), P)$ constitutes a discrete probability space for which the probability is the mapping defined according to (q4), (q5) and (3) and such that:

$$P(E(i)) = \sum_{e \in E(i)} p(e) \qquad (12)$$

If $i = e \in E$, then the probability of an answer is none other than $p(e)$. If $i \in F$, then the probability associated with i will be the probability $p(E_1)$ of an event $E_1 \in \mathfrak{P}(E)$.

Property 6 We can build a discrete probability space $(E, \mathfrak{P}(E), P)$ on any arborescent questionnaire Ω in which E is the set of answers and P the valuation of the answers, that is the mapping $P: E \to\ ]0,1[$. The answers of the questionnaire are the eventualities of $(E, \mathfrak{P}(E), P)$.

Let $x_T \in F$ be a vertex all of whose successors are answers;

---

[1] Whence the name "eventualities" often given to the answers of an arborescent questionnaire.

there exists at least one such vertex because the support of Q is a finite graph (arborescence). The probability of the event $E(x_T)$ is

$$P(E(x_T)) = \sum_{e \in \Gamma x_T} p(e) .$$

Noting that for any $i \in F$, the formula

$$p(i) = \sum_{j \in \Gamma i} p(j) = \sum_{e \in \Gamma i \cap E} p(e) + \sum_{j \in \Gamma i - E} (\sum_{e \in \Gamma j \cap E} p(e) + \sum_{k \in \Gamma j - E} p(k))$$

can be developed in a finite number of terms, it is possible to define by proceeding, step by step, from E to α, the valuation of i by

$$p(i) = \sum_{e \in E(i)} p(e) . \qquad (13)$$

It is then possible to identify $p(i)$ and $p(E(i))$ for any $i \in X$; in particular we recover the formula $p(\alpha)=1$.

Property 7 The valuation $p(i)$ defined at any vertex of an arborescent questionnaire is the probability of the event $E(i)$ formed by the subset $\hat{\Gamma} i \cap E$.

If $i \in E$, the event is elementary: it is the probability of the answer i. Otherwise we will say that $p(i)$ is the probability of asking the question i. In terms of a transportation network we will say that $p(i)$ is the probability of *transiting through i*.

From (9) and (13), taking into account that $E(j) \subseteq E(i)$, if $i\Gamma j$, and putting $p(i,j) = P(E(i) \cap E(j))$ we deduce:

Property 8 $p(i,j)$ can be interpreted as the probability that a path linking α to E contains the arc (i,j).

We will also say that $p(i,j)$ is the *probability (of transit) of the arc* (i,j). It is the probability of the occurrence of both the events $E(i)$ and $E(j)$. The quotient $\dfrac{P(E(j))}{P(E(i))}$ will then be called the *conditional probability* of j *relative to i*, $p(j|i)$, or the *conditional probability of the arc* (i,j) *relative to i*, $p((i,j)|i)$ and we note that

$$p(i) = \sum_{j \in \Gamma i} p((i,j)|i) p(i)$$

and

$$p(i,j) = p((i,k)|i) p(i) .$$

Remark Although every vertex i defines an event $E(i) \in \mathfrak{P}(E)$, we must note that, in any given arborescent questionnaire, not every event $A \in \mathfrak{P}(E)$ can be associated with a question.

The number of questions $\sum_{k \in K} q_k$ given by (1) is actually less than the number of elements of $\mathfrak{P}(E)$ containing more than one answer which is $2^N - (N+1)$. In fact there is equality only if $N=2$ and $|F|=1$.

It is precisely to make possible a choice of questions considered as elements of $\mathfrak{P}(E)$ that the theory of questionnaires has been introduced.

4.3 Routing

Definition 8 The *routing of a latticoid* Λ is the set of distinct paths from the root to the terminal vertices of Λ. The *routing of a questionnaire* Q is the routing of its support Λ.

We will denote the set of distinct paths from the root to the terminal e by $\mathcal{C}_e$ and the routing will be denoted by
$$\mathcal{C} = \bigcup_{e \in E} \mathcal{C}_e.$$

4.3.1 The Arborescence of paths in a latticoid questionnaire

Definition 9 The *arborescence of paths in a latticoid* $\Lambda = (X, \Gamma)$ with root α is the latticoid $A = (T(X), T(\Gamma))$ with root ω in which, corresponding to any path $\mu_s = [\alpha i]$ of Λ, there is exactly one vertex i^s in A and exactly one path $[\omega i^s]$ called the *image of* μ_s *under the transformation* T.

This transformation T which is a bijection from the routing of Λ to the routing of A, but which is not in general a mapping from Λ to A, facilitates the discussion of the routing of Λ.

The set $T(i)$ of vertices i^s defined by T from the set of paths from α to $i \in X$ is called the *image of* i. $|T(i)|=1$ if and only if there is only one path $[\alpha i]$ in Λ.

Let a be the outdegree of i. Then any path $[\alpha i]$ can be extended in a distinct ways to a path having one arc more. Consequently for all μ_s, i^s has the same outdegree a as does i. We will first construct ω as the unique vertex of $T(\alpha)$, then the images of the successors of ω, and so on.

If there are c paths $\mu_1, \mu_2, \ldots, \mu_s, \ldots, \mu_c$ from α to i in Λ, then $T(i)$ will consist of c vertices $i^1, i^2, \ldots, i^s, \ldots, i^c$ such that there are unique paths from ω to each of i^s.

If $T(i)$ has only one element i^T, then the images in A of the arcs $(i, j_1), \ldots, (i, j_a)$ of Λ will be a arcs coming out of i^T. Otherwise the images of these arcs will consist of a arcs coming out of each element of i^T, that is ca arcs in all.

This method of constructing the graph of paths coming out of the root of a latticoid is exhaustive. Since each vertex of A is situated along a unique path from ω, A is indeed an arborescence.

Note that the transformation

$$T^* : A \to \Lambda$$

is onto Λ (surjective), mapping vertices and arcs of A onto those of Λ, in such a way that the transformation from the paths of A to the paths of Λ is by definition a bijection. We write $T^*(i')=i$ the vertex of Λ defined by T^* for $i^s \in A$.

We shall distinguish three cases:

(a) If Λ is arborescent, then the arborescence A of paths of Λ is isomorphic to Λ and T is a bijection from A onto Λ:

$$|T(i)| = 1 \text{ for all } i.$$

Thus in this case T^* is the inverse mapping of T.

(b) If Λ is Dedekindian, there will be two kinds of question.

- If a question i is linked to α by a unique path, then any arc (vertex) or $[\alpha i]$ has an image in A consisting of only one arc (vertex) as in case (a).

- Suppose the question $\xi \in X$ with rank $r \geq 2$ is linked to α by c paths, $c \geq 2$, of the same length r:
  $[\alpha x_i u_i \ldots v_i \ldots w_i \xi]$, where the v_i (i: 1,...,c) have the same rank and where $\exists i,j$ such that $v_i \neq v_j$. Then the image of ξ in A will consist of c vertices $\xi^1, \xi^2, \ldots, \xi^c$; the image of the a arcs coming out of ξ will consist of c.a arcs coming out from the ξ^i (i:=1,...,c) and entering into the c.a distinct vertices which will be of rank r+1.

(c) If Λ is a latticoid which is not Dedekindian then in the same way any vertex ξ linked to α by c paths of the same length will have as image c vertices with the same rank. But if a vertex ξ is linked to α by at least two paths of different lengths, such as $[\alpha \ldots u \ldots \xi]$ of length r and $[\alpha \ldots v \ldots \xi]$ of length r+s, then the image of ξ will consist of at least two vertices having respective ranks r and r+s.

We will call a *maximal sub-arborescence of* A any sub-arborescence all of whose terminals are terminals of A.

If Λ is not an arborescence, then the distinct vertices belonging to the image of a non-terminal vertex ξ, with more than one entering arc are the origins of maximal sub-arborescences which are isomorphic and which belong to the image of the sub-latticoid with root ξ.

Let A be an arborescence in which there exist two distinct isomorphic maximal sub-arborescences with roots η and ζ. If η and ζ have a common predecessor $T(x)$ and if there is no other isomorphism of maximal sub-arborescences of A, then the maximal sub-arborescences issuing from η and ζ cannot constitute the image of one and the same sub-graph of a latticoid (otherwise η and ζ would constitute the image of a successor of x linked to x by two arcs and Λ would be a 2-graph). If

η and ζ have no common predecessor, then η and ζ can be considered as belonging to the image of a vertex of a latticoid questionnaire having at least one non-terminal vertex with indegree greater than 1.

Property 9 A necessary and sufficient condition for an arborescence A to be the arborescence of the paths of a non-arborescent latticoid admitting at least one non-terminal vertex with indegree strictly greater than 1 is that A contains at least two isomorphic maximal sub-arborescences the roots of which do not have the same predecessor.

Let Λ be a latticoid having N terminal vertices, $|e|=V$ paths coming out of α and entering into E, and d arcs entering into E.

It is possible to determine d and V by summation of certain parameters related to any question $i \in F$. Let a_i, b_i, be the outdegrees and indegrees of i and c_i the number of paths entering into i.

We put
$$d(i) = a_i - b_i, V(i) = (a_i - 1)c_i, \text{ if } i \in \overline{F}$$
and
$$d(i) = V(i) = a_i, \text{ if } i = \alpha.$$

If there is no path of length 2 coming out of i, then contraction of the set $A = \{i\} \cup \Gamma i$ creates a unique terminal and

reduces d by $d(i)$,   if $a_i > b_i$

preserves d,   if $a_i = b_i$

increases d by $|d(i)|$, if $a_i > b_i$.

This contraction reduces the number of paths by $(a_i-1)c_i=V(i)$, if $i \in \overline{F}$ and by a_d, if $\overline{F}$ is empty.

But such a contraction can be effected step by step to any vertex of Λ by starting from a question all of whose outcomes are answers. It follows that

$$d = \sum_{i \in F} d(i) \quad \text{and} \quad V = \sum_{i \in F} V(i),$$

That is,
$$d = \sum_{i \in F} (a_i - b_i)$$
and
$$V = a_d + \sum_{i \in \overline{F}} (a_i - 1)c_i .$$

But d and N are related in any case by
$$d = N + \sum_{e \in E} (b_e - 1) .$$

Hence we deduce that

$$V = N + \sum_{e \in E}(b_e - 1) + \sum_{e \in \overline{F}}\{(a_i - 1)c_i - (a_i - b_i)\}$$

and we have:

Property 10 The number of paths $|\mathcal{C}| = V$, and the number of terminals N of a latticoid Λ are connected by the formula

$$V = N + \sum_{i \in E \cup \overline{F}}(b_i - 1) + \sum_{i \in \overline{F}}(a_i - 1)(c_i - 1)$$

and $V = N$ if and only if Λ is arborescent.

If Λ is not arborescent, $V > N$ because

$$b_e \geq 1, \quad \text{for all } e \in E$$
$$c_i \geq b_i \geq 1, \ a_i > 1, \quad \text{for all } i \in \overline{F}$$

and one of the inequalities for b_e or b_i is strict for at least one vertex of E or of $\overline{F}$.

On the other hand let m_Λ and m_A be the number of arcs of Λ and A. Any question $i \in \overline{F}$ which is the extremity of c_i paths issuing from α and the origin of a_i arcs, possesses an image in A consisting of c_i vertices from which there are $c_i \cdot a_i$ arcs coming out. Hence we have:

Property 11 The number of arcs of A is

$$m_A = m_\Lambda + \sum_{i \in \overline{F}}(c_i - 1)a_i ,$$

$m_A = m_\Lambda$ if and only if there is only one arc entering into each internal question of Λ.

4.3.2 Compatible Arborescent Questionnaires

We have just seen that the arborescence A of the paths of a latticoid questionnaire is such that its routing is in one-to-one correspondence with the routing of Λ.

However we must also valuate A by a flow in such a way as to extend the discussion of the routing to the weighted paths of the latticoid. Thus we are led to construct an arborescent questionnaire with support A.

Definition 10 An *arborescent questionnaire compatible with the questionnaire Q* is any arborescent questionnaire C satisfying the following three conditions:

1. The support is the arborescence A of the paths of Λ.

2. The image of a vertex i of Λ, having b_i arcs

$$(h_t, i) , \ t = 1, 2, \ldots, b_i$$

GENERAL PROPERTIES OF QUESTIONNAIRES

and c_i paths

$$\mu_s = [\alpha \ldots h_s i] \quad , \quad s \in S = \{1, 2, \ldots, c_i\}$$

entering into i and a_i arcs

$$(i, j_u) \quad , \quad u \in U = \{1, 2, \ldots, a_i\}$$

coming out of i, consists of c_i vertices i^s with outdegrees a_i. We will write simply a, b and c when there is no risk of ambiguity. (Figure 4.7)

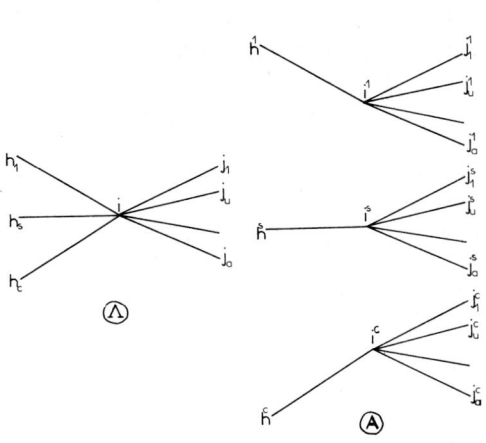

Figure 4.7

3. The valuations $p_C(i^s, j_u^s)$ of the arcs (i^s, j_u^s) in the image of the arc (i, j_u) are such that

$$\sum_{s \in S} p_C(i^s, j_u^s) = p(i, j_u) \quad (14)$$

$$\sum_{u \in U} p_C(i^s, j_u^s) = p(h_s, i) \quad (15)$$

Remark c_i distinct paths enter into i by b_i different arcs and $b_i \leq c_i$. From the point of view of routing, a vertex h predecessor of i and situated on two paths

$$[\alpha \ldots x_1 \ldots h] \text{ and } [\alpha \ldots x_2 \ldots h]$$

will be represented by two distinct vertices h_s and h_t situated on two paths $[\alpha i]$. Since the different vertices $h^s, h^t \ldots$ belonging to the image of h and $i^1, i^2, \ldots i^{c_i}$ forming the image of i are determined successively from α, it is legitimate to consider h_s and h_t as distinct predecessors of i while constructing $T(i)$. That is why we wrote (15) in the way we did; this amounts to reasoning locally as if i had c_i predecessors instead of b_i which reflects the privileged role played by the paths coming out of α; the flows of h_s and h_t are such that $p_C(h_s) + p_C(h_E) \leq p(h)$, and there is equality if h is transformed into only two vertices.

The first condition of compatibility (14) requires the image of any arc (i,j) coming out of i to be a set of arcs in which the sum of the flows is equal to that of (i, j_u).

The second condition of compatibility (15) requires that in the sub-arborescence with root i^s, the sum of the flows of the arcs coming out of i^s is equal to $p(h_s, i)$.

Since every vertex i^s of A is the terminal extremity of a unique arc (h^s, i^s),

$$p_c(i^s) = p_c(h^s, i^s) = p(h_s, i)$$

and it follows that

$$\sum_{s \in S} p_c(i^s) = p(i) . \qquad (16)$$

Summing the two sides of (14) over u and the two sides of (15) over s we then have

$$\sum_{u \in U} \sum_{s \in S} p_c(i, j_u) = \sum_{s \in S} \sum_{u \in U} p_c(i^s, j_u^s) = p(i) . \qquad (17)$$

The construction of an arborescent questionnaire C compatible with Q requires the determination of $c_i a_i$ flows coming out of the image of every $i \in \overline{F}$ where there are c_i paths entering from α. If $c_i = 1$, there are no indeterminates but if $c_i > 1$, we must determine $c_i a_i$ flows with the aid of the a_i flows of the arcs coming out of i, that is a_i relations of the type (14). Moreover we have c_i relations of type (15) expressing the valuation of the vertices of the image of i, but, because of (16), there are only c_i-1 independent relations of type (15).

There are therefore

$$(c_i-1) a_i - (c_i-1) = (c_i-1)(a_i-1)$$

indeterminates which we must eliminate in order to construct any new question of a compatible questionnaire C.

Thus to determine one of the questionnaires C, it is necessary to introduce a set of parameters related to the questions of C and of which

$$\sum_{i \in \overline{F}} (c_i-1)(a_i-1)$$

will be arbitrary. There will be finally a choice to be made which will correspond to different semantic interpretations of the same latticoid questionnaire. Let us introduce the strictly positive parameters $\lambda_u^s(i)$ by writing:

$$p_c(i^s, j_u^s) = \lambda_u^s(i) p(i, j_u) \frac{p_c(i^s)}{p(i)} , \quad (\lambda_u^s(i) > 0) . \qquad (18)$$

We can sum the two sides of (18) over s or over u which gives, taking into account (14) and (15)

$$\sum_{s \in S} p_c(i^s, j_u^s) = p(i, j_u) \sum_{s \in S} \lambda_u^s(i) \frac{p_c(i^s)}{p(i)}$$

or

$$\sum_{s \in S} \lambda_u^s(i) p_c(i^s) = p(i) \qquad (19)$$

and

$$\sum_{u \in U} p_c(i^s, j_u^s) = p_c(i^s) \sum_{u \in U} \lambda_u^s(i) \frac{p(i, j_u)}{p(i)}$$

or
$$\sum_{u \in U} \lambda_u^S(i) p(i,j_u) = p(i) . \qquad (20)$$

Finally we have a system of a + c equations with a.c unknowns because there are c arcs entering into i and a arcs coming out of i, thus a.c parameters $\lambda_u^S(i)$ and only a equations of type (19), c equations of type (20) — which we will write in the homogeneous form:

$$\begin{aligned} \sum_{s \in S} p_c(i^s)(\lambda_u^S(i)-1) &= 0 \\ \sum_{u \in U} p(i,j_u)(\lambda_u^S(i)-1) &= 0 \end{aligned} \qquad (21)$$

and such that
$$\lambda_u^S(i) \in \ ]\, 0, \min\bigl(\tfrac{p(i)}{p_c(i^s)}, \tfrac{p(i)}{p(i,j_u)}\bigr)\, [$$

Thus for any question i of the latticoid questionnaire Λ we have a.c-(a+c)+1 = (a-1)(c-1) arbitrary parameters $\lambda_u^S(i)$ facilitating the realization of a questionnaire compatible with Λ.

We see that

(α) If a = 1 or c = 1, $\lambda_u^S(i)$ takes exclusively the value 1: in this case nothing is arbitrary. This corresponds in particular to the case when Λ is an arborescence (c = 1 for all i) or when there is only one arc entering into i. A general latticoid can have a = 1 but axiom (q_3) for questionnaires requires a > 1.

(β) If a = c = 2, then it is possible to choose arbitrarily one and only one parameter, for example $\lambda_1^1(i)$ and to deduce from this all the $p_c(i^s, j_u^s)$

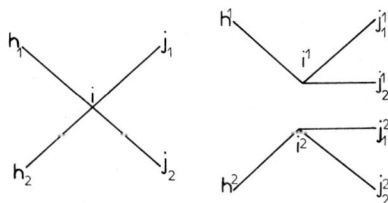

Figure 4.8

$$p_c(i^1, j_1^1) = \lambda_1^1(i) p(i,j_1) \frac{p_c(i^1)}{p(h_1,i)+p(h_2,i)}$$

$$p_c(i^1, j_2^1) = p(h_1,i) - p_c(i^1, j_1^1)$$

$$p_c(i^2, j_1^2) = p(i,j_1) - p_c(i^1, j_1^1)$$

$$p_c(i^2,j_2^2) = p(h_2,i) - p_c(i^2,j_1^2) = p(i,j_2) - p_c(i^1,j_2^1).$$

(γ) Lastly the system of equations (21) naturally has the trivial solution

$$\lambda_u^s(i) = 1 \quad (\forall i,s,u).$$

This solution corresponds to the particular case when the flows of the arcs coming out of i^s are proportional to those of the arcs coming in (Fig. 4.9):

$$p_c(i^s,j_u^s) = \frac{p(i,j_u)}{p(i)} p_c(i^s). \tag{22}$$

Definition 11 The compatible arborescent questionnaire obtained by the trivial solution of (21) for every question i of Λ will be referred to as the questionnaire B.

The questionnaire B satisfies what we call:

The Hypothesis of Proportionality of Flow

Whatever the vertices i^s and i^t belonging to the image of i in a compatible questionnaire and whatever the successor j_u of i, they satisfy the equation

$$\frac{p_c(i^s,j_u^s)}{p_c(i^s)} = \frac{p_c(i^t,j_u^t)}{p_c(i^t)} \tag{23}$$

Conversely, if a compatible arborescent questionnaire satisfies (23), then we can write

$$\frac{p_c(i^s,j_u^s)}{p_c(i^s)} = \frac{\sum_s p_c(i^s,j_u^s)}{\sum_s p_c(i^s)} = \frac{p(i,j_u)}{p(i)}$$

which implies $\lambda_u^s(i) = 1$ ($\forall i,s,u$).

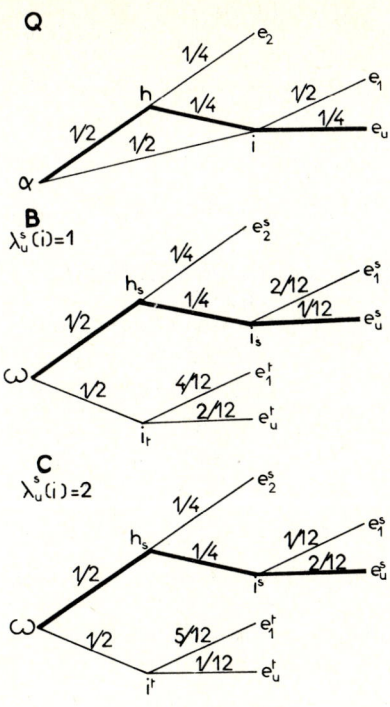

Q: Latticoid questionnaire
B,C: arborescent questionnaires compatible with Q $\lambda_u^s(i) \in \,]0,3[$

Property 12 The compatible questionnaire B for which $\lambda_u^s(i)=1$ ($\forall \, i,s,u$) is the only compatible questionnaire satisfying the hypothesis of proportionality of flow (23) at every question i.

Figure 4.9

4.4 Probabilities in a Latticoid Questionnaire

4.4.1 Probabilities of the vertices

The probabilistic discussion made in the arborescent case in §4.2.4 will be extended to the latticoid case by means of the arborescent questionnaires B and C.

Let $Q = (X,\Gamma,P_\Gamma)$ be a latticoid questionnaire having N terminal vertices into which there are V distinct paths entering from the root α. Any compatible questionnaire C with support equal to the arborescence $A=(Y,\Delta)$ of the paths of (X,Γ) has V terminal vertices consisting of a set E_C such that there exists a surjection $E_C \to E$ preserving the number of elementary paths linking the root to the terminals. By properties 6, 7 and 8, it is possible to construct a discrete probability space $(E_C, \mathfrak{P}(E_C), P_C)$ on C, where P_C is the valuation of the V answers of $C: v \to P_C(v)$.

The valuation $P_C(i^s)$ of a question i^s of C is equal to the probability $P_C(E_C(i^s))$ of the event formed by the subset $\Delta_i^s \cap E_C$ and the valuation of the arc $(i^s, j_u^s) \in \Delta$ is the probability that a path linking ω to E_C contains the arc (i^s, j_u^s).

Let T(e) be the image of $e \in E$ in E_C. Then the valuation

$$p(e) = \sum_{v \in T(e)} p_e(v) \quad (\forall e \in E)$$

is the probability of the event $\bigcup_{v \in T(e)} E_c(v) \subset \mathfrak{P}(E_c)$ and is therefore

$$p(e) = P_c(\bigcup_{v \in T(e)} E_c(v)). \tag{24}$$

We remark that this event of $(E_c, \mathfrak{P}(E_c), P_c)$ is not in general associated with a vertex of C but with an answer of Q. The eventualities of $(E_c, \mathfrak{P}(E_i), P_c)$ are the events $E_c(v)$ where $\in E_c$, whereas the answers of Q form the eventualities of this space only if they are the extremities of only one path coming out of α.

If Q is arborescent, the answers $e \in E$ constitute a complete system of eventualities because

$$0 < p(e) < 1 \quad \text{and} \quad \sum_{ve \in E} p(e) = 1$$

Otherwise it is the answers $V \in E_c$ which form a complete system of eventualities whereas the answers $e \in E$ constitute a complete system of events (exclusive and of which the sum of probabilities, all non-zero, is 1).

We can also consider that $p(e) = P(E(e))$ if $e \in E$, by using the probability space $(E, \mathfrak{P}(E), P)$. But we cannot in general write an analogous equation for a question.

If every path linking α to the descendants of i passes through i, then

$$p(i) = \sum_{e \in E(i)} p(e) .$$

But if there exists an arc $u = (x,y)$ with flow $p(u)$ entering into a descendant of i and such that $x \notin \Gamma i$, then

$$\sum_{e \in E(i)} p(e) = p(i) + p(u)$$

(see for example figure 4.2. There the question i has two entering arcs of flows 0.2 and 0.4 and is such that $\sum_{e \in E(i)} p(e) = 0.7$.)

For a latticoid questionnaire the inequality

$$p(i) \leq \sum_{e \in E(i)} p(e) \tag{25}$$

prevents us from associating the valuation $p(i)$ with the probability $P(E(i))$ because this valuation is not a measure on $(E, \mathfrak{P}(E))$.

We now study the union of the events associated with the

vertices $i^s \in T(i)$ for $i \in F$. According to the properties of probability spaces, we have

$$P_C(\bigcup_{i^s \in T(i)} E_C(i^s)) = \sum_{i^s \in T(i)} \sum_{v \in E_C(i^s)} P_C(v).$$

But the first summation has already been carried out in (13) for arborescent questionnaires, such as C, that is:

$$P_C(\bigcup_{i^s \in T(i)} E_C(i^s)) = \sum_{i^s \in T(i)} p_C(i^s)$$

and finally by (16):

$$p(i) = P_C(\bigcup_{i^s \in T(i)} E_C(i^s)) \quad (\forall i \in F) \qquad (26)$$

and in particular $p(\alpha) = P_C(E_C) = 1$.

The formulae (24) and (26) lead to:

Property 13 The valuation $p(i)$ defined at every vertex of a latticoid questionnaire Q is the probability of the event $E_C(i) = \bigcup_{i^s \in T(i)} E_C(i^s)$ defined in the probability space $(E_C, \mathfrak{P}(E_C), P_C)$ built on a compatible arborescent questionnaire C.

The importance of this property is connected on the one hand with the probabilistic interpretation of $p(i)$ in a latticoid questionnaire and on the other with the indifference to the choice of values of the arbitrary parameters $\lambda_u^s(i)$; this independence of $p(i)$ as regards C really results from the compatibility relations (14) and (15) which have made possible the construction of the probability space having E_C for eventualities: the choice of the $\lambda_u^s(i)$ modifies the mapping P_C from E_C onto $]0,1[$ and influences the union $\bigcup_{i^s \in T(i)} E_C(i^s)$ but not the probability of this event.

We can now speak of the probabilities of the vertices of a latticoid questionnaire by noting that the definition will concern $(E_C, \mathfrak{P}(E_C))$ and not $(E, \mathfrak{P}(E))$, these two probabilizable spaces coincide or at least are isomorphic if and only if Q is arborescent.

Definition 12 The *probability of a vertex* x (question or answer) of a questionnaire is its valuation $p(x)$.

4.4.2 Probabilities of the arcs and conditioning

The probabilistic interpretations of $p_C(v)$, $p_C(i^s)$ and $p_C(i^s, j_u^s)$ on the one hand and of $p(e)$ and $p(i)$ on the other will enable us to determine for any arc the unconditional or conditional probabilities relative to the initial extremity or the path entering into this initial extremity.

By (26) we can write

$$p(j) = P_c(\bigcup_{i^s \in {'T'}(j)} E_c(j^s) \bigcup_{j^t \in T(j)-T'(j)} E_c(j^t))$$

where $T'(j) \subset T(j)$ is the subset of the image of j consisting of those vertices which are the terminal extremities of arcs coming out of $T(i)$.

The event

$$E_{IJ} = (\bigcup_{i^s \in T(i)} E_c(i^s)) \cap (\bigcup_{j^t \in T(j)} E_c(j^t))$$

is such that

$$(\bigcup_{i^s \in T(i)} E_c(i^s)) \cap (\bigcup_{j^t \in T(j)-T'(j)} E_c(j^t)) = \emptyset;$$

and thus

$$E_{IJ} = (\bigcup_{i^s \in T(i)} E_c(i^s)) \cap (\bigcup_{i^s \in T'(j)} E_c(j^s)) .$$

But $j^s \in T'(j)$ if and only if $j^s \in \Delta i^s \cap T(j)$ so that by (14) and Property 8 we obtain

$$p(i,j) = \sum_{i^s \in T(i)} P_c(i^s, j^s) = P_c(\bigcup_{i^s \in T(i)} (E_c(i^s) \cap E_c(j^s)))$$

and consequently $p(i,j) = P_c(E_{IJ})$.

Property 14 The valuation $p(i,j)$ defined on any arc of a questionnaire is the probability

$$p(i,j) = P_c([\bigcup_{i^s \in T(i)} E_c(i^s)] \cap [\bigcup_{j^t \in T(j)} E_c(j^t))] .$$

By (3) we have

$$\sum_{j \in \Gamma i} \frac{p(i,j)}{p(i)} = 1 ,$$

and the quotient $\frac{p(i,j)}{p(i)}$ is then:

$$\frac{P_c([\bigcup_{i^s \in \Gamma(i)} E_c(i^s)] \cap [\bigcup_{j^t \in T(j)} E_c(j^t))]}{P_c(\bigcup_{i^s \in T(i)} E_c(i^s))}$$

which is a conditional probability on

$$(E_c, \mathcal{P}(E_c), P_c)$$

We will also introduce the probability of the arc (i, j_u) conditioned by a particular path $\mu_s = [\alpha i]$. The image of μ_s is the path $\nu_s = [\omega i^s]$ of C and we know that $P_c(j_u^s) = P_c(i^s, j_u^s)$. Consequently

$$p_c(i^s, j_u^S) = P_c(E_c(j_u^S)),$$

that is that the event of $(E_c, \mathfrak{P}(E_c), P_c)$ determined by the vertex j_u^S must be associated with the arc (i, j_u) when this arc is conditioned by the path μ_S.

Definition 13 In a latticoid questionnaire the *probability of the arc* (i,j) is its valuation $p(i,j)$; the *probability of* (i,j) *conditioned by the vertex i* is

$$p((i,j)|i) = \frac{p(i,j)}{p(i)}$$

the *probability of* (i,j) *conditioned by the path* $\mu_S = [\alpha i]$ is

$$p((i,j)|\mu_S) = p_c(j_u^S)$$

where the last arc of the image of μ_S ends in i^s a predecessor of j_u^S in C.

These probabilities are connected by the equations

$$p((i,j_u)|i) = \frac{p(i,j_u)}{\sum_{j_u \in \Gamma i} p(i,j_u)} \qquad (27)$$

and

$$\sum_{s \in S} p((i,j_u)|\mu_s) = \sum_{s \in S} p_c(i^s, j_u^S) = p(i,j_u)$$

that is

$$p((i,j_u)|i) = \frac{\sum_{s \in S} p((i,j_u)|\mu_s)}{\sum_{u \in U} p(i,j_u)} \qquad (28)$$

Now let μ_σ be a path of Q which we will denote by

$$[x_0 x_1 \ldots x_{k-1} x_k] \qquad \text{with } x_0 = \alpha, \ x_{k-1} = i, \ x_k = j_u$$

and its image ν_σ in C:

$$[y_0 y_1 \ldots y_{k-1} y_k] \qquad \text{with } y_0 = \omega, \ y_{k-1} = i^s, \ y_k = j_u^S.$$

Let μ_S (resp. ν_S) be the restriction of μ_σ (resp. ν_σ) to its (k-1) first arcs. The parameters $\lambda: (.)$ relative to the arcs of μ_σ will be denoted by $\lambda_h^S(h-1)$, for h: = 1,2,...k.

Then we can write successively:

$$p_c(y_k) = \frac{p_c(y_{k-1}, y_k)}{p_c(y_{k-1})} \, p_c(y_{k-1})$$

and for any h:

$$p_c(y_h) = \frac{p_c(y_{h-1}, y_h)}{p_c(y_{h-1})} p_c(y_{h-1})$$

so that
$$p_c(y_k) = \prod_{h:=1}^{k} \frac{p_c(y_{h-1}, y_h)}{p_c(y_{h-1})}$$

Therefore, taking into account (18) and (27):

$$p((i,j)|\mu_s) = \prod_{h:=1}^{k} \lambda_u^s(h-1) p((x_{h-1}, x_h)|x_{n-1}) \qquad (29)$$

If the compatible arborescent questionnaire is B then

$$p((i,j_u)|\mu_s) = \prod_{h:=1}^{k} p((x_{h-1}, x_h)|x_{h-1}) \qquad (30)$$

Since
$$p(j_u) = \sum_{y_k \in T(j_u)} p_c(y_k)$$

and
$$p(j_u) = \sum_{s \in S} p((i,j_u)|\mu_s) ,$$

we can write
$$p(j_u) = \sum_{s \in S(x_{h-1}, x_h)} \lambda_h^s(h-1) p((x_{h-1}, x_h)|x_{h-1}) ,$$
$$\mu_\sigma$$

a formula which enables us to evaluate the probability of any vertex j_u of a latticoid questionnaire by using exclusively the conditional probabilities of the arcs situated along a path $[\alpha j_u]$ as well as the parameters $\lambda_h^s(h-1)$. But because $p(j_u)$ is independent of these parameters, we can further write according to (30)

$$p(j_u) = \sum_{s \in S(x_{h-1}, x_h)} \prod_{\in \mu_\sigma} p((x_{h-1}, x_h)|x_{h-1}) . \qquad (31)$$

However the formulae (29) and (30) remain distinct.

Property 16 Let μ_s be a path linking the root to i and μ_σ a path formed by the concatenation of μ_s with the path $\{(i,j)\}$. The probability $p((i,j)|\mu_s)$ is the product of the probabilities of the arcs of the path μ_σ conditioned by their origins if the path μ_σ is unique or if the compatible arborescent questionnaire is B. Otherwise $p((i,j)|\mu_\sigma$ is given by (29).

Remark Since any path μ_σ ending at j is formed by the concatenation of a path μ_s ending at i with an arc ending at j we can speak of the probability of j conditioned by μ_s and we will write:

GENERAL PROPERTIES OF QUESTIONNAIRES   107

$$p(j|\mu_s) = p((i,j)|\mu_s) \qquad \text{when} \qquad i\Gamma j$$

In the same way we will write

$$p(j|i) = p((i,j)|i) \qquad \text{when} \qquad i\Gamma j$$

We can then generalize (29) and (30) to the paths τ_s ending at i but having as initial extremity an internal question x_{h_0}:

$$p((i,j)|\tau_s) = \prod_{h:=h_0}^{k-1} \lambda^s_{h+1}(h) p((x_h, x_{h+1})|x_h) p(x_{h_0})$$

and speak of $p(j|x_{h_0})$ when $x_{h_0}\hat{\Gamma}j$ and $x_{h_0}\hat{\gamma}j$; if $x_0\hat{\gamma}j$, then $p(j|x_0)=0$.

4.4.3   Example of a Semantic

The two questionnaires B and C associated with the latticoid questionnaire Q (cf. above Fig. 4.9) differ in the choice of the parameter $\lambda^s_u(i)$ which takes the values 1 and 2 respectively. This choice is arbitrary at this point of the theory where the questionnaire Q is determined by the axioms (q_1) to (q_3) governing the support and (q_4) and (q_5) governing the flow along the arcs and thus the probabilities of the vertices (cf. Property 13).

However, the use of a questionnaire requires the introduction of a semantic enabling us to give a significance to the different vertices. If the questionnaire is arborescent, we can consider that the root α is a question causing a partition of E into a_α disjoint sub-sets $E_1, E_L, \ldots, E_{a\alpha}$ and in the same way each question causes a partition of a sub-set of E. Finally the answers appear as atoms or eventualities of E.

If the questionnaire is latticoid, the root can always have the same significance but as soon as a question is the terminal extremity of many paths there is a regrouping (union) of as many subsets as there are distinct paths and everything takes place as if the questioner had lost track of some answers.

Moreover, the significance of the answers or the partition of E depends directly on the formulation of a question and we will say that there is a constraint if there are limitations on the formulation. In the case of the game of heads or tails only two answers are possible: H and T. In the case of a computer the formulation is most often realized by a comparison of type > or ⩾ or = with two possible outcomes; sometimes (Fortran language) there are three answers: <, =, >.

As an example of a semantic enabling us to give an interpretation to the questionnaires Q, B and C, we will consider the problem of establishing a partition of the total orderings of four distinct numbers α, β, γ, ∂ knowing that $\alpha<\beta$. There are $\frac{4!}{2}=12$ total orderings compatible with $\alpha<\beta$. If we have an arborescent questionnaire in which the answers have probabilities:

$(\frac{3}{12}, \frac{2}{12}, \frac{1}{12}, \frac{4}{12}, \frac{2}{12})$ or $(\frac{3}{12}, \frac{1}{12}, \frac{2}{12}, \frac{5}{12}, \frac{1}{12})$

that is a questionnaire B or C, it is possible to determine the semantic, in order that this questionnaire makes it possible to partition the set of twelve orders into five classes to be specified.

α β γ δ	a	γ α β δ	a'
α β δ γ	b	γ α δ β	b'
α γ β δ	c	γ δ α β	c'
α γ δ β	d	δ α β γ	d'
α δ β γ	e	δ α γ β	e'
α δ γ β	f	δ γ α β	f'

The latticoid questionnaire Q only allows a partition into three classes (Fig. 4.9).

In Figures 4.10 and 4.11 the sign + indicates the arcs corresponding to an affirmative outcome to the question asked in a block; the eventualities are indicated by the names of the orders which occur as solutions.

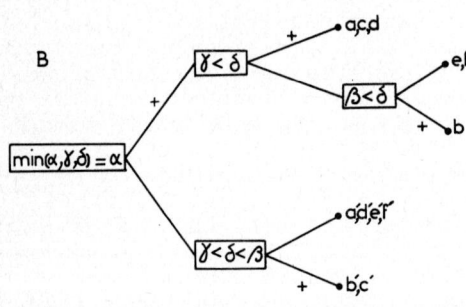

Figure 4.10

The contraction of questionnaire B to a latticoid Q_B (Fig. 4.12) leads to the formulation of a question involving the connectives *and*, *or*

$\{\beta<\delta<\gamma$ or $\gamma<\delta<\beta\}$?

and having for outcome +, the orders $\{b,b',c'\}$ the probability of which is $\frac{1}{4}$.

Similarly for C (Fig. 4.13) we will formulate the question

$\{(\alpha<\delta$ and $\beta<\gamma)$ or $(\gamma<\delta<\alpha)\}$?

having for outcome +, $\{b,e,c'\}$ the probability of which is $\frac{1}{4}$.

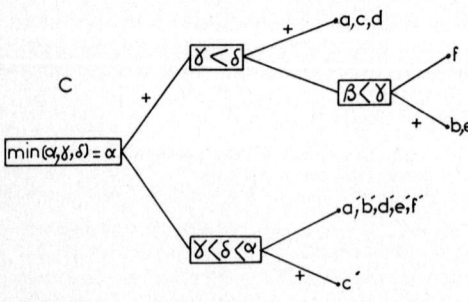

Figure 4.11

The information (in the naive sense) obtained by the answer to the first question (root) is partially lost because

the question with two entering arcs operates a partition on
{b,e,f}∪{a',b',c',d',e',f'} which are primitively disjoint.

However the information obtained by the second question
($\gamma<\delta$?) is not lost and this allows the elimination of case d
for which the double inequality $\gamma<\delta<\beta$ is true.

Remarks

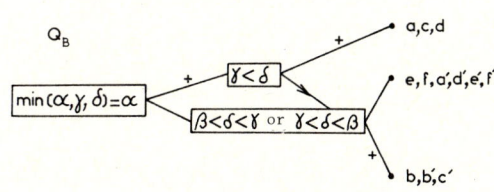

Figure 4.12

1. An arborescence can be the arborescence of the paths of many latticoids (cf. Property 9).

2. It is possible to associate with any latticoid questionnaire Q, a family (C) of compatible arborescent questionnaires all having the same support and to adjoin to each of the elements of (C) a semantic. (C) is determined by the range of variation of the different parameters $\lambda_u^s(j)$ attached to each question where there is more than one path entering from the root of Q. This range is an open set interior to a parallelepiped of

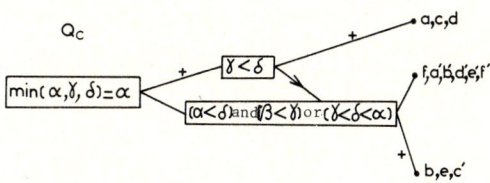

Figure 4.13

R^k, where $k = \sum\limits_{i \in F'} a_i c_i$, F' being the set of such questions
($F' \subset \overline{F}$) and each of the edges of the parallelepiped being

$$[\,0,\ \min\,(\frac{p(i)}{p_c(i^s)},\ \frac{p(i)}{p(i,j_u)})\,]\ \subset\ [\,0,\ \frac{p(i)}{p(i,j_u)}\,]\ ;$$

this last evaluation is independent of the paths μ_s.

Finally each element of C may correspond to a different utilisation of the same questionnaire Q the support and the valuations of which remain fixed.

4.4.4 A restriction of the theory

We make a modification in the system of axioms and show that it leads to a restriction of the questionnaires compatible only with the questionnaires B corresponding to the trivial solution of the system (21) for any i. To this end we substitute two new axioms (b_1) and (b_2) in place of (q_4) and (q_5).

Definition 14 A questionnaire K is a valued l.q.s.c. graph
$K = (X,\Gamma,P_1,P_2)$ such that the set of its vertices admits the
partition $X = E \cup F$ satisfying the axioms (q_1) to (q_3), (b_1) and

(b_2):

(b_1) there exists a mapping from the set of arcs Γ into $[0,1]$ $P_1:(i,j) \mapsto p_1(i,j)$ such that $\sum_{j \in \Gamma i} p_1(i,j) = 1$ for all $i \in F$.

(b_2) there exists a mapping P_2 of the set of vertices X into $]0,1[$, $P_2: i \mapsto p_2(i)$ such that if $\Gamma^{-1}i$ is non-empty, then

$$p_2(i) = \sum_{h \in \Gamma^{-1}i} p_1(h,i) p_2(h)$$

and if $\Gamma^{-1}i$ is empty, then $p_2(i) = 1$.

The axiom (b_1) says that, with each question i we can associate a complete system of elementary events $E(j)$ determined by each $j \in \Gamma i$ and we will associate with i an event $E(i) = \bigcup_{j \in \Gamma i} E(\gamma)$ in such a way that $(E(i), \mathcal{P}(E(i)), P_1)$
constitutes a probability space with probability $P_1(i,j)$. Only the events $E(i)$ will in fact be considered. Since this space is related to question $i \in F$, we will speak of $P_1(i,j)$ as the probability of the arc (i,j) conditioned by the vertex i and this probability is independent of the paths linking α to i.

Consider a questionnaire Q. We shall show that it is possible to associate a questionnaire K with it.

For, it is possible to write (q_4) in the form

$$p(i) = \sum_{h \in \Gamma^{-1}i} \frac{p(h,i)}{p(h)} p(h) \quad \text{if} \quad \Gamma^{-1}i \neq \emptyset .$$

Therefore putting

$$p_2(h) = p(h),$$

$$p_1(h,i) = \frac{p(h,i)}{p(h)},$$

we obtain one and only one questionnaire K with the same support; $p_2(i)$ is then a probability in the sense of Property 13.

Conversely, let K be a questionnaire with root α. Then it is possible to calculate the valuations of the questions, in K, by a relation analogous to (30) and valid for any path μ_s of K:

$$p_2(x_k) = \sum_{\mu_s \in \mathcal{C}_h} \prod_{(x_{u-1}, x_u) \in \mu_\sigma} p_1(x_{u-1}, x_u)$$

by using again the notation of §4.4.2 and noting that $p_2(\alpha)=1$. We can then construct a questionnaire Q by forming the probabilities of the vertices using this formula and calculating the valuations of the arcs from

$$p(i) = p_2(i)$$
$$p(i,j_u) = p_1(i,j)p(i) .$$

But then whatever the path entering into i, the equation

$$\frac{p_c(i^s, j_u^s)}{p_c(i^s)} = \frac{p(i,j_u)}{p(i)}$$

will be satisfied by the compatible questionnaire which by Property 12, is the questionnaire B.

Consequently we have:

Property 16 There is an equivalence between the questionnaire Q (Definition 1) and the questionnaire K (Definition 14) if and only if the hypothesis of proportionality of the flow is satisfied.

The theory of the questionnaires K will continue as a special case of the theory determined by the axioms (q_1) to (q_5). The questionnaire K no longer admits multiple interpretations and applications, unlike those which we met in §4.4.3 where the inequality C≠B allowed a different use of the same questionnaire Q.

We remark that the choice of the parameters determining C is possible only for the questions forming a set F' where there is more than one distinct path coming from α. Thus the questionnaires Q and K are identical when F' is empty. Consequently we have:

Property 17 An arborescent or latticoid questionnaire in which the root is linked by a unique path to each of the other questions is determined indifferently by Definitions 1 and 14.

4.5 Routing length

Definition 15 The *routing length* $L(Q)$ of a questionnaire Q is the mathematical expectation of the length of the paths linking the root to the answers of Q.

Let D be a questionnaire, arborescent or Dedekindian. The unweighted length of the path or of the paths linking the root to an answer is the rank $r(e)$ (cf. Definition 2.9). The weighted length of it is $p(e)r(e)$ and because the answers constitute a complete system of events:

$$L(D) = \sum_{e \in E} p(e)r(e) . \qquad (32)$$

Now consider a latticoid questionnaire Q. If Q is not Dedekindian we can no longer speak of the rank of an answer e, but of the length of a path μ_σ linking α to e. This path will be formed by concatentation of a path $\mu_S = [\alpha i]$ with the arc (i,e) which may be common to other paths entering into e. We will denote by $r_{\mu_s}(e)$, the length of the path μ_σ. We use the notation of §4.4.2 and of its final remark: $p(e|\mu_s)$ is the probability of e conditioned by the path μ_s, that is

$$p(e|\mu_s) = p((i,e)|\mu_s)$$

is given by (29). Therefore let $\mathcal{C}_e$ be the set of distinct paths linking α to e. Since the study of the weighted paths of a latticoid questionnaire implies a choice of the $\lambda:(.)$, we will try to evaluate $L(Q_C)$, the routing length of Q relative to a certain compatible questionnaire C such that

$$L(C) = \sum_{v \in E_C} p_C(v) r_C(v) .$$

Noting

$$\sum_{\mu_\sigma \in \mathcal{C}_e} p(e|\mu_s) = \sum_{i \in \Gamma^{-1}e} \sum_{\mu_s \in \{[ai]\}} p((i,e)|\mu_s)$$

or

$$\sum_{\mu_\sigma \in \mathcal{C}_e} p(e|\mu_s) = \sum_{i \in \Gamma^{-1}e} p(i,e) = p(e)$$

and thus that

$$\sum_{e \in E} \sum_{\mu_\sigma \in \mathcal{C}_e} p(e|\mu_s) = 1 ,$$

we find that the mathematical expectation of the length of the paths is

$$L(Q_C) = \sum_{e \in E} \sum_{\mu_\sigma \in \mathcal{C}_e} p(e|\mu_s) r_{\mu_s}(e) \qquad (33)$$

But there exists $v \in T(e)$ such that

$$p(e|\mu_s) = p_C(v) \quad \text{and} \quad r_{\mu_s}(e) = r_C(v) ,$$

thus

$$L(Q_C) = \sum_{e \in E} \sum_{v \in T(e)} p_C(v) r_C(v) ,$$

that is

$$L(Q_C) = \sum_{v \in E_C} p_C(v) r_C(v)$$

that is

$$L(Q_C) = L(C) . \qquad (34)$$

Let $A_C(v)$ be the set of proper ancestors of v.

We can write $r_C(v) = \sum_{t^s \in A_C(v)} 1$ and

$$L(C) = \sum_{v \in E_C} \sum_{i^s \in A_C(v)} p_C(v) .$$

Setting $l_v(i^s) = 1$, if $i^s \in A_c(v)$

$\qquad\qquad\quad\; 0$, if $i^s \in F_c - A_c(v)$,

where F_c is the set of questions of C; we have

$$L(C) = \sum_{v \in E_c} \sum_{i^s \in A_c(v)} l_v(i^s) p_c(v)$$

and

$$L(C) = \sum_{v \in E_c} \sum_{i^s \in F_c} l_v(i^s) p_c(v) \ .$$

But by (13) $\sum_{v \in \hat{\Delta}i^s \cap E_c} p_c(v) = p_c(i^s)$ so that

$$L(C) = \sum_{i^s \in F_c} \sum_{v \in \hat{\Delta}i^s \cap E_c} p_c(v) = \sum_{i^s \in F_c} p_c(i^s) \ ,$$

$$L(C) = \sum_{i \in F} \sum_{i^s \in T(i)} p_c(i^s)$$

and finally:

$$L(Q_c) = \sum_{i \in F} p(i) \ .$$

But this last summation only concerns the probabilities of questions of Q, independently from the compatible questionnaire C used. Consequently

$$L(Q) = L(C) \qquad\qquad (35)$$

for any C compatible with Q.

Hence

$$L(Q) = \sum_{i \in F} p(i) \ . \qquad\qquad (36)$$

This final formula is always valid whether the support of Q is a latticoid, a Dedekindian or an arborescence.

Property 18 The routing length of a questionnaire is equal to the sum of the probabilities of its questions.

Property 19 All the arborescent questionnaires compatible with a given latticoid questionnaire have the same routing length. Sometimes we will say that (32) (resp. 33) is the *first form of the routing length* for Dedekindians (resp. for latticoids) and that (36) is the *second form of the routing length*.

Remarks

1. In an earlier work essentially concerning arborescent questionnaires, a *mean length*, weighted by the probabilities of the lengths of the paths was introduced:

$$L_\eta(Q) = \sum_{e \in E} \frac{p(e)}{|\mathcal{C}_e|} \sum_{\mu_s \in \mathcal{C}_e} r_{\mu_s}(e) . \qquad (37)$$

If the questionnaire is arborescent or Dedekindian, then $r_{\mu_s}(e) = r(e)$ ($\forall$ B) so that $L_\eta(D) = L(D)$ and (37) can be substituted for (33) for Dedekindian questionnaires.

If the questionnaire is latticoid and such that with B compatibility for example we have

$$p(e|\mu_s) = \frac{p(e)}{|\mathcal{C}_e|} \quad (\forall e, s)$$

then $L_\eta(Q) = L(Q)$. Otherwise $L_\eta(Q)$ can be greater or less than $L(Q)$ (Fig. 4.14).

2. Removal of axiom (q_3) from Definition 1 leads to the concept of a "quasi-questionnaire" (see further §8.3) as a valuated graph in which the vertices of F can have only one successor. If Γ_i reduces to a unique vertex j, then we note from (q_4) that $p(j) = p(i)$ and thus

$$E_c(j) = E_c(i) .$$

It will be possible to recover and to extend the general properties of questionnaires for this type of generalization. In particular we can use without modification the concepts of probability, routing and routing length.

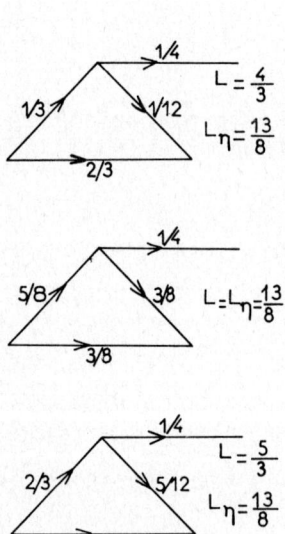

Figure 4.14

EXERCISES

Consider the latticoid questionnaire Q shown in Fig. 4.15.

1. Determine the range of variation of the arc $\lambda_u^s(i)$; Evaluate $L(Q)$.

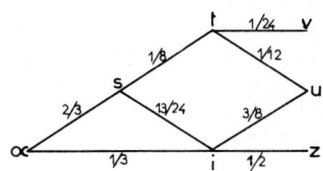

Figure 4.15

2. Determine the probabilities of the images of u and z in the compatible arborescent questionnaires corresponding to the choices of $\lambda_u^s(i)$:

$\lambda_A = 0,5; \lambda_B = 1; \lambda_C = 1,5.$

3. For these same choices, determine the probabilities of the arcs (i,u) and (i,z) conditioned by the paths [αsi] and [αi], and by the vertex i.

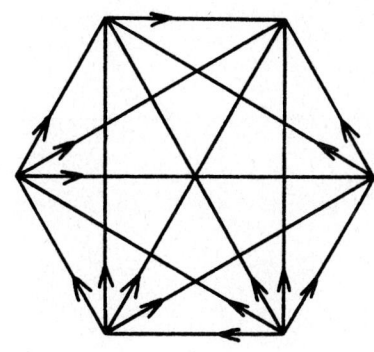

Chapter Five

THE CONSTRUCTION OF QUESTIONNAIRES

This chapter is intended to introduce a set of tools for the construction of questionnaires satisfying the axioms (q_1) to (q_5) independently of any semantic consideration.

We will use operations or transformations of graphs (cf. Chapter 3) and we will consider certain problems concerning valuations, in particular the relationship between those of the answers and those of the arcs.

We will define optimality of routing and having shown that in every case an optimal solution can be an arborescent questionnaire, we will study some properties of this type of questionnaire in relation to the support or to the probability distribution on the set of answers.

Representation From here on, any arborescences which we draw in diagrams will usually carry some indication of the questions posed.

As an aid to clarity, the vertices of the questionnaire will usually be represented as follows:

- questions by a circle or a rectangle containing the question posed but without the question mark

- answers by a point with the coding or label displayed alongside it.

Arcs drawn without arrows will be understood to have their origins on the left and their terminal extremities on the right. In the case when $a = 2$, arcs going upwards will represent the outcome "yes" (or 1) and arcs going downwards will represent the outcome "no" (or 0). Finally the probability of a vertex will be shown, except possibly for a scaling factor, by a value attached to the arc inwardly adjacent to the vertex (that is, to the left of the vertex).

5.1 Operations on Questionnaires

5.1.1 Definitions

Since a questionnaire Q is a valuated graph, it is possible to define an operation on Q by effecting a modification of the valuation P_Γ with or without preservation of the support; any modification of the support leads necessarily to a modification of the paths and sometimes to a modification of the probability space (E_C, $\mathfrak{P}(E_C)$, P_C) built from the valuated answers of a compatible questionnaire C. We will denote by Q^T the result of an operation T on Q.

Definition 1 A *downstream to upstream* process (respectively *upstream to downstream*) is any algorithm processing the elements of the questionnaire, vertices, arcs and valuations from the answers to the root, (resp. from the root to the answers) in such a way that if $i\Gamma j$, then j is processed before i (resp. after i).

Examples

1. The formation of the compatible questionnaire is an upstream to downstream process.

2. The determination of $p(i)$ by the formulae (4.3) is an upstream to downstream process when we write $p(i) = \sum_{h \in \Gamma^{-1}i} p(h,i)$ and we know the $p(h,i)$; it is a downstream to upstream process when we use $p(i) = \sum_{j \in \Gamma i} p(i,j)$ and we know the $p(i,j)$.

Certain operations preserving the support of a questionnaire Q can lead to choices: this is the case for the formation of the compatible questionnaires B or C when the arborescence of the paths A has already been obtained. These choices can often be determined by making a hypothesis analogous to that of proportionality of the flow, but where the flow would be obtained downstream to upstream. Let $p(i) = \sum_{h \in \Gamma^{-1}i} p(h,i)$ be the probability of a vertex of Q and $p^T(i)$ the probability of this vertex in Q^T obtained from Q by a downstream to upstream process. The admissible hypothesis

$$p^T(h,i) = p(h,i) \frac{p^T(i)}{p(i)} \quad (\forall h) \tag{1}$$

allows a step by step determination of the probabilities of the entering arcs and the ancestor vertices of i. However this hypothesis is not necessary and we can do without it by introducing parameters analogous to $\lambda:(.)$ of Chapter 4. To simplify the exposition, we will often make implicit use of this hypothesis.

The simplest transformation we can perform on P is the following named *reallocation of probabilities*.

Let $Q = (X, \Gamma, P_\Gamma)$ be a questionnaire admitting e_1 and e_2 as answers with probabilities $p(e_1)$ and $p(e_2)$.

Let $Q^T = (X, \Gamma, P^T)$ be the transform given by

$$p^T(e_j) = p(e_j), \text{ if } e_j \in E \text{ and } e_j \neq e_1 \text{ or } e_2, \tag{2}$$

$$p^T(e_1) = p(e_2)$$

and $\quad p^T(e_2) = p(e_1)$.

If (X, Γ) is an arborescence, then those vertices which are ancestors of both e_1 and e_2 and those which are ancestors of neither e_1 nor e_2 all keep the same probability. The probabilities of those vertices j which are ancestors of one of e_1 or e_2 may be determined by a straightforward downstream to upstream process.

If (X,Γ) is a latticoid, then again those vertices which are ancestors of both or of neither of e_1 and e_2 all keep the same probability. However for the probabilities of the other vertices, if there are several paths entering into either e_1 or e_2, it will be necessary to introduce parameters attached to any arc which has its terminal extremity in common with other arcs or to use hypothesis (1). A downstream to upstream process will then give a determination of Q^T (see further Problem 1 and Theorem 1, §5.2).

Definition 2 A *sub-questionnaire*

$$Q_\sigma = (X', \Gamma', P')$$

of a questionnaire $Q = (X, \Gamma, P_\Gamma)$ is a questionnaire in which

a. the support is a sub-latticoid of (X,Γ) with root σ such that only the arcs of Γ entering into X' all have terminal extremity σ and the questions of Q_σ have the same basis in Q as in Q_σ;

b. the valuation is determined by $p'(i,j) = \dfrac{p(i,j)}{p(\sigma)}$ for any $(i,j) \in \Gamma'$.

The probabilities of the vertices of Q_σ are then $p'(i) = \dfrac{p(i)}{p(\sigma)}$ ($\forall i \in X'$) and in particular $p'(\sigma) = 1$.

If $\sigma = \alpha$, the sub-latticoid will be obtained by suppression of a certain number of terminals and, possibly, some internal vertices.

If $\sigma \neq \alpha$ the terminals of the sub-questionnaire may or may not be coincident with those of Q.

In particular a sub-questionnaire generated by a question and its outcomes can be considered as an elementary questionnaire the arcs of which have probability $p'(\sigma,j) = p((\sigma,j)|\sigma)$.

When σ is a terminal we will consider the graph $(\{\sigma\}, \emptyset)$ valuated by $p'(\sigma) = 1$ as a *degenerate sub-questionnaire*. A sub-questionnaire such that $1 < |X'| < |X|$ will be called a *proper sub-questionnaire*.

Contraction at a subset A of X, where the sub-graph generated by A is connected will be used especially in the following cases.

R1 There exists a vertex ξ of outdegree 1 and $\Gamma\xi = \{\zeta\}$. We put $A = \{\xi, \zeta\}$ and the graph (X^T, Γ^T) obtained after contraction is then such that $X^T = X - \{\xi\}$, while the new valuation is:

$$p^T(h,\zeta) = p(h,\xi) , \quad p^T(\zeta) = p(\zeta)$$

and for the other arcs and vertices common to Q and Q^T

$$p^T(i,j) = p(i,j) , \quad p^T(i) = p(i) .$$

If several vertices $\xi_1, \xi_2, \ldots, \xi_p$, of degree 2 are situated consecutively on the same path, then A will consist of $\{\xi_1, \xi_2, \ldots, \xi_p, \zeta\}$, if (ξ_p, ζ) is an arc of (X, Γ).

R2 Suppose that question i is such that all the paths linking α to e, $(\forall e \in \hat{\Gamma} i)$ pass through i. Then the subgraph generated by $A = \hat{\Gamma} i$ is a latticoid and the only arcs entering into A all have i as terminal extremity: $p^T(i) = p(i)$.

Then contraction at A reduces the number of answers by $|\hat{\Gamma} i \cap E| - 1$, i becoming an answer of Q^T.

The valuation of i and of the vertices of $X - \hat{\Gamma} i$ remains unchanged. This type of contraction will be used especially when all the successors of i are answers.

Substitution of arcs, that is the transformation substituting an arc (x,b) in place of the arc (a,b) will be effected when the only arc entering into $\hat{\Gamma}_b$ has b as its terminal extremity: $p^T(x,b) = p(a,b)$ (Fig. 5.1). Substitution of arcs allows us to increase the outdegree of x and to reduce that of a.

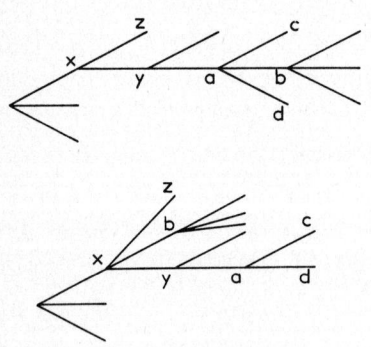

Figure 5.1

Exchange of arcs will be effected in the same way on the two sub-latticoids with roots b and y when the only two arcs entering into $\hat{\Gamma}_b$ or $\hat{\Gamma}_y$ have respectively b or y as terminal extremity. The sub-latticoids with root b and y will then have their valuation unchanged whereas the ancestors either of b or of y will have a modified valuation.

The term *transfer of arborescence* will mean an exchange of arcs transforming an arborescent questionnaire $Q = (X, \Gamma, P_\Gamma)$ into another arborescent questionnaire $Q^T = (X^T, \Gamma^T, P^T)$.

We will say that we have *exchanged the arcs* (x,y) and (a,b) or that we have *transferred the sub-arborescences with roots* y and b such that $\hat{\Gamma}_y \cap \hat{\Gamma}_b = \emptyset$ (Fig. 5.2).

The valuations will be related by

$$p_T(j) = p(j) \quad \text{if} \quad j \in \hat{\Gamma}y \cup \hat{\Gamma}b \qquad (3)$$

and for $h, k \neq b$ and $h, k \neq y$:

$$p_T(h) = p(h) + p(y) - p(b) \quad \text{if} \quad h\hat{\Gamma}b$$

and thus if $h\hat{\Gamma}^T y$,

$$p_T(k) = p(k) + p(b) - p(y) \quad \text{if} \quad k\hat{\Gamma}y \qquad (4)$$

and thus if $k\hat{\Gamma}^T b$.

If the sub-questionnaires with roots y, b are isomorphic or degenerate this operation is equivalent to one or more re-allocations of probabilities.

It is possible to define an operation of transfer of sub-graphs in latticoid questionnaires by making one or more exchanges of arcs; special care needs to be taken when $\hat{\Gamma}y \cap \hat{\Gamma}b \neq \emptyset$ (Ghersi).

The product of questionnaires and the restricted product of questionnaires are the valuated extensions of the latticoid product and the restricted latticoid product (Definitions 3.11 and 3.12).

Let $Q = (X,\Gamma,P_\Gamma)$ and $R = (Y,\Delta,P_\Delta)$ be two latticoid questionnaires with roots a, b and answers $E=\{\xi_1,\ldots,\xi_T\}$, $D=\{\eta_1,\ldots,\eta_{T'}\}$ respectively. Then the product of the questionnaires will be

$$QR = (Z,\Lambda,P_\Lambda),$$

where

$$(Z,\Lambda) = (X,\Gamma)(Y,\Delta)$$

and

$$p_\Lambda(ib,jb) = p_\Gamma(i,j) \text{ for } (i,j) \in \Gamma$$

$$p_\Lambda(\xi_h k, \xi_h \ell) = p_\Gamma(\xi_h) p_\Delta(k,\ell)$$
$$\text{for } (k,\ell) \in \Delta$$

Consequently

$$j \in X \rightarrow p_\Lambda(jb) = p_\Gamma(j)$$

$$k \in Y \rightarrow p_\Lambda(\xi_h k) = p_\Gamma(\xi_h) p_\Delta(k)$$

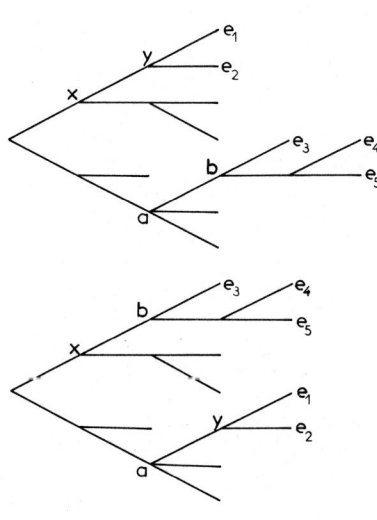

Figure 5.2

so that

$$p_\Lambda(ik) = p_\Gamma(i) p_\Delta(k) \qquad (5)$$

with

$$i \in X - E \quad \text{and} \quad k = b \quad \text{such that} \quad p_\Lambda(b) = 1$$

or
$$i \in E \quad \text{and} \quad k \in Y .$$

In particular QR will have $|E| \cdot |D|$ answers with probability:

$$p_\Lambda(\xi_h \eta_j) = p_\Gamma(\xi_h) p_\Delta(\eta_j) \quad (\xi_h \in E, \eta_j \in D) ,$$

the indices Γ, Δ, Λ being relative to the valuations of the arcs P_Γ, P_Δ, P_Λ or to those which are induced for the vertices.

The arborescence of the paths of QR will have $V_Q \cdot V_H$ terminals (where V_G is the number of paths of the questionnaire G) and this facilitates the construction of the probability space related to a questionnaire compatible with QR.

A sub-questionnaire of QR with root $\xi_h b$ will be such that

$$p'(\xi_h k) = \frac{p_\Lambda(\xi_h k)}{p_\Lambda(\xi_h b)} = p_\Delta(k) , \tag{6}$$

and it will therefore be isomorphic to R and the bijection acting on the arcs preserves the valuation P_Δ.

The contractions of all the sub-questionnaires with root $\xi_h b$ (for all $\xi_h \in E$) leads to a questionnaire isomorphic to Q.

The contraction of one or more sub-questionnaires, the roots of which are the vertices $\xi_h b$, where $\xi_h \in E' \subset E$ gives a questionnaire Q◊R, a valuated extension of the restricted latticoid product.

If $|E'|=1$, the restricted latticoid product Q◊R is a questionnaire isomorphic to that obtained by concatenating the path of Q ending at ξ_1 (by setting $E'=\{\xi_\Lambda\}$) and the paths of R coming out of its root b. Thus if $|E'|=1$, the restricted latticoid product appears as the operation inverse to that consisting in forming a sub-questionnaire: R is isomorphic to the sub-questionnaire of QR whose root is $\xi_1 b$ and whose terminals form the set $\{\xi_1 \eta_j | \eta_j \in D\}$.

In the same way that the latticoid product is associative, we can show that the product of questionnaires is also associative. On the other hand we can obtain the product of questionnaires QR by means of $|E|$ restricted products of questionnaires Q and R the restriction bearing successively on each of the answers of Q.

5.1.2 Operations and Routing Length

The principal properties of routing in connection with operations on questionnaires can be established either by the global formula (4.36) or by starting from (4.32) for arborescent questionnaires.

Re-Allocation of Probabilities: We calculate

$$L(Q^T) = L(Q) + \sum_{i \in F^X} (p_T(i) - p(j)) \quad (7)$$

where $p_T(i)$ is the probability of question i in Q^T and F^X is the set of questions not having the same probability in Q as in its transform Q^T.

Contraction R1 at $A = \{\xi, \zeta\}$, where ξ is of outdegree 1:

$$L(Q^T) = L(Q) - p(\xi) . \quad (8)$$

Contraction R2 at X' where X' generates a sub-questionnaire with root σ: we will denote by $\overline{F}_\sigma$ the set of internal questions of the sub-questionnaire

$$L(Q^T) = L(Q) - \sum_{i \in \overline{F}_\sigma} p(i) . \quad (9)$$

Sub-questionnaires Q_σ *with root* σ :

Let F_σ be the set of questions of Q_σ. Then

$$L(Q_\sigma) = \sum_{i \in F_\sigma} p'(i) , \quad (10)$$

$p'(i)$ being the valuation of the sub-questionnaire. Since this valuation is related to the questionnaire according to Definition 2, we obtain

$$L(Q_\sigma) = \frac{1}{p(\sigma)} \sum_{i \in F_\sigma} p(i) . \quad (11)$$

If the sub-questionnaire has only one question then

$$L(Q_\sigma) = 1 .$$

Transfer of arborescences with roots b and y:

Put

$$B = E \cap \hat{\Gamma}b = \{u\}$$

and

$$Y = E \cap \hat{\Gamma}y = \{v\} .$$

Then

$$L(Q) = \sum_{e \in E-B-Y} p(e)r(e) + \sum_{u \in B} p(u)r(u)$$

$$+ \sum_{\in Y} p(v)r(v)$$

$$L(Q^T) = \sum_{e \in E-B-Y} p(e)r(e)$$

$$+ \sum_{u \in B} p(u)[r(u) - r(b) + r(y)]$$

$$+ \sum_{v \in Y} p(v) [r(v) - r(y) + r(b)]$$

and since $\sum_{u \in B} p(u) = p(b)$, $\sum_{\in Y} p(v) = p(y)$ it follows that

$$L(Q^T) = L(Q) + [p(b) - p(y)] [r(y) - r(b)] \quad . \quad (12)$$

This formula has been generalised by Ghersi for the transfer of sub-latticoids. He defines for that case the *equivalent rank* $R(y)$ as the expectation of the rank of the image $T(y)$ in the compatible questionnaire B; the substitution of $[R(y) - R(b)]$ for $[r(y) - r(b)]$ produces a generalization of (12).

Substitution of an arc with terminal extremity b, root of a sub-questionnaire the set of questions of which is F_b:

Let the arc $(x,b) \in Q^T$ be substituted in place of $(y,b) \in Q$: Then the ranks of b in Q and in Q^T are $r(y) + 1$ and $r(x) + 1$. The preceding calculation leads to

$$L(Q^T) = L(Q) + (\sum_{i \in F_b} p(i)) [r(x) - r(y)] \quad .$$

Denoting by $L(Q_b)$ the length of the sub-questionnaire with root b, we find:

$$L(Q^T) = L(Q) + p(b)L(Q_b) [r(x) - r(y)] \quad . \quad (13)$$

Product of questionnaires

If Q and R are arborescent, we will denote by $r_\Gamma(\xi_h)$ the rank of ξ_h in (X,Γ) etc.

$$L(QR) = \sum_{\xi_h \in E} \sum_{\eta_j \in D} p_\Lambda(\xi_h \eta_j) r_\Lambda(\xi_h \eta_j)$$

$$= \sum_{\xi_h \in E} \sum_{\eta_j \in D} p_\Gamma(\xi_h) p_\Delta(\eta_j) (r_\Gamma(\xi_h) + r_\Delta(\eta_j))$$

$$= \sum_{\xi_h \in E} p_\Gamma(\xi) r_\Gamma(\xi_h) + \sum_{\eta_j \in D} p_\Delta(\eta_j) r_\Delta(\eta_j)$$

that is

$$L(QR) = L(Q) + L(R) \quad (14)$$

If Q and R are latticoids we will make an analogous calculation for the questionnaires of type B for example, compatible respectively with Q, R and QR. We find the same relationship which may be called commutativity - in the sense of routing length for the latticoid product.

In the case of the *restricted product* we find

$$L(Q \diamond R) = \sum_{\xi_i \in E-E'} p_\Gamma(\xi_i) r_\Gamma(\xi_i)$$

$$+ \sum_{\xi_h \in E'} \sum_{\eta_j \in D} p_\Lambda(\xi_h \eta_j) r_\Lambda(\xi_h \eta_j)$$

$$= \sum_{\xi_h \in E} p_\Gamma(\xi_h) r_\Gamma(\xi_h)$$

$$+ \sum_{\xi_h \in E'} p(\xi_h) \sum_{\eta_j \in D} p_\Delta(\eta_j) r_\Delta(\eta_j)$$

Putting $p(E') = \sum_{\xi_h \in E'} p(\xi_h)$ we have

$$L(Q \diamond R) = L(Q) + p(E') L(R) \ . \qquad (15)$$

If $E = \{\xi_1\}$, then

$$L(Q \diamond R) = L(Q) + p(\xi_1) L(R) \ .$$

Let Q be an arborescent questionnaire

$$Q = (E \cup F, \Gamma, P_\Gamma) \ .$$

Now let us construct a questionnaire step-by-step as follows:

1. Make a series of $\bar{F}$ contractions at each of the internal questions by a downstream to upstream process. We thus obtain an elementary questionnaire with root α, answers Γ_α and routing length $L(Q^{(o)}) = 1$.

2. Make a series of $\bar{F}$ restricted products in inverse order to the preceding operations in such a way as to reconstitute by an upstream to downstream process the initial questionnaire Q. The ith restricted product will operate on a questionnaire $Q^{(i-1)}$ having i questions and on an elementary questionnaire Q_i formed by the ith internal question σ_i and its outcomes and arcs coming out, and we will have

$$L(Q^{(i)}) = L(Q^{(i-1)}) + p(\sigma_i) L(Q_i) \ .$$

The first step starts with $Q^{(o)}$ and the last causes the reconstruction of Q so that:

$$L(Q) = 1 + \sum_{\sigma_i \in \bar{F}} p(\sigma_i) L(Q_i) \ .$$

This formula which is evidently equivalent to (4.36), because of (11), shows that the routing length of a questionnaire is the sum of the products $p(\sigma_i) L(Q_i)$ calculated for all questions $\sigma_i \in F$.

We can say that $p(\sigma_i) L(Q_i) = p(\sigma_i)$ is the *contribution to the routing length* of the question σ_i.

Let $Q_\sigma = (E' \cup F', \Gamma', P')$ be a sub-questionnaire of Q such that $E' = E(\sigma)$ then $\Gamma_\sigma = E' \cup F^+_\uparrow$ and let α be the basis of σ. Then the routing length of Q_σ is

$$L(Q_\sigma) = \sum_{e \in E(\sigma)} (r(e) - r(\sigma)) \frac{p(e)}{p(\sigma)} .$$

Suppression of σ transforms the arborescence $(E' \cup F', \Gamma')$ into a forest comprising a arborescences possibly degenerate. Consider the a corresponding sub-questionnaires they have routing length for $j \in \Gamma\sigma$:

$$L_j = \sum_{e \in E(j)} (r(e) - r(\sigma) - 1) \frac{p(e)}{p(j)}$$

and if further $j_o \in E \cap \Gamma_\sigma$, then we find that $L_{j_o} = 0$.

We can then write

$$L(Q_\sigma) = \sum_{j \in \Gamma\sigma} \sum_{e \in E(j)} [\frac{p(e)}{p(\sigma)} + (r(e) - r(\sigma) - 1)\frac{p(e)}{p(\sigma)}]$$

that is

$$L(Q_\sigma) = 1 + \sum_{j \in \Gamma\sigma} \frac{p(j)}{p(\sigma)} L_j . \qquad (16)$$

In particular

$$L(Q) = 1 + \sum_{h \in \Gamma} p(h) L_h$$

If it is possible to calculate L_j for all the outcomes of a question σ, and afterwards in the same way, for all $\sigma \in F$, then we can evaluate $L(Q)$ by recursive use of (16), downstream to upstream.

5.2 Valuations on the Answers and the Arcs

Let $\mathcal{F}$ be a family of questionnaires whose supports all have the same number of terminals $|E|=N$ and for which the set $\mathcal{P}$ of probabilities of the answers is given;

$$\mathcal{P} = \{p_1, p_2, \ldots, p_N\} ,$$

where $p_k > 0$ for all k and $\sum_{k=1}^{N} p_k = 1$.

The questionnaires with given support (X, Γ) and the same set $\mathcal{P}$ but distinct because of the allocation of the p_k to the answers $e_j(k) \in E$ are determined by the permutations

$$\begin{pmatrix} 1 & 2 & \ldots & N \\ j(1) & j(2) & & j(N) \end{pmatrix}$$

and constitute a sub-family of $\mathcal{F}$ containing at most $N!$

elements. There are less than N! elements if the p_k are not all distinct or if there exist certain isomorphisms of subgraphs containing terminals.

We take one element of this sub-family in order to mark or "label" each terminal. For economy of writing we will choose to use the identity permutation and in that case we have a mapping
$$P_E : E \to \,]0,1[$$
such that $P_E(e_j) = p_j$ for all $e_j \in E$.

We may now pose the following construction problem.

Problem 1 Suppose $\Lambda = (X,\Gamma)$ is a latticoid and P_E is a valuation on its terminals. Then the problem is to construct a questionnaire (X,Γ,P_Γ) such that the law of the probabilities of the answers induced by P_Γ is P_E.

It is possible to consider Problem 1 again for each of the valuations which may be formed from (X,Γ) and $\mathcal{P}$, and this amounts to solving the problem generalized to "non-labelled" answers.

Note that, for any vertex $e \in E$ with only one entering arc (i,e), we have $p(i,e) = p_E(e)$ and this will mean that the solution of Problem 1 is trivial in the first of the cases to follow.

Since the valuation p_E is the restriction to E of the mapping $p: X \to \,]0,1[$ defined in §4.22, we may omit the indices from the notation p_E.

1. Λ *is an arborescence*

Here every vertex $i \in X-\{\alpha\}$ is the terminal extremity of only one arc. Hence as noted above

(i) for $i \in \Gamma^{-1}e_j$ and $e_j \in E$, we put $p(i,e_j) = p(e_j)$
and
(ii) if $p(i,j)$ has been determined for all $j \in \Gamma i$, then we put $p(i) = \sum_{j \in \Gamma i} p(i,j)$ and then $p(h,i) = p(i)$ as soon as $i \in \Gamma h$.

Recursive use of these rules leads to a description of the arborescence by a process acting from upstream to downstream and facilitates a unique determination of $p(i)$ and $p(h,i)$ for all vertices $i \in X-\{\alpha\}$ and all arcs $(h,i) \in \Gamma$.

Thus we find a unique mapping P_Γ with $p(\alpha)=1$.

2. Λ *is a Dedekindian or a latticoid*

a. It is possible to construct the arborescence A of the paths in Λ and with the help of certain degrees of freedom to reduce this case to the preceding one.

If A has V terminal vertices, then we know the image $T(e) \subset A$ of each $e \in E$.

If $|T(e)|=1$, say $T(e)=\{e_c\}$, then

$$P_C(E_C(T(e))) = p(e)$$

whatever the arborescent questionnaire based on A and compatible with Λ.

If $|T(e)|=c_e \geq 2$, say $T(e) = \{v_1, v_2, \ldots, v_{c_e}\}$; and we must realize $p_C(v_k) > 0$ for all $k = 1, 2, \ldots, c_e$ and

$$\sum_{k=1}^{c_e} p_C(v_k) = p(e) \tag{17}$$

Thus we have one equation with c_e unknowns which gives $c_e - 1$ degrees of freedom. The same is true for any vertex $e \in E$ so that by means of

$$\sum_{e \in E} (c_e - 1)$$

arbitrary parameters (strictly positive and such that (17) is satisfied) we will have determined a mapping $P_{E_C}: E_C \rightarrow]0,1[$, E_C being the set of terminals of A such that the arborescent questionnaire is compatible.

A downstream to upstream process consisting in a sequence of contractions at $T(i)$ for any $i \in X - \{\alpha\}$ then achieves the construction of a questionnaire solution of Problem 1.

b. At any vertex $i \in X$ with only one entering arc (h,i) it is possible to determine the probability of this arc starting from the vertex because $p(h,i) = p(i)$.

If a vertex i (question or answer of the questionnaire under discussion) has b_i entering arcs, then the probabilities of these arcs are such that

$$\sum_{k=1}^{b_i} p(h_k, i) = p(i) \tag{18}$$

The solution of this equation requires $b_i - 1$ arbitrary strictly positive parameters.

Operating once more downstream to upstream - the graph is finite - we must have recourse to

$$\sum_{i \in E \cup \overline{F}} (b_i - 1)$$

arbitrary parameters to determine the probabilities of the arcs enabling us to ensure that $p_E: E \rightarrow]0,1[$ in conformity with the data.

The determination of a compatible questionnaire then requires

$$\sum_{i \in \overline{F}} (a_i-1)(c_i-1)$$

other parameters of the type $\lambda:(.)$.

Thus in conclusion, when (X,Γ) is latticoid, we can choose between two algorithms:

a. Determine directly a compatible questionnaire and then regroup the elements belonging to the image of a vertex or an arc of Λ.

b. Determine progressively $p(i)$ and $p(h,i)$ for all $i \in X$ and all $(h,i) \in \Gamma$ starting from the arcs entering into E and going up to the arcs coming out of the root.

From the point of view of solving Problem 1 this second algorithm is much quicker than the first because it requires only the determination of the probabilities of the $m(\Lambda)$ arcs of Λ, whereas (a) also requires that of the probabilities of the $m(A)$ arcs of A. However solution (a) offers the advantage of forming an arborescent compatible questionnaire C which is necessary for the construction of the probability space $(E_C, \mathcal{P}(E_C), P_C)$.

Algorithm (a) requires

$$\sum_{e \in E} (c_e-1)$$

arbitrary parameters to determine C before determining P_Γ and no further arbitrary parameters to determine P_Γ from C.

Algorithm (b) requires

$$\sum_{i \in E \cup \overline{F}} (b_i - 1)$$

arbitrary parameters to determine P_Γ and then (cf. §4.3.1)

$$\sum_{i \subset \overline{F}} (a_i-1)(c_i-1)$$

arbitrary parameters are necessary to determine C.

The two methods are equivalent when obtaining an arborescent questionnaire C compatible with a questionnaire (X,Γ,P_Γ) in which P_E is imposed. In both cases the probabilities of $m(A) + m(\Lambda)$ arcs are to be determined.

The number of arbitrary parameters required by (a) and (b) is in fact the same because to write

$$\sum_{e \in E} (c_e-1) = \sum_{i \in E \cup \overline{F}} (b_i-1) + \sum_{i \in \overline{F}} (a_i-1)(c_i-1)$$

amounts to

$$\sum_{e \in E} (c_e - b_e) = \sum_{i \in \bar{F}} \{(a_i-1)(c_i-1) + (b_i-1)\} \tag{19}$$

which is equivalent to Property 4.10.

The solution of Problem 1 can now be summarized as:

Theorem 1 The solution of Problem 1 is unique if and only if the support Λ is an arborescence. If Λ is a latticoid the valuation P_Γ depends on $\sum_{i \in E \cup \bar{F}} (b_i-1)$ parameters, where b_i is the number of arcs entering into the vertex i distinct from the root.

5.3 L-Optimal Supports

Definition 3 A questionnaire in which the routing length is minimal among the routing lengths of a family of questionnaires is called a *questionnaire with optimal routing* or an *L-optimal questionnaire*.

Given a graph Λ and a mapping P_E, it is possible to construct at least one questionnaire such that P_E is the restriction E of the valuation of X. We propose to determine the nature of the support of an L-optimal questionnaire when the set of bases $\mathcal{A}$, the number of answers N and the distribution $\mathcal{P}$ are all fixed.

5.3.1 Homogeneous Questionnaires

Problem 2 Determine what type of support a homogeneous questionnaire with basis a must have in order that the routing length may be minimal when $\mathcal{P} = \{p_1, p_2, \ldots, p_N\}$ is given.

We need to determine $Q_o \in \mathcal{F}$ such that

$$L(Q_o) = \min_{Q \in \mathcal{F}} L(Q), \tag{20}$$

noting that we can impose any restriction which is not contradictory with Definition 4.1. In particular the events $E_C(i)$ associated with the questions i of the arborescent questionnaire compatible with Q can be arbitrary within the limits defined by

$$|E_c(i)| \geq a \quad \text{and} \quad E_c(i) \subset E_c .$$

The solution of Problem 2 is given by

Theorem 2 There exists an arborescent questionnaire which is L-optimal in the family of homogeneous questionnaires with basis a in which the probability distribution $\mathcal{P}$ of the answers is fixed.

The proof will be given in two steps corresponding to the cases in which the relation of strict compatibility (4.2) of arborescent questionnaires is and is not satisfied respectively.

5.3.1.1   (a-1) is a divisor of (N-1)

Let $\Lambda = (E \cup F, \Gamma)$ be a latticoid having N terminals and in which the non-terminal vertices all have the same outdegree $a > 1$. Γ necessarily has $|F| \geq \frac{N-1}{a-1}$ non-terminal vertices, with equality if and only if Λ is acyclic. Let us make the follow- hypothesis: (H) Λ is not an arborescence and the questionnaire Q_0 (a solution of Problem 2) has support Λ and valuation of the terminals P_E, determined by $\mathcal{P}$ and a permutation of the indices.

We will show by a sequence of transformations that (H) leads to a contradiction.

T.1 Let S be the set of arborescences with exactly N terminals and which are partial sub-graphs of Λ and let $A_1 \in S$ be the arborescence possessing property P1:

P1 Whatever the terminal e of A_1 and of Λ, the path [αe] of A_1 is the path admitting the least number of arcs among the paths linking α to E in Λ.

Let us first show that A_1 exists and can be constructed by an algorithm c:

1. For each $e \in E$ determine one of the smallest non-weighted paths [αe] by using a classical algorithm. Let μ_e be such a path.

2. Let $\mathcal{C}_1 = \{\mu_e | e \in E\} \subset \mathcal{C}$ be the set of those paths. Determine the graph A_1 of which $\mathcal{C}_1$ is the routing.

Now A_1 is a partial sub-graph of Λ having N terminals each being linked to α by a unique path. Hence A_1 is an arbores- cence satisfying P_1.

We remark that A_1 may not be a partial graph of Λ. If for example there exists a vertex k of Λ such that the paths linking α and E and passing through k are all longer than the paths not passing through k (Fig. 5.3), then the algorithm constructed in the preceding way does not give any path passing through k.

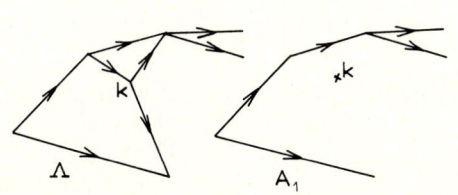

Figure 5.3

Moreover some paths of A_1 may have only one successor, which violates (q3); respec- ting the final remark of §4.5 we will nevertheless still speak of the question- naire Q_1 with support A_1, since $L(Q_1)$ is defined with- out any difficulty.

The quasi-questionnaire Q_1 whose terminals have the same

valuation P_E as in Q_0 is easily constructed. The routing lengths are:

$$L(Q_0) = \sum_{e \in E} \sum_{v \in T(e)} p_C(v) r_C(v)$$

and

$$L(Q_1) = \sum_{e \in E} p(e) r_1(e)$$

with

$$p(e) = \sum_{v \in T(e)} p_C(\ ) \quad \text{and} \quad r_1(e) = \min_{v \in T(e)} r_C(v)$$

C being an arborescent questionnaire compatible with Ω_0.

Consequently

$$p(e) r_1(e) \leq \sum_{\in T(e)} p_C(v) r_C(v) ,$$

with equality if and only if

$$r_C(v) = \text{cst.} \quad (\forall v \in T(e)) .$$

From this it follows that

$$L(Q_1) \leq L(Q_0) , \tag{21}$$

and there would be equality if and only if all the discarded paths did not possess more arcs than the paths with the same terminal extremity which have been preserved in A_1 (case where Λ is Dedekindian).

T.2 The quasi-questionnaire Q_1 is not necessarily homogeneous. In particular it may contain vertices ξ with basis $a_\xi=1$ which are transmitters (cf. §1.2) and which cannot be preserved later on.

Any vertex ξ such that $a_\xi=1$ will be removed from the graph by an operation of contraction at $A \supset \xi$ (cf. §3.3) as shown in Fig. 5.4.

If A_1 has h transmitters, then h contractions will transform A_1 into a new arborescence A_2 in which all the outdegrees are strictly greater than 1. In the questionnaire Q_2 with support A_2 the answers which are terminal extremities of paths through the transmitters of A_1 have rank $r_2 < r_1$, so that

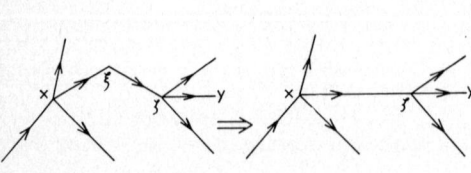

Figure 5.4

$$L(Q_2) \leq L(Q_1) \tag{22}$$

with equality if and only if all the vertices i of A_1 have a basis $a_i > 1$.

T.3 If in A_2, there exists a vertex η with basis a_η such that $1 < a_\eta < a$, then we will transform A_2 in such a way as to obtain a homogeneous arborescence if possible. Since $a - 1/N - 1$, there exists a strict polychotomic arborescence with basis a having N terminals.

Let J and H be two non-empty families of indices and $\{\eta_j | j \in J\}$, $\{\zeta_h | h \in H\}$ be a partition into two classes of the vertices of A_2 with basis less than a such that the ranks are:

$$r_2(\eta_j) \leq r_2(\zeta_h) \quad (\forall j, h) .$$

Make the substitution of arcs between some ζ_h and the η_j in such a way as to obtain in the transformed arborescence A_3 (Fig. 5.5) vertices η_j with basis a and vertices ζ_h with basis 1.

The ranks of a terminal vertex descending from ζ_h in A_2, $r_2(e)$ and from η_j in A_3, $r_3(e)$ will be such that

$$r_2(e) \leq r_3(e)$$

with equality if and only if

$$r_2(\eta_j) = r_2(\zeta_h) .$$

The ranks of the answers which do not descend from some ζ_h will be left unchanged in such a way that the arborescent questionnaire Q_3 with support A_3 and with valuation P_E is such that:

$$L(Q_3) \leq L(Q_2) \tag{23}$$

with equality if and only if all the vertices with basis less than a are of the same rank.

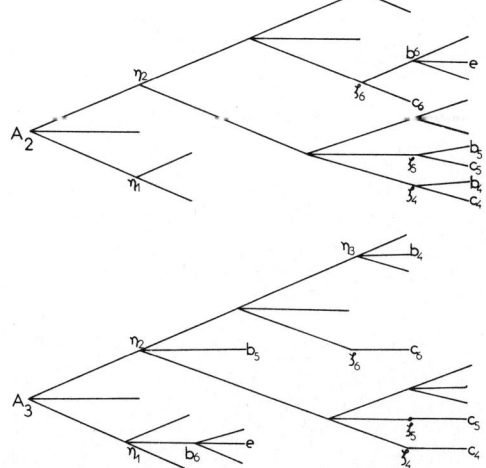

Figure 5.5

T.4 The arborescence A_3 therefore has internal vertices with bases equal to a or 1. Making contractions as in T.2 we obtain a polychotomic questionnaire Q_4 with support A_4 such that
$$L(Q_4) \leq L(Q_3) \tag{24}$$
with equality if and only if A_2 is polychotomic.

Finally
$$L(Q_4) \leq L(Q_o)$$
and by (21) to (24) there is equality if and only if Q_o is Dedekindian and Q_1 is homogeneous with basis a.

But, according to formula 4.2 a homogeneous Dedekindian possesses at least one vertex more than any homogeneous arborescence having the same terminals. The search for the partial sub-graph of Λ reduces to suppressions of arcs and vertices. If all the vertices are preserved there exists at least one with basis smaller in A_1 than in Λ so that (22), (23) or (24) is a strict inequality. If a vertex k is suppressed, then the arcs entering into k are also suppressed, hence the predecessors of k (in Λ) have a basis which is smaller in A_1 than in Λ. If all the ancestors of k were suppressed, A_1 would not have the same root as Λ, which is absurd. Consequently if Q_o is Dedekindian, Q_1 has at least one vertex x with basis a_x which is non-zero and less than the outdegree of x in Λ and Q_1 is not homogeneous.

It follows that one of the inequalities (22) to (24) is necessarily strict and therefore:
$$L(Q_4) < L(Q_o) \tag{25}$$
which implies the impossibility of (H). This completes the proof of Theorem 2, when $a-1 | N-1$ and also gives the sharper result:

Property 1 If $a-1$ is a divisor of $N-1$, then no homogeneous Dedekindian questionnaire is L-optimal for fixed a and $\mathcal{P}$.

5.3.1.2   (a-1) is not a divisor of (N-1)

Put $N = v(a-1) + \beta + 1$, where v and β are integers and $0 < \beta < a-1$; v is the number of questions of a polychotomic questionnaire having $v(a-1)+1$ answers.

If the support of the optimal questionnaire had at least $v+2$ questions, then a succession of transformations analogous to T_1, T_2, T_3, T_4 would reduce this number to $v+1$ and produce a homogeneous arborescent questionnaire in the broad sense.

Suppose therefore that Q_1 is a polychotomic questionnaire containing $v+1$ questions and N_1 answers such that
$$N_1 = (v+1)(a-1) + 1$$

Then the $N_1-N = a-1-\beta$ supplementary answers are obtained by the systematic valuation $\frac{Pm}{a-\beta}$ and by substitution of the probability $\frac{Pm}{a-\beta}$ for the probability Pm of an element of $\mathcal{P}$. The optimal questionnaire Q having N_1 answers is indeed arborescent according to Property 1 and each of its $\nu+1$ questions possesses a outcomes.

Making a contraction at $A \subset E$ by regrouping the $a-\beta$ answers with probability $\frac{Pm}{a-\beta}$, the new vertex s deduced from A will have probability

$$p(s) = \sum_{e \in A} p(e) = p_m .$$

1. If these answers are the outcomes of $a-\beta$ distinct questions with the same rank, then the questionnaire obtained will be Dedekindian, that is Q_2.

2. If these answers are the outcomes of b questions, distinct and of the same rank, where $1<b \leqslant a-\beta$, then the contraction will lead to an s-graph having cycles. Suppression of the multiple arcs without modification of the valuation of the answers will lead to a Dedekindian having at most $a-\beta$ questions with basis less than a. We will still use Q_2 to denote this Dedekindian questionnaire.

3. If the answers are the outcomes of b distinct questions ($1<b \leqslant a-\beta$) not all of the same rank, then the contraction will lead to a non-Dedekindian latticoid ($b=a-\beta$) or to an s-graph with cycles which does not satisfy the Jordan-Dedekind condition. Suppression of the multiple arcs and the arcs with origins not of the smallest rank produces a Dedekindian questionnaire which we will still denote by Q_2 and which will admit at most $a-\beta$ questions with basis less than a.

From this Dedekindian questionnaire Q_2, it is possible to deduce a homogeneous arborescent questionnaire in the broad sense, Q_3, by making at most $a-\beta$ suppressions of arcs and such that $L(Q_3)=L(Q_2)$.

If the question with basis less than a has a rank less than that of some other question, then substitutions of arcs will allow a reduction of the routing length as in the transformation T3 of §5.3.1.1. Let Q_4 be the new polychotomic questionnaire in which the question with basis $\beta+1$ is of maximal rank (among questions).

Any Dedekindian questionnaire of type Q_2 is such that

$$L(Q_2) \geqslant L(Q_4) \tag{26}$$

and there is no questionnaire which has fewer questions than Q_4: therefore there exists a questionnaire having as support a homogeneous arborescence in the broad sense which is a solution of Problem 2. Thus in all cases there exists a polychotomic questionnaire, which is a solution of Problem 2 and if the transformations 2 and 3 are made the Dedekindian is no longer homogeneous in the strict sense.

This completes the proof of Theorem 2 and we get:

Property 2 There exists a Dedekindian questionnaire solution of Problem 2, when a-1 is not a divisor of N-1 if and only if a polychotomic questionnaire solution has at least a-β questions the outcomes of which have rank equal to the height of the arborescence.

5.3.2 Heterogeneous questionnaires

The extension of Theorem 1 to the case of heterogeneous questionnaires requires certain precautions. If the maximum number of quesions was not fixed for each basis, it is clear that substitution of arcs coming out of vertices having as basis the greatest possible basis (replacing arcs coming out of smaller bases) would allow the creation of supports with vertices having only one arc coming out. Contractions will reduce the routing length and we would be brought back to the homogeneous problem with a fixed basis which, as we are able to show, is equal to the greatest one. If the number of questions in each basis is fixed and satisfies the compatibility relation (4.1), then the only admissible solution, for N fixed, is an arborescent questionnaire. That is why the extension of Problem 2 takes the form:

Problem 3 Determine the type of the support of L-optimal heterogeneous questionnaires having a maximum of q_k questions with basis a_k where k belongs to a non empty set of indices K and

$$\mathcal{P} = \{p_1, p_2, \ldots p_N\}$$

is fixed.

For the problem always to have a solution, we also need the hypothesis that the number q_1 of questions with base $a_1=2$ is unlimited.

We know that a dichotomic questionnaire exists for all N because (a_1-1) is always a divisor of N and this hypothesis ensures only that we are able to construct at least one questionnaire for any given N.

Supposing that the support of a solution Q_0 of Problem 3 is latticoid, we perform transformations analogously as in the case of strict homogeneity.

T_1 and T_2 allow the construction of an arborescence A_1 as a partial subgraph of the latticoid by eliminating the transmitters while forming an arborescence A_2.

T_3 will be somewhat modified. Let J and H be disjoint subsets of K; The vertices of the first subset will be denoted by η_j and will have basis $a_j, j \in J$; Similarly for ζ_h with basis a_h

and we will suppose that $a_j > a_h (\forall j, h)$. A substitution of arcs will be made between the ζ_h and the η_j in such a way as to create the maximal number of vertices η_j with basis as great as possible among the a_k which are not saturated, and the maximum number of vertices ζ_a with basis 1 or 2. A reduction of L follows.

After an operation T_4 of contraction made with a view to eliminating the vertices with only one arc coming out, we obtain a questionnaire Q_4 in which the number of questions with basis 2 is

$$q_2 = N - 1 + \sum_{\substack{k \neq 2 \\ k \in K}} q_k - \sum_{\substack{k \neq 2 \\ k \in K}} a_k q_k \qquad (27)$$

The routing length of Q_4 is

$$L(Q_4) \leq L(Q_o) \qquad (28)$$

with equality if and only if Q_o is Dedekindian and if the intermediate inequalities concerning $L(Q_i)$, for $i = 1, 2, 3$ are equalities.

In the case when the relation (4.1) can be satisfied with saturation of the greatest bases, at least one of these inequalities is strict and the only solution of Problem 3 is an arborescent questionnaire; the Dedekindian will only be a solution in a case where the questions of basis 2 are necessary without saturation of the number of questions of greater basis.

Note that any homogeneous questionnaire in the broad sense can be considered as a heterogeneous questionnaire with $|K|=2$.

Theorem 3 There exists an arborescent questionnaire which is L-optimal in the family of heterogeneous questionnaires with basis $a_k (k \in K)$ when the maximal number q_k of questions with basis a_k and the probability distribution $\mathcal{P}$ of answers are fixed.

5.4 Properties of Arborescent Questionnaires

Theorems 2 and 3 establish the existence of arborescent questionnaires possessing the property of L-optimality. Before trying to build them directly it is useful to determine some properties related to an arborescent support.

5.4.1 The Number of Vertices and Notation

Property 3 In a polychotomic questionnaire in the strict sense, the number of vertices of any rank is a multiple of a.

$Q = (X, \Gamma, P_\Gamma)$ possesses M-1 internal vertices and (a-1)M+1 terminal vertices and there are therefore aM proper descendants of α, a of which are successors. The vertices with rank r possess a successors if they are questions and 0 successors otherwise. Consequently for any r, the number of vertices with rank r+1 is a multiple of a as long as r+1 ≤ h, the height of Q.

If the questionnaire is polychotomic in the broad sense, then there exists a question having β+1 outcomes only.

Property 3 then becomes:

Property 4 In a polychotomic questionnaire in the broad sense the number of vertices of any rank except 1 is a multiple of a.

Any vertex z of a polychotomic questionnaire in the strict sense will be referred to by two indices [r,s], where r is the rank of z (0 ≤ r ≤ R) and s specifies the position of z among the vertices of rank r; s is an r-digit number expressed to the base a. The digit of greatest weight (to the left) will be 0, 1, 2 ...... or (a-1) according as the path linking α to z will have passed along the arc linking α to $\Gamma_0 \alpha, \Gamma_1 \alpha, \ldots$ or $\Gamma_{a-1}\alpha$; the following digit will also be 0, 1, 2 ...... or (a-1) according to the arc taken by this path between rank 1 (second question) and rank 2 (third question). Similarly for all the digits of s, the rth digit of s, that of smallest weight, placed to the right, indicating the arc taken after the rth question linking a vertex of rank r-1 to the vertex z with rank r.

s may be expressed in a different system of numeration without giving rise to any confusion with the help of the index r; for example the number expressed in decimal by s and in the system with base a by αβγ...σ will be part of the labelling indices of the vertices:

r	αβγ...σ	or	[r , s] ,
r + 1 ,	0αβγ...σ	or	[r + 1,s] ,
r + 2 ,	00αβγ...σ	or	[r + 2,s] ,
r + 3 ,	000αβγ...σ	or	[r + 3,s] ,

This notation makes it possible to write very quickly all the successors of any question x ∈ F. For let x = [r,s] be a vertex of F. Then the successors of x are obtained by multiplying the number s by a (that is to say in the system with base a by making a shift to the left and adding a zero on the right) then by adding a unit of least weight corresponding to the index, comprised between 0 and a-1, of the arc taken at the exit of x. The successors of x are then

[r+1,as] , [r+1,as+1] ,..., [r+1,as+(a-1)].

Let k_0 by the minimal rank of an answer; then s will take all the values between 0 and a^r-1 for all $r \le k_0$. There are therefore

$$X_{k_0} = 1 + a + a^2 + a^3 + \ldots + a^{k_0-1} = \frac{a^{k_0}-1}{a-1} \qquad (29)$$

questions of rank less than the rank of any answer.

If the answers are not all of the same rank, between k_0 and h the values taken by s will be distributed in sequences of a consecutive values comprised between 0 and a^r-1.

Every vertex of a polychotomic questionnaire in the broad sense or of a heterogeneous arborescent questionnaire will be specified in an analogous way. We must first determine the maximum value of a_m, that is a_M, then express the vertices with the aid of two indices $[r,s]$ where s is an r-digit number expressed to the base a_M. The system of the s thus formed is sparse if only a few questions are of basis a_M because few vertices will have a numeral taking the value a_M-1 but there will be no ambiguity. If the vertex $[r,s]$ possesses c outcomes ($c \leq a_M$), its successors will be denoted by

$$[r + 1, a_M s], [r + 1, a_M s + 1] ,...,$$
$$[r + 1, a_M s + i] ,..., [r + 1, a_M s + c - 1]$$

and will obviously be distinct from the successors of $[r,s-1]$ and of $[r,s+1]$ if they exist because the nearest values of the indices s of these vertices will be at worst:

$$a_M(s - 1) + a_M - 1 < a_M s$$
$$\text{and} \quad a_M(s + 1) > a_M s + c - 1 .$$

Remark These notations may also be used to describe the paths of Dedekindians and latticoids, but we must adjoin two rules to denote the vertices of these graphs:

1. The code $[r,s]$ of a vertex i with several entering paths will have index of rank equal to the smallest index r of the paths entering into i.

2. If several paths denoted by $[r,s_j]$ ($j \in J$ where J is a family of indices), enter into a vertex i and all have the same length r, then i will be denoted by $[r, \min_{j \in J} s_j]$.

The coding of the vertices obtained by application of these rules corresponds to a unique arborescence, being a partial graph of the latticoid. However in the second rule we can substitute a different citerion of choice in place of $\min_{j \in J} s_j$.

5.4.2 Arborescences of minimal height

Problem 4 Construct a homogeneous arborescence with basis a having N terminals and for which the height is minimal.

We distinguish three cases according as N is a power of k and then according, as a-1 is a divisor of N-1 or not.

a. $N = a^k$:

By (4.2) the number of questions is:

$$M = \frac{N-1}{a-1} = \sum_{r:=0}^{k-1} a^r. \qquad (30)$$

In view of Property 3, it is possible to construct an arborescence A whose N terminals are of rank k: for any rank r there are a^r questions.

If a terminal was of rank r<k, then there would be at most $a^{r+1}-a$ vertices of rank r+1 so that such an arborescence with basis a having a^k terminals cannot be of height less than k.

b. $\quad N = a^k + \alpha(a-1), \ 0 < \alpha < a^k$ (¹)

Consider the arborescent solution of Problem 4 for $N'=a^{k+1}$.

Any contraction of a set A consisting of question i and its a successors reduces the number of terminals by a-1 and creates a terminal with rank k.

Let us make $(a^k-\alpha)$ such contractions of sets the terminals of which are all of rank k+1. Then we obtain an arborescence A' having

$$a^{k+1} - (a^k - \alpha)(a - 1) = N$$

terminals, that is an arborescence of height k+1 and of minimal rank k.

c. $\quad N = a^k + \alpha(a - 1) + \beta, \quad 0 \leqslant \alpha < a^k$ and $0 < \beta < a - 1$.

Start with an arborescent solution of Problem 4 for

$$N'' = a^k + (\alpha + 1)(a - 1)$$

A contraction of a terminal of rank k+1 and its predecessor preserves the number of terminals with rank k and reduces by 1 the number of terminals with rank k+1.

We make a-(β+1) such contractions of the sets {i,j}, where $j \in \Gamma i$. This is always possible because i has a successors in the homogeneous arborescence in the strict sense and a-1>β.

This allows the production of a homogeneous arborescence in the broad sense having

$$a^k + (\alpha+1)(a-1) - [a - (\beta+1)] = N$$

vertices which are of height k+1 as in the preceding case. This gives

Property 4 Whatever a and N, there exists a homogeneous arborescence in the strict sense or in the broad sense for which the

___

(¹) We will take care not to make any confusion between the integer α and the root a.

difference between the ranks of the terminals is at most equal to 1 and the height is minimal.

Definition 4 An arborescence, homogeneous or heterogeneous in which the ranks of the terminals differ by 1 at most is called a *balanced arborescence*.

If $N = a^k$ we have seen that the arborescence of minimal height is obtained when all the terminals are of rank k; otherwise, if some terminal is of rank k-1, then there are necessarily a of them for which the rank is k+1.

It is not always the same when

$$N = a^k + \alpha(a-1) + \beta$$

and $\alpha\beta > 0$.

The contraction of the set formed by one question of rank k-1 and its successors (terminals) followed by a restricted product bearing on an arborescence with N-(a-1) terminals and an arborescence with a terminals leads to an arborescence having:

1 terminal with rank k-1

$a^k - (\alpha+1) - (a+1)$ terminals with rank k

$(\alpha+1)a+\beta+1$ terminals with rank k+1

if
$\quad \alpha + a + 2 \leq a^k$,

and requires a question of rank k+1 if

$$\alpha + a + 2 > a^k .$$

If the inequality $\alpha+a+2 \leq a^k$ holds, then there exists at least one arborescence of height k+1, one terminal of which has a rank less than k. Otherwise it is left to the reader to study the contractions bearing on the question with basis $\beta+1$. Such an arborescence is a solution of Problem 4 which is a minimax problem (that is, minimum of the maximal rank) whereas the balanced arborescence is another solution of the more precise problem of maximin (that is maximum of minimal rank).

Property 5 Any arborescence of height k+1 and having $N=a^k+\alpha(a-1)+\beta$ terminals is a solution of a minimax problem whereas only balanced arborescences are solutions of the maximin problem.

Among the balanced arborescences we distinguish the class where the question with basis $\beta+1$ is of rank k or of rank k-1 and we shall see that the minimal routing length is obtained when this question is of rank k.

5.4.3 Questionnaires with Balanced Support

Here we will construct questionnaires in which the support is a balanced arborescence.

If the probability distribution is uniform:

$$\mathcal{P} = \left\{ \frac{1}{N}, \frac{1}{N}, \ldots, \frac{1}{N} \right\}$$

that is if $p(e) = \frac{1}{N} (\forall e \in E)$, then we can determine the routing length of a balanced arborescence.

According to the values of α and β we find:

- $N = a^k \quad L = K$

- $N = a^k + \alpha(a-1) \quad 0 \leqslant \alpha \leqslant a^k$

$$L = \frac{1}{N} \{ (a^k - \alpha)k + a\alpha(k+1) \}$$

that is

$$0 \leqslant \alpha \leqslant a^k : L = k + \frac{a\alpha}{N} \qquad (31)$$

- $N = a^k + \alpha(a-1) + \beta \quad 0 \leqslant \alpha < a^k$ and $0 < \beta < a - 1$

$$L = \frac{1}{N} \{ [a^k - (\alpha+1)] k + [a\alpha + (\beta+1)] (k+1) \}$$

that is

$$0 \leqslant \alpha < a^k \text{ and } 0 < \beta < a-1 : L = k + \frac{a\alpha + (\beta+1)}{N} \qquad (32)$$

where we have written $\beta+1$ in brackets as a reminder that this last relation must not be applied when $\beta=0$.

If the probability distribution is not uniform, then L remains equal to k when $N=a^k$.

Since $k \leqslant r(e) \leqslant k+1$ in any questionnaire in which the support is a balanced arborescence it follows from the formula for the routing (4.32) that

$$k < L \leqslant k + 1$$

when there exist terminals with rank $k+1$.

In the equiprobable case these inequalities are strict, escept for extremal values for which, either $\alpha=\beta=0$, or:

$$\alpha = a^k - 1 \quad \text{and} \quad \beta > 0 ,$$

that is

$$N = a^k + (a^k - 1)(a - 1) + \beta ,$$

because in this case

$$L = k + \frac{a(a^k-1) + \beta + 1}{N} = k + 1 ;$$

but then all the terminals are of rank k+1.

Now consider the case of balanced arborescences for

$$N = a^k + \alpha(a-1) + \beta$$

with

$$0 \leq \alpha < a^k - 1 , \quad 0 \leq \beta < a - 1$$

or

$$\alpha = a^k - 1 \quad \text{and} \quad \beta = 0 ,$$

for which there exist terminals with ranks k and k+1

We show that we can find probability distributions such that L-k and k+1-L can be made as small as we please.

Let

$$\mathcal{P} = \{1-\varepsilon, \frac{\varepsilon}{N-1}, \frac{\varepsilon}{N-1}, \ldots, \frac{\varepsilon}{N-1}\} .$$

If the answer e_1, with probability $p_1 = 1-\varepsilon$, is of rank k, then

$$L = (1-\varepsilon)k + \frac{\varepsilon}{N-1} [a^k - (\alpha+2)] k$$
$$+ \frac{\varepsilon}{N-1}(a\alpha + \beta + 1)(k+1)$$

that is

$$L = k + \frac{a\alpha + (\beta+1)}{N-1} \varepsilon$$

which tends to the limit k as $\varepsilon \to 0$.

If the answer e_1 is of rank k+1, then

$$L = \frac{\varepsilon}{N-1}[a^k - (\alpha+1)] k$$
$$+ \frac{\varepsilon}{N-1}(a\alpha+\beta)(k+1) + (1-\varepsilon)(k+1)$$

or

$$L = k + 1 - (1 - \frac{a\alpha+\beta}{N-1})\varepsilon ,$$

which tends to the limit k+1 as $\varepsilon \to 0$.

Questionnaires with balanced arborescent support, of height k+1 and of minimal rank k, have a routing length satisfying

$$k < L < k + 1 . \tag{33}$$

5.4.4 Arborescences and Questionnaires of Maximal Height

Problem 5 Construct a homogeneous arborescence with basis a having N terminals for which the height is maximal.

Suppose there is a homogeneous arborescence in which:

(1) each question except one has a-1 terminals and one question as its successors and (2) the exceptional question has a terminals as its successors. Then there is a vertex $x \in F$ and $(a-1)$ vertices $y \in E$ of all ranks ranging from 1 to h-1, that is M=h. The height h is related to N by the equation

$$h = \frac{N-1}{a-1} \qquad (34)$$

which characterizes a homogeneous arborescence A_M with N terminals having the greatest possible rank. A vertex of F cannot have rank $\frac{N-1}{a-1}$ because it would then have only one descendant contrary to axiom (q_3).

If the N answers are equiprobable,

$$p(e_j) = \frac{1}{N} \quad (\forall j),$$

then A_M is the support of a questionnaire in which the routing length is

$$L_M = \frac{1}{N} \left\{ [1+2+3+\ldots+(\frac{N-1}{a-1})] \ (a-1) + \frac{N-1}{a-1} \right\},$$

that is

$$L_M = \frac{N-1}{N} \frac{N+a}{2(a-1)}.$$

A_M has the following properties:

1. L_M increases approximately in proportion with N.

2. A homogeneous arborescence in the strict sense with N terminals cannot have a height greater than $\frac{N-1}{a-1}$.

If the arborescence is homogeneous in the broad sense, then $N = a^k + \alpha(a-1) + \beta$ $(\beta > 0)$ and the question with basis $\beta+1$ can be of rank r, such that $0 \leq r \leq h-1$. Putting $N'' = N+a-(\beta+1)$, we find that the height is

$$h = \frac{N''-1}{a-1} \qquad (36)$$

and the routing length, in the case of a uniform distribution when the question with basis $\beta+1$ is of rank h-1, is

$$L = \frac{N''-1}{N} \frac{N+\beta+1}{2(a-1)},$$

If the question with basis $\beta+1$ is of rank r, then h remains unchanged whereas the routing length will be increased by

$$\delta = [\frac{N''-1}{a-1} - (r+1)] \frac{a-(\beta+1)}{N}.$$

Note that the factor in the brackets is an integer.

The maximal value of L obtained when the question with basis $\beta+1$ is the root, so that r=0, is

$$L_M = \frac{N''-1}{N} \frac{N+\beta+1}{2(a-1)}$$
$$+ \frac{1}{N} \frac{[N - (\beta+1)] \quad [a - (\beta+1)]}{a - 1} \qquad (37)$$

For non-uniform distributions, we can obtain an upper bound for the routing length of the polychotomic questionnaire in the broad sense.

Taking an answer with probability $1-\varepsilon$ at rank h and the question with basis $\beta+1$ at rank 0 we find

$$L = \frac{N''-1}{a-1} - 0(\varepsilon) . \qquad (38)$$

5.4.5 Extremal properties of the supports

It is possible to resolve some other problems which demonstrate properties of duality between (non-arborescent) latticoids and arborescences.

According to relations 4.2 and 4.2' we see that arborescences are the solutions of

Problem 6 Construct a latticoid in which the set of non zero outdegrees $\mathcal{A}$ is given and maximising the number of terminals N.

In the homogeneous case and when $a-1|N-1$ this problem has the same solutions as

Problem 7 Construct a homogeneous latticoid with N terminals having the minimal number of questions.

We can show that certain arborescences, solutions of Problem 7, are also solutions of

Problem 8 Construct a homogeneous latticoid with minimal height for given N and a when $a-1|N-1$.

We will see that latticoids which are not solutions of Problem 7 are however solutions of Problem 8 if N is not a power of a.

The dual of Problem 7 can be put in the form:

Problem 9 Construct a homogeneous latticoid with M questions having the minimal number of terminals.

A more interesting problem may be formulated by using maximal paths, that is paths which can not be extended (cf. §1.6.1).

We know that the number of maximal paths of a latticoid is $V=|\mathcal{C}|$, the number of paths linking the root to the terminals; moreover $V \geqslant N$ and equality holds if and only if the latticoid is an arborescence.

Problem 10 Construct a homogeneous graph of outdegree a and without circuits having the maximal number of distinct maximal paths. We ask that only one arc enters into any terminal.

The solution of Problem 10 is a particular latticoid.

The number of "questions" having one terminal among their successors is a and there are

$$a + (a-1) + \ldots + 1 = a\frac{a+1}{2}$$

terminals.

The construction of this latticoid, which we will call the *Fibonacci latticoid*, is made downstream to upstream. For $1 \leq i \leq a$, the ith question constructed has i-1 questions and a-i+1 terminals as its successors. The other questions admit as successors the a questions constructed last: the a+hth question has as successors the questions constructed in the

$$h^{th}, h+1^{st}, \ldots, h+(a-1)^{st}$$

place (Figures 5.6 and 5.7).

The number V of maximal paths is related to the number of questions M by

$$V(M) = \sum_{m=1}^{M-1} V(M) + a - (M-1),$$
$$\text{if } M \leq a \quad (39)$$

with $V(1) = a$, and

$$V(M) = \sum_{m=M-a}^{M-1} V(m),$$
$$\text{if } M > a. \quad (40)$$

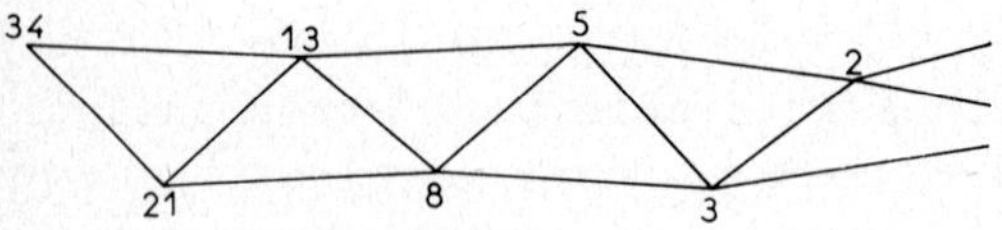

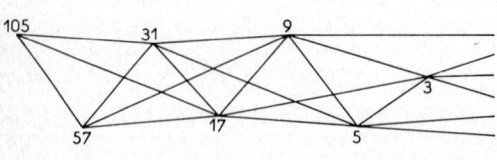

Figure 5.6

Figure 5.7

When a=2, the elements of the sequence V(M) are the same as those of the Fibonacci series and this justifies the name given to these latticoids.

We remark that Chaty obtained the same generalized Fibonacci sequence with a=3 while searching for the number of maximal paths (Hamiltonian) of strongly connected graphs having the minimal number of circuits.

THE CONSTRUCTION OF QUESTIONNAIRES

EXERCISES

1. Using a language such as ALGOL for example, write detailed programs for the algorithms (a) and (b) §5.2 to construct (X,Γ,P_Γ) from given (X,Γ) and P_E.

2. (i) Using algorithm (a) on the one hand and algorithm (b) on the other, show that the arborescent questionnaires compatible with the latticoid questionnaires with the support shown in Figure 5.8 have 3 degrees of freedom.

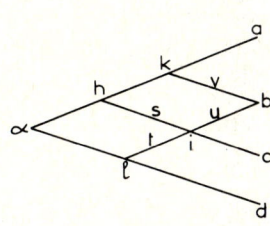

Figure 5.8

(ii) For this graph and the equiprobable distribution $\mathscr{P} = \{\frac{1}{4}, \frac{1}{4}, \frac{1}{4}, \frac{1}{4}\}$ take

$$\frac{p(v)}{p(u)} = \frac{1}{2}, \quad \frac{p(t)}{p(s)} = \frac{3}{2}, \quad \frac{p(b \;|[\alpha h i])}{p(b \;|[\alpha \ell i])} = 2$$

and determine P_Γ, then a compatible arborescent questionnaire by using algorithm (b). Determine then the choice of free parameters of algorithm (a) which produces the same compatible arborescent questionnaire as well as the value of $\lambda_b^h(i)$.

3. It is proposed to construct a latticoid questionnaire Λ with minimal routing length when (X,Γ) and P_E are fixed.

Determine P_Γ according to the algorithm of Problem 1 in such a way as to render L minimal and examine the two cases: Dedekindian and latticoid but not Dedekindian. Show that we can reduce L only in the second case.

Show that in the Dedekindian case we can however form certain valuated graphs.

$$G = (X,\Gamma,P_\Gamma),$$

having an arc with zero flow and which do not then satisfy axiom (q_4). Deduce from this by transformations a questionnaire

$$Q = (Y,\Delta,P_\Delta),$$

P_Δ inducing the distribution P_E on the set of answers

$$E \subset X \cap Y.$$

Show that $L(Q) < L(\Lambda)$.

As an example, examine the graph shown in Figure 5.9 with the distribution P_E or P_E'.

P_E	P'_E
$p(e_1) = 0.4$	$p(e_1) = 0.399$
$p(e_2) = 0.3$	$p(e_2) = 0.3$
$p(e_3) = p(e_4) = p(e_5) = 0.1$	$p(e_3) = p(e_5) = 0.1$
	$p(e_4) = 0.101$.

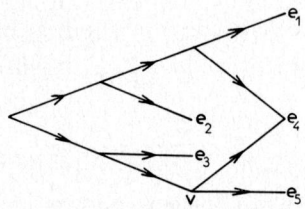

Figure 5.9

Do these results still hold if we impose the extra condition $p(v, e_4) \leq 0.1$?

Chapter Six

OPTIMAL ROUTING

6.1 Determination of an L-optimal questionnaire

Consideration of Problems 2 and 3 of the preceding chapter has shown that there always exists an arborescent questionnaire among the questionnaires of optimal routing determined by a pair $(\mathcal{A}, \mathcal{P})$ where $\mathcal{A}$ is a set of bases of the questions and $\mathcal{P}$ is a probability distribution of the N answers. We propose to study and to construct these optimal arborescent questionnaires, noting that the latticoid questionnaires with the same routing length can then be deduced when the conditions of Property 5.2 and §5.3.2 are satisfied.

We will establish necessary conditions and a sufficient condition for L-optimality and will give a unique construction algorithm which is applicable in the heterogeneous as well as the homogeneous case.

The construction of an arborescent questionnaire belonging to the family determined by $(\mathcal{A}, \mathcal{P})$ may be undertaken with the aid of the following two techniques:

- Starting from an arbitrary questionnaire in the family, perform a series of transformations with a view to decreasing the routing length to its minimal value.

- Supposing that the answers $p_1,\ldots,p_a$ are the outcomes of question i, group them together into one new answer and put $p_i' = p_1+\ldots+p_a$ and $p_j' = p_j$, if $j \neq 1,\ldots,a$ or i, and thereby produce a distribution $\mathcal{P}'$ of $N-(a-1)$ probabilities. With $\mathcal{A}' = \mathcal{A} - \{a\}$, we then have a questionnaire with $N-(a-1)$ answers defined by $(\mathcal{A}', \mathcal{P}')$.

The first of these techniques will be used to establish necessary conditions for L-optimality and the second to obtain a sufficient condition.

6.2 Necessary conditions for L-optimality

Suppose that Q is an arborescent questionnaire with N answers belonging to the family $\mathcal{F}$ defined by $(\mathcal{A}, \mathcal{P})$ in which the number of questions with basis 2 satisfies (5.27) in such a way that the compatibility condition (4.1) is also satisfied.

Then we will apply the various transformations to Q and study their effects on the routing length.

6.2.1 Substitutions of arcs

Case (1) There exist two elements a_x and a_y of $\mathcal{A}$ such that

$a_x = a_y - 1$. Let $x \in Q$ be a vertex with basis a_x and $y \in Q$ a vertex with basis a_y. Substitution of the arc (x,b) for the arc (y,b) leads to a new questionnaire Q^T of $\mathcal{F}$ in which the routing length is given by 5.13:

$$L(Q^T) < L(Q) \quad \text{if and only if} \quad r(x) < r(y) .$$

In Q^T, the vertex x has basis a_y and the vertex y has basis a_x.

Case (2) There exist two elements a_x and a_y in $\mathcal{A}$ such that

$$a_x = a_y - h$$

Let $b_1, \ldots, b_h$ be the successors of vertex y, with basis a_y, in Q. Then h substitutions of arcs

$$(y, b_i) \rightarrow (x, b_i) \quad \text{for} \quad i := 1, 2, \ldots, h$$

produce a new questionnaire Q^T in $\mathcal{F}$ such that $L(Q^T) < L(Q)$ if and only if $r(x) < r(y)$.

Such substitutions of arcs which reduce the routing length of Q are possible as long as there exists a pair of vertices x, y such that

$$a_x < a_y , \quad r(x) < r(y) . \tag{1}$$

Substitution of arcs can be interpreted as a global displacement of a vertex b and the sub-arborescence with root b, the arc entering into b being (y,b) in Q, and (x,b) in Q^T. We thus have:

Rule Ro *(Ordering of the bases)* The routing length of a questionnaire can be reduced by making substitutions of arcs as long as any vertex of given rank has basis greater than a vertex of smaller rank.

Whence:

Property 1 The questions of a heterogeneous L-optimal questionnaire ordered by non-decreasing ranks have non-increasing bases.

This result generalizes the condition obtained in §5.3.1.2 concerning the question with basis $\beta + 1 < a$ in a polychotomic questionnaire in the broad sense to the case of heterogeneous questionnaires. This is logical because the former questionnaire may be considered to be heterogeneous, the set of bases being reduced to

$$\mathcal{A} = \{a, \beta + 1\} .$$

6.2.2 <u>Transfers of arborescences</u>

We have already established equation (5.12)

$$L(Q^T) = L(Q) + [p(b) - p(y)] \, [r(y) - r(b)]$$

relating two questionnaires Q and Q^T obtained by the permutation of arborescences with roots b and y.

If b and y are answers then this operation reduces to a simple re-allocation of probabilities and then

$$L(Q^T) < L(Q)$$

if and only if
$$[p(b)-p(y)] \quad [r(y)-r(b)] < 0$$

which leads to

Rule R1 (*Permutation of answers* $b,y \in E$) The routing length of an arborescent questionnaire can be reduced without modifying the support by allocating the greatest probability to the answer of smallest rank and then by allocating the probabilities in a non-increasing order, to answers of non-decreasing ranks, the smallest probabilities being applied to the answers of maximal rank.

If b and y are questions, the transfer of sub-arborescences with roots b and y will give a reduction of the routing length according to the same principle. If b is a question and y an answer we will consider that y is the root of a trivial arborescence and it will be required to make a transfer according to the sign of
$$[p(b)-p(y)] \quad [r(y)-r(b)] .$$

Thus:

Rule R2 (*Permutation of sub-arborescences*) The routing length of an arborescent questionnaire can be reduced by modifying the structure of the arborescence by allocation of the greatest probability to the vertex, question or answer, with the smallest possible rank.

Whence:

Property 2 The vertices of an L-optimal questionnaire ordered according to non-decreasing ranks are allocated with non-increasing probabilities.

Consider in an arborescent questionnaire B a vertex U with rank r_0 and probability p_0 and a vertex V with rank $\geq r_0+2$ and probability p_0. All the vertices with rank r_0+1 must have a probability not greater than $p(U)$ and not smaller than $p(V)$ therefore equal to p_0. But the ancestor of V with rank r_0+1, that is W, has a probability strictly greater than $p(V)$. Thus $p(W) > p(U)$ and $r(W) > r(U)$. The routing of B is therefore non-optimal and some operation of transfer will produce an arborescent questionnaire B^T with smaller routing length. Consequently we have:

Property 3 In an L-optimal questionnaire the difference between the ranks of vertices with the same probability is never greater than 1.

Iterated application of rules R0, R1 and R2 allows the construction of a questionnaire with routing length smaller than that of the given arborescent questionnaire. If it is not possible to pursue the reduction by a transfer operation of type R1 or R2, we must perform a transfer of sub-arborescences with roots z_a and z_b of the same rank $r_a = r_b$.

Suppose the questionnaire is homogeneous of basis a.

Then the routing in Q^T will have the same length as in Q. However the type of transfer just mentioned allows a re-allocation of the probabilities in such a way that, if there are at least 2a vertices (questions or answers) of rank r, then we can arrange these vertices as follows: The a vertices $U_0, U_1, U_2,...,U_{a-1}$ with the least probabilities are the successors of the same vertex $\overline{U}$, each of the following groups of a vertices consists of the successors of some vertex and the a vertices $V_0, V_1, V_2,...,V_{a-1}$ with the greatest probabilities are the successors of the vertex $\overline{V}$.

These vertices of rank r are then given by

$$\Gamma\overline{U} = \{U_i\} \quad (i,j: = 0,1,2,...,a-1)$$

$$\Gamma\overline{V} = \{V_j\}$$

$$r(\overline{U}) = r(\overline{V}) = r - 1$$

$$p(U_0) \leq p(U_1) \leq ... \leq p(U_{a-1}) \leq ...$$

$$\leq p(V_0) \leq p(V_1) \leq ... \leq p(V_{a-1}).$$

If there exists at least one vertex, say $W = V_{a-1}$, such that

$$p(W) > \sum_{i:=0}^{a-1} p(U_i),$$

that is, if
$$p(W) > p(\overline{U}) \quad \text{with} \quad r(W) = r(\overline{U}) + 1,$$

then the arborescence can still be modified by application of R1 or R2 to produce a reduction in the routing length.

When Q is heterogeneous the new ordering of the vertices of the same rank must be carried out with the greatest of care.

Let $\overline{U}$ and $\overline{V}$ be questions of Q with rank r-1 having respectively the smallest basis, a_U, and the largest basis a_V.

The regrouping of the vertices of rank r with the smallest probabilities is made in such a way that in Q^T these vertices form the a_U descendants of $\overline{U}$; in the same way the a_V descendants of $\overline{V}$ will be the vertices of rank r with the greatest probabilities.

If after this modification, the probabilities of $\overline{U}$ and of $\overline{V}$ are such that

$$p^T(W) > p^T(\overline{U}) \quad \text{with} \quad r(W) = r(\overline{U}) + 1,$$

then the questionnaire can be modified by R1 or R2 to produce a new reduction of the routing length.

Whence we have:

Rule R3 (Transfers preserving L) When the rules R1 and R2 do not allow a reduction of L, it is necessary to reorder the vertices of the questionnaire in such a way that for each rank r, the vertices are grouped together by increasing order of probabilities, the vertices with the smallest probabilities being outcomes of the question of rank r-1 with smallest basis and the vertices of the greatest probabilities being outcomes of the question of rank r-1 with greatest basis.

6.2.3 Sub-Questionnaires

Among the transformations of questionnaires, contractions play a predominating role because they allow a reduction in the order of the support with preservation of some of the answers and substitution of an answer for a sub-arborescence.

F. Dubail has studied sub-questionnaires in which the answers are also answers in the initial questionnaire. S. Petolla has studied the sub-questionnaires obtained by contraction of the set consisting of a question and all its descendants.

Without making any hypotheses concerning the partitions of E authorized by the first of these authors or concerning the valuation of the questions by costs as by the other, we will use absolutely arbitrary sub-questionnaires which will lead us to a property including both the results of F. Dubail and S. Petolla.

Let Q be an L-optimal questionnaire and Q^σ a sub-questionnaire of Q with root σ and answers consisting of vertices of Q. The support (X,Γ) of Q admits a decomposition into various sub-arborescences of which it can be considered as being the union. Q will be formed (except for an isomorphism) by restricted products of questionnaires generated by these subgraphs. By making restricted products from left to right in the following expression we will write (Fig. 6.1)

$$Q = Q_0 \diamond Q^\sigma \diamond Q_1 \diamond Q_2 \diamond \ldots \diamond Q_t .$$

. Q_0 is obtained from Q by the contraction at $A=\hat{\Gamma}_\sigma$ and Q_0 is empty if σ coincides with the root α of Q. The terminals of Q_0 distinct from σ form a subset $E_0 \subset E$.

. The root of Q^σ is σ and its answers are a sub-set $E' \subset X$. Let $e_1,\ldots,e_x$ be the elements of $E' \cap E$.

The sub-questionnaires $Q_1,\ldots,Q_t$ have roots equal to the terminals $x_1, x_2,\ldots, x_t$ of Q which are not in E, that is the elements of $E'-E$; the terminals of $Q_1,\ldots,Q_t$ form the set $E-E'-E_\sigma$.

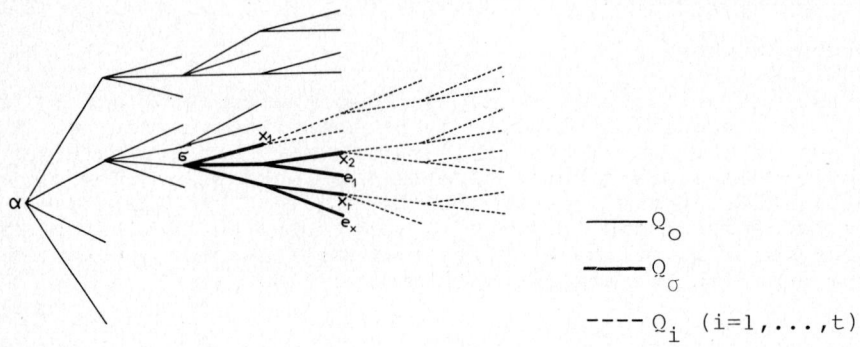

Figure 6.1

The routing length of Q can then be written as:

$$L(Q) = L(Q_o) + p(\sigma)L(Q^\sigma) + \sum_{i:=1}^{t} p(x_i)L(Q_i) \qquad (4)$$

The questionnaire Q^T has $|E'|$ answers the probabilities of which are

$$\frac{p(x_1)}{p(\sigma)}, \frac{p(x_2)}{p(\sigma)},\ldots, \frac{p(x_t)}{p(\sigma)}, \frac{p(e_1)}{p(\sigma)},\ldots,\frac{p(e_x)}{p(\sigma)}$$

Suppose that Q^T is not L-optimal. Then there exists at least one L-optimal questionnaire in the family defined by the pair $(\mathcal{A}', \mathcal{P}')$ of bases of Q and the probability distribution defined on E'. Let Q^T and $L(Q^T)$ be this L-optimal questionnaire (or one of these optimal questionnaires) and its routing length:

$$L(Q^T) < L(Q^\sigma)$$

The questionnaire defined by the restricted products

$$Q^R = Q_o \diamond Q^T \diamond Q_1 \diamond Q_2 \diamond \ldots \diamond Q_t$$

then has routing length

$$L(Q^R) < L(Q)$$

and consequently Q is not L-optimal contrary to hypothesis. This contradiction leads to a new necessary condition:

Property 4 Every sub-questionnaire of an L-optimal questionnaire is L-optimal.

6.3 A sufficient condition for L-optimality

Whereas transfers performed according to rules R0, R1 and R2 and the sub-questionnaires have enabled us to obtain the 4 necessary conditions for L-optimality (Properties 1 to 4), it is rule R3 which will enable us to construct an L-optimal questionnaire, and then to show that all L-optimal questionnaires can be generated. For R3 imposes that inside each rank the probabilities will have to be allocated in non-increasing order of vertices by sequences of a vertices ordered according to increasing indices s.

If after this modification there exists a pair of vertices satisfying inequality (2), then the arborescence obtained, A, is not yet optimal. But if this inequality is not satisfied for every pair of vertices of the arborescence A, then A possesses an optimal routing.

For in this arborescence the comparisons required by R3 are more forced than in other arborescences of the same routing length and the irreducibility of A implies that of the others.

The probability distribution $\mathcal{P}$ of answers being given by hypothesis, the a_i answers of smallest probability will have maximum rank and the comparison of their sum with the probability which is immediately greater will indicate the relative rank of certain other answers. Then the contraction of a_i answers and their common antecedent i (with determined probability) allows us to repeat the same discussion but with $N-a_i+1$ answers. Such transformations allow the construction of A, starting from the vertices with the smallest probability, and thus with the greatest rank, and progressing towards the vertices (answers and questions) of greater probabilities. The values of the probability of the questions are obtained in a unique manner as soon as R0, R1, R2 and R4 are satisfied because of the order adopted to classify the vertices of the same rank. Since the routing length is determined by the probabilities of the questions only (cf. Property 4.18) we will deduce from these that the questionnaire thus constructed is L-optimal. We now introduce some new definitions:

Definition 1 Given a pair of bases and probabilities of answers ($\mathcal{A}, \mathcal{P}$), we say that a *question σ is minimal* if $a_\sigma = \min_{a \in \mathcal{A}} a$ and if p_σ is the sum of the a_σ smallest probabilities of $\mathcal{P}$.

Note that a questionnaire can have no minimal question; none of the questions with basis $a_\sigma = \min_{a \in \mathcal{A}} a$ can have its outcomes equal to the answers with the a_σ smallest probabilities in $\mathcal{P}$. If $\mathcal{A}$ has two questions with minimal basis a_σ and $\mathcal{A}$ has at least $2a_\sigma$ smaller elements with the same value p_1, then we can construct a questionnaire having two minimal questions with probability $a_\sigma \cdot p_1$ and we may work equally well with either.

These questions are called minimal because it is not possible to construct on $(\mathcal{A}, \mathcal{P})$ a question with smaller probability.

From Definition 1 and Properties 1 and 2 we have:

Property 5 If a questionnaire Q is L-optimal and has a minimal question with rank $r(\sigma)$ and no minimal question with rank $r(\sigma)+1$, then the height of Q is $h=r(\sigma)+1$.

Definition 2 In a questionnaire with height h, a minimal question is said to be of *maximal rank* if it is of rank h-1.

Definition 3 When a questionnaire Q admits a minimal question σ of maximal rank, the contraction at $\hat{\Gamma}_\sigma$ is called *condensation*.

Let Q be a questionnaire and $Q^{③}$ a questionnaire obtained from Q by a rearrangement of the vertices as defined by R3:

$$L(Q^{③}) = L(Q).$$

Property 6 If it is impossible to form a questionnaire Ω' such that $L(\Omega')<L(Q)$ by making a substitution or an exchange of arcs according to rules R0, R1, R2 and R3, then $\Omega^{③}$ has at least one minimal question of maximal rank.

For if $Q^{③}$ had no minimal question it would be possible to make one of the transfers R0 to R3 which is contrary either to the definition of $Q^{③}$ or to the hypothesis on Ω'; if the minimal question σ of $Q^{③}$ had rank r and if a question q of $Q^{③}$ had rank r+1, then the exchange of arcs ending in σ and in q would produce a questionnaire Ω' such that $L(\Omega') < L(\Omega)$.

Now let Q be an L-optimal questionnaire having a minimal question σ. Then every sub-questionnaire of Ω is L-optimal and in particular the sub-questionnaire obtained by the contraction at $\hat{\Gamma}_\sigma$.

Property 7 A sufficient condition for Q to be L-optimal is that every sub-questionnaire obtained from Q by a series of condensations possesses a minimal question with maximal rank.

It is to be agreed that the question σ must be minimal relative to the pair $(\mathcal{A}', \mathcal{P}')$ formed starting from the bases and the probabilities of the sub-questionnaire.

Proof Consider a questionnaire Q admitting M questions and such that every sub-questionnaire obtained by a series of condensations has a minimal question of maximal rank. Let $Q_1, Q_2, \ldots, Q_M$ be a sequence of sub-questionnaires of Q with the same root defined by

$Q_1 = Q$

Q_i is the condensation of Q_{i-j}

Q_M has the root for unique question (and is optimal).

Q_{M-1} has a minimal question with rank 1 from which are issued the a_q answers (among the $a_\alpha + a_q - 1$) with smallest probabilities.

Any questionnaire built on the same bases a_α, a_q and having the same probability distribution has routing length at least equal to $L(Q_{M-1})$; therefore Q_{M-1} is optimal.

Suppose Q_i is optimal and Q_{i-1} is not. Let K_{i-1} be an optimal questionnaire of the same family ($\mathcal{A}'$, $\mathcal{P}'$) as Q_{i-1}. Then $K^{\circled{3}}_{i-1}$ obtained from K_{i-1} according to rule R3 is also optimal.

By Property 6, $K^{\circled{3}}_{i-1}$ has at least one minimal question σ with maximal rank and probability $p(\sigma)$. Let K_i be the condensation of $K^{\circled{3}}_{i-1}$. Then:

$$L(K_i) = L(K_{i-1}) - p(\sigma)$$

and in the same way

$$L(Q_1) = L(Q_{i-1}) - p(\sigma) .$$

Hence $L(K_i) < L(Q_i)$, which is absurd.
Q_{i-1} is therefore optimal and this proves Property 7.

6.4 Huffman's Algorithm

In this section we carry out a direct construction of a questionnaire satisfying the sufficient condition and show how it is possible to generate all the L-optimal questionnaires. The algorithm is a generalisation (Picard, 1963) to the heterogeneous case of the one formulated by Huffman as early as 1952 for codes (homogeneous case). The questionnaire thus formed will be called *a Huffman questionnaire*.

Algorithm

1. Order the N answers according to decreasing probabilities

$$p_1 \geq p_2 \geq \cdots \geq p_N$$

and the M bases according to decreasing order:

$$a_1 \geq a_2 \geq \cdots \geq a_M .$$

2. Take the a_M smallest probabilities p_j and form their sum

$$p := \sum_{j=N-a_M+1}^{N} p_j .$$

Memorize the values of p and the p_j thus chosen. Remove the p_j from the list of answers and insert p instead to constitute a list of $N-a_M+1$ answers ordered according to decreasing probabilities. Make $M := M-1$.

3. If M=0, go to 4.

    If M>0, return to 2.

4. End of the formation of questions. Go to 5.

Comments The a_1 answers of the elementary questionnaire which it is possible to constitute at the end of the last phase 2 are vertices of the optimal questionnaire. The vertices of rank r ($r \geq 2$) are those whose probabilities have been summed to form the questions of rank $r-1$. All the vertices of maximal rank are answers.

5. Determine the rank of a vertex x by counting the number of times the questions have been formed with the aid of its probability $p(x)$, from the root to its predecessor.

6. Codify the indices s in such a way that the vertices with greatest probabilities have, at given rank, the smallest possible indices s.

7. Form the sum of the probabilities of the questions, that is L_H, the routing length of the questionnaire.

8. End.

Definition 4 Any questionnaire constructed according to Huffman's algorithm is called a *Huffman questionnaire*.

Any such questionnaire will be denoted by Ω_H.

Example Consider the pair ($\mathcal{A}$, $\mathcal{P}$) (Fig. 6.2)

$\mathcal{A} = \{4, 3, 3, 3, 3, 2, 2\}$

$\mathcal{P} = \{30, 25, 9, 8, 7, 4, 4, 3, 2, 2, 2, 2, 1, 1\}$

The probabilities are shown multiplied by 100. The indices [r,s] of the vertices recall the paths: r is the rank; s = 310 is relative to the path in which the first arc is indexed 3, the second 1, the third 0: the successors of [3,310] are [4,3100] and [4,3101].

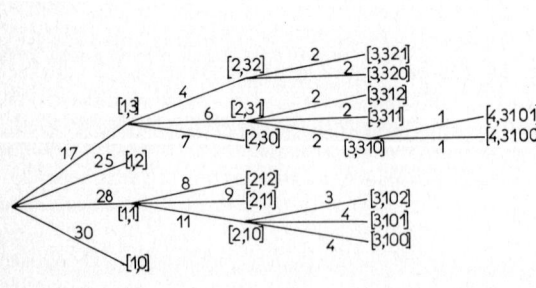

Figure 6.2

This algorithm calls for the following

Remarks

1. It is a recursive process operating downstream to upstream

by successive formation of the M questions; each one of them is constituted, in phase 2, in such a way that it is a minimal question for the pair ($\mathcal{A}'$, $\mathcal{P}'$) formed after the suppression of answers and insertion of questions made in the course of preceding applications of phase 2. This question is a minimal question of maximal rank in the optimal questionnaire built on ($\mathcal{A}'$, $\mathcal{P}'$). Any pair ($\mathcal{A}'$, $\mathcal{P}'$) deduced from ($\mathcal{A}$, $\mathcal{P}$) in the course of the algorithm has as Huffman questionnaire a sub-questionnaire of the questionnaire Q_H($\mathcal{A}$, $\mathcal{P}$).

2. The formation of a question implies that all its descendants are part of the set on which has been made at least one contraction in the course of one or more applications of phase 2. Formed with the aid of its successors indexed essentially by their probabilities, a question can be considered as the outcome of another question which will itself be formed in the course of a further application, immediate or not, of phase 2. If all the phases 2 allow an immediate use of the questions, then Q_H is always determined by extending an arborescence (of maximal height, for fixed $\mathcal{A}$). If two questions k and ℓ are the outcomes of the same one question i, then the formation of ℓ will have been accomplished independently of that of k and (for example) afterwards. In the course of elaboration of the questionnaire we will have realized two disjoint sub-questionnaires k and ℓ. Consequently Huffman's algorithm may lead to the formation of sub-graphs of Q_H being either arborescences (immediate reutilisation of every question) or forests.

The graphical construction of the arborescent support of Q_H can be realized downstream to upstream in parallel with phase 2 but every time the support is a forest we can not be sure of the rank of the new question: the rank will be less than or equal to the rank which has already intervened (and which is not yet determined); then either the index s will be smaller or else it will be a question of rank smaller by 1, with the greatest index s for this new rank (Fig. 6.2).

That is why the algorithm requires, as described above, that all the passages through phase 2 be made before the synthetic construction of the support realized in phase 5 by an upstream to downstream process; it is this method which is used in the algorithms of programming.

3. Phase 6 is a means, among others, of encoding the vertices, that is, it enables us to label vertices of the same rank. Several processes are equivalent as to the L-optimality of Q_H but the method proposed here allows the indices s to be increased as the vertices are formed (phase 5) by allocating progressively the successors to each question and by then respecting the decreasing order of the bases and the probabilities as their intervention takes place.

4. In the course of phase 2 the insertion of probability p in the list of the $N-a_M+1$ probabilities may require a choice in the case of equality. An interesting strategy consists in substituting p+ε for p (where ε is arbitrarily small and

such that no vertex has a probability belonging to the open interval $]p, p+\varepsilon[$), then to restore p at the moment of memorizing. One example of this strategy is given in Figure 6.3. By operating in such a manner any question will have a rank at most equal to that of the answer of the same probability. This labelling strategy also facilitates the solution of the following problems:

Auxiliary Problem 1 Among the L-optimal questionnaires built on the same pair ($\mathcal{A}, \mathcal{P}$), construct a questionnaire of minimal height.

Auxiliary Problem 2 Among the L-optimal questionnaires built on the same pair ($\mathcal{A}, \mathcal{P}$), construct a questionnaire minimizing $S = \sum_{e \in E} r(e)$.

For two Huffman questionnaires, distinct by the choice of rearrangements of a question with basis a having the same probability as another vertex, will have the same value for L; the one in which the choice is fixed according to the method given here will correspond to a value of S which is either equal to or less than the other by a-1.

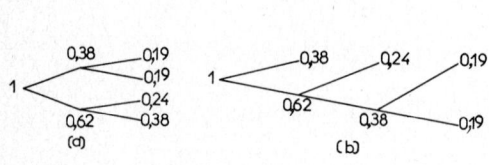

Figure 6.3

Repetition of such choices by means of this strategy gives the solution of auxiliary Problem 2 and implies that of auxiliary Problem 1 (Schwartz).

5. To emphasise the fact that the suppression of the labelling performed at phase 6 of Huffman's algorithm does not destroy L-optimality, we will still call any questionnaire built according to phase 1 to 5 of the algorithm a Huffman questionnaire.

6. The number of non-isomorphic Huffman questionnaires constructed on a pair ($\mathcal{A}, \mathcal{P}$), depends on this pair, but the labelling strategies such as that used at phase 6 of the algorithm and at remark 4 may be used to determine a unique class of isomorphic arborescences.

Phase 5 of Huffman's algorithm facilitates the construction of the graph Q_H upstream to downstream by substitution for an answer of a question obtained by the transformation inverse to condensation. Consequently the sufficient condition (Property 7) is satisfied.

Theorem 1 Every Huffman questionnaire is L-optimal.

In fact Q_H satisfies a stronger condition than the given sufficient one: Let Q_a be a sub-questionnaire of Q_H with height h_a, and let k and ℓ be questions in Q_a of the same rank and with bases a_k and a_ℓ such that $a_k \geq a_\ell$. According to the algorithm, the probability of every vertex $k' \in \Gamma k$ is at least equal to the probability of every vertex $\ell' \in \Gamma \ell$.

Within any rank the question with minimal basis has a probability which is minimal (for the rank) and at most equal to that of any vertex with lower rank. This is true for the greatest possible rank, that is $h_a - 1$. Thus we obtain:

Property 8 In a Huffman questionnaire any sub-questionnaire has a minimal question with maximal rank.

It follows also that

$$Q^{\circledS}_a = Q_a \qquad (6)$$

so that it is now possible to group together the necessary properties (1 to 5) and the sufficient properties (7 and 8).

Theorem 2 A questionnaire Q is optimal if and only if the application of rules R0, R1, R2, R3 preserves the routing length.

The condition is necessary because of Properties 1 to 4. The preservation of the routing length by application of R0 and R3 implies the existence of a minimal question with maximal rank. Moreover the rearrangement imposed by R3 is not modified by the condensation operations. Hence any sub-questionnaire obtained from Q by a sequence of condensations satisfies the condition of Property 7, and this completes the proof of Theorem 2.

It follows that a construction technique for an optimal questionnaire may be obtained independently of Huffman's algorithm by starting from any questionnaire and making substitutions or exchanges of arcs according to rules R0, R1, R2, R3. The preservation of the routing length by R3 will then serve as a stopping criterion.

Thus it is possible to generate the set of L-optimal questionnaires on $(\mathcal{A}, \mathcal{P})$ by performing transfers while preserving L by starting from some one given questionnaire, Q_H for example: one needs to make a transfer of arborescence if and only if R0, R1 and R2 are respected. If the ranks of the roots are the same, then there is preservation of the height and the probabilities of these roots can be distinct.

If the probabilities of the roots are the same the ranks can be distinct and the height can be modified by the transfer. Permutation of two isomorphic sub-arborescences will not create a new questionnaire as long as no semantic is used and will lead to a simple permutation of labels.

We will now describe an algorithm which will enable us to generate the set of questionnaires with minimal routing length on $(\mathcal{A}, \mathcal{P})$. The motivation for doing this is related to the semantic aspect which is adjoined to the study of a questionnaire for practical use, because, an operation such as the exchange of arcs modifies completely the events associated with the vertices which are descendants of the vertices belonging either to one or to the other of the transposed subarborescences. The semantic will generally admit some of these events but will have to reject others.

Remark Formula 5.16 can be applied for minimization, and F. Dubail has effectively established an algorithm based on dynamic programming to form an L-optimal questionnaire by looking for a questionnaire Q as a solution of

$$L(Q) = 1 + \min \{ \sum_{h \in \Gamma \alpha} p(h) L_h \}$$

where the minimum is taken on the set of questionnaires built on $(\mathcal{A}, \mathcal{P})$. F. Dubail's algorithm is necessary to construct the L-optimal arborescent questionnaire when the semantic forbids the use of Huffman's questionnaire. Except for this case we can say that the property of Huffman's algorithm is to furnish at each phase, in the construction of any question, local minimization as well as global minimization. G. Petolla has studied in a more general case the problem of families of optimal sub-graphs and he has considered in particular the "forest questionnaires" formed by the forest sub-graphs which may be encountered in Huffman's construction.

6.5 Questionnaires and Coding

The problem of the homogeneous questionnaire with N answers and optimal routing is equivalent to the following coding problem: transmit a series of messages with the aid of a characters in such a way that the time of transmission, which is directly proportional to the number of characters used is minimal.

The characters will be associated by groups to form the code, the whole of which consists of N symbols or words.

Each of the N symbols is a succession of characters which can take the values 0, 1, 2... or a-1. A message will consist of a set of symbols written in succession without any space. The frequency of use of each of the N words constituting the vocabulary of the message expressed with the aid of a characters is characterized by giving a priori the appearance probabilities of each word $p(y_j)$ ($j:=1,2,\ldots,N$).

To transmit a message in the minimum mean time amounts to giving a code with minimal redundancy; the solution of this problem is equivalent to that of the optimal questionnaire because we have an isomorphism between the set of words and the set of answers.

This coding problem has been the subject of numerous works: in particular the codes proposed by Shannon and Fano have been studied with the aim of using, on average, the minimum number of characters with an alphabet of two characters, to transmit messages using N symbols y_j (j=1,2,...,N) with frequency $p(y_j)$. However the codes of Shanon and Fano were conceived by starting from a principle which does not exactly correspond to optimal coding:

Distribute the symbols in a classes of probabilities as nearly equal as possible, as long as this is possible, that is: distribute the successors of questions of the same rank in such a way that the probabilities are as nearly equal as possible up to the greatest rank.

This principle of separation is often admitted as one of the best heuristic guides to constructing a code or a questionnaire. However even if this procedure can lead to an optimal code for certain probability distributions, such as those of the Shannon-Fano code shown in Fig. 6.4, it does not *guarantee* that the code obtained is optimal.

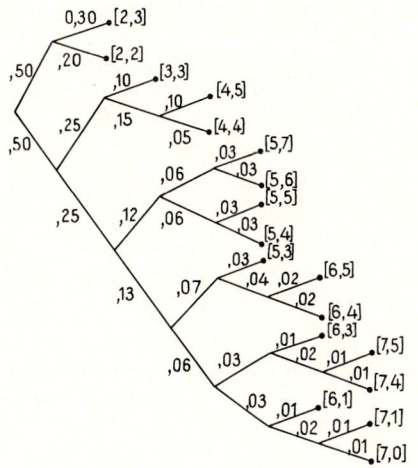

Figure 6.4

Shannon-fano Code The 18 answers are indexed by [r,s] the probabilities are shown on the arcs incident inwardly to the questions and to the answers.

With a set E consisting of eight answers with probabilities

$p(y_j)$: = 0 50-0.12-0.12-0.10-0.07-0.05-0.03-0.01, 0.50 etc.

the "fifty-fifty" separation gives rise to the arborescence I of Fig. 6.5. Along this arborescence the separation into two classes of equal probabilities, that is, into two sub-arborescences with roots having equal probabilities, is realized for the ranks 0 and 1. This arborescence satisfying rules R1 and R2 is not optimal, because of R3.

The arborescence II of the same figure is optimal because of Property 7. However in II the "fifty-fifty" separation has only been realized at rank 0.

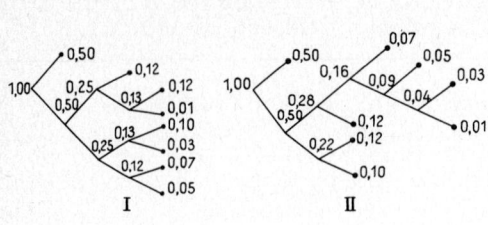

Figure 6.5

The lengths of routing $L = \sum_{x_i \in Q} p(x_i)$ are: for the "fifty-fifty" arborescence

$$L_1 = 2.38 \quad (h = 4)$$

and for the optimal arborescence:

$$L_2 = 2.29 \quad (h = 6).$$

As a matter of fact the L-optimal homogeneous questionnaires are equivalent to the codes with minimal redundancy studied by Huffman.

Without exhibiting rule R3, but using it implicitly, Huffman has obtained the optimal code. As Huffman is not interested in the probabilities of the questions, his aim being to form the set of N words with the aid of a characters with the most economical notation, his theory presents two differences from that of questionnaires:

1. Huffman calculates L with the aid of the probabilities and the ranks of the answers, whereas the calculation of L can be done much more rapidly (without multiplication) with the aid of the probabities of the questions which in any case Huffman uses implicitly in his algorithm.

2. The notational system of Huffman's answers has only one index s identical with that defined above and expressed to base a; for, two answers y_A and y_B do not have the same index s because the paths linking the roots to two answers are at least different starting from the rank of one of them.

This economy in notation does not make it possible to index the questions nor to express the indices s in a basis different from a as does the notation [r,s].

Huffman's theory based on two hypotheses (a) and (b), equivalent to Properties 2 to 7 leads to defining three properties (c), (d) and (e) of codes with minimal redundancy, equivalent to polychotomic questionnaires in the broad sense.

Expressed from the point of view of questionnaires, they are:

(a) *Property 9* The indices s and s' of the answers [r,s] and [r',s'] are represented to base a by two different numbers.

(b) *Property 10* We may write the digits to base a of the indices s of a sequence with repitition, of k answers $e \in E$ without separating them out and without risk of confusion.

OPTIMAL ROUTING   165

(d) *Property 11*   The indices s of at least two and at most a answers with maximal rank h differ only in their last digits to base a.

(e) *Property 12*   In a sequence of indices s, each sequence of h digits represents an answer of rank h or else contains in its digits of strongest weight (left) those of an answer of rank r<h.

(c) is Property 2.

(a) (Huffman's hypothesis) has been proved above.

(b) and (e) follow from Theorem 5.2 of the notational system [r,s] and from (a); h is the height of Huffman's question- naire.

(d) is a consequence of Property 5.4 of polychotomic question- naires.

6.6 Equiprobable Polychotomic Questionnaires

An L-optimal equiprobable polychotomic questionnaire may be obtained when it is no longer possible to make any transfer of sub-arborescences to reduce the routing length. Any two answers then have the same rank or have only one rank differ- ence (Property 3) and the L-optimal questionnaire has a balanced arborescence as its support.

The probability of any answer is $p(e) = \frac{1}{N}$ and the routing length of the L-optimal questionnaire has been determined by (5.3.1) and (5.3.2).

$$\alpha = \beta = 0 \Rightarrow L = k, \tag{7}$$

$$\left\{ \begin{array}{c} 0 \leq \alpha \leq a^k \\ \beta = 0 \end{array} \right\} \Rightarrow L = k + \frac{a\alpha}{N} \tag{8}$$

$$\left\{ \begin{array}{c} 0 \leq \alpha < a^k \\ 0 < \beta \leq a-1 \end{array} \right\} \Rightarrow L = k + \frac{a\alpha + \beta + 1}{N} \tag{9}$$

6.6.1   A characteristic property of homogeneous balanced arborescences

According to Definition 5.4 this is an arborescence A_0 for which the difference of ranks of the various terminals is never greater than 1.

1. Let
$$N = a^k + \alpha(a-1)$$

Let $A(z_i)$ be the sub-graph of A_0 generated by a vertex z_i with rank i and its descendants.

Let $\mathcal{C}_{z_i}$ be the set of paths of $A(z_i)$ coming out of z_i and having for terminal extremity a vertex y of E.

$A(z_i)$ is itself an arborescence.

N-1 being divisible by a-1 the number of elements of e_{z_i} satisfies the inequalities

$$a^{k-1} \leq |e_{z_i}| \leq a^{k-i+1} \tag{10}$$

for all i; the equalities correspond to the cases when the terminal vertices are all of rank k or all of rank k+1.

$|e_{z_i}|$ is the number of terminal vertices descending from z_i.

We can partition e_{z_i} into a sub-sets $e_{z_i}^j$ (j=0,1,2,...,a-1) such that $e_{z_i}^j$ is the set of paths whose first arc is $(z_i, \Gamma_j z_i)$, if $z_i \in F$ and

$$a^{k-i-1} \leq |e_{z_i}^j| \leq a^{k-i}$$

$$\sum_{j=0}^{a-1} |e_{z_i}^j| = |e_{z_i}| . \tag{11}$$

If i=k, the paths coming out of vertices $z_k \in F$ are of length 1:

$$|e_k^j| = 1 (\forall j) \text{ and } |e_{z_k}| = a .$$

If i=k or i=k+1 and $y_i \in F$ we make the convention that $|e_{y_i}|=1$ because only one path passes through y.

The inequalities (10) are then satisfied by any

$$z_i \in E \cup F \; (i := 0, 1, \ldots, k, k+1) .$$

These inequalities indicate that the number of paths starting from any vertex of rank i is enclosed between the same consecutive powers of a related to the rank.

In particular the ratio of the numbers of paths in the two sub-arborescences satisfies the following inequalities for all i:

$$\frac{1}{a} \leq \frac{|e_{z_j}|}{|e_{z_i}|} \leq a$$

that is

$$\frac{a-1}{a} + 1 \leq \sum_{\substack{h:=0 \\ h \neq j}}^{a-1} \frac{|e_{z_i}^h|}{|e_{z_i}^j|} + 1 \leq a(a-1) + 1$$

and

$$\frac{1}{a(a-1)+1} \leq \frac{|e_{z_i}^j|}{|e_{z_i}|} \leq \frac{a}{2a-1} . \tag{12}$$

The equiprobability hypothesis for the answers implies that the probabilities of the vertices of any rank i lie between two fractions in which the numerators are two consecutive powers of a and the denominators are equal to the number of answers.

$$\frac{a^{k-1}}{N} \leq p(x_i) \leq \frac{a^{k-i+1}}{N} \quad ;$$

if i=k+1:
$$p(y_{k+1}) = \frac{1}{N}$$

if i=k:
$$p(y_k) = \frac{1}{N} < p(x_k) = \frac{a}{N} \, .$$

The ratio of the probabilities of two vertices of the same rank is then

$$\frac{1}{a} \leq \frac{|e_{z_i}^h|}{|e_{z_i}^j|} = \frac{p(i+1,h)}{p(i+1,j)} \leq a$$

and the conditional probability of a vertex of rank r+1:

$$\frac{1}{a(a-1)+1} \leq \frac{p(r+1,as+h)}{p(r,s)} \leq \frac{a}{2a-1} \tag{13}$$

Among all the arborescences with root x_o and N answers, relation (10) characterizes balanced arborescences and implies (13).

We will show that conversely arborescences satisfying relation (10) are balanced.

For let C be an arborescence with maximal rank k+1 the vertices of which satisfy the double inequality

$$a^{k-i} \leq |e_{z_i}| \leq a^{k-i+1}$$

with the convention that

$$z_i = y_i \in E \Rightarrow |e_{y_i}| = 1 \, .$$

The answers with rank k+1 and k satisfy the equations

$$|e_{y_{k+1}}| = a^{k-(k+1)+1} = 1$$
$$|e_{y_k}| = a^{k-(k)} = 1 \, .$$

If there exists an answer with rank r less than k, then there corresponds to it on the one hand,

$$|e_{y_r}| = 1$$

and on the other

$$|e_{y_r}| = a^{k-r} \geq a,$$

which is absurd. Consequently the answers of C are of rank k, or k+1 only, which establishes the converse.

2. Let
$$N = a^k + \alpha(a - 1) + \beta \quad (0 < \beta < a - 1).$$

Then there exists at least one question q_1 with rank h possessing less than a outcomes and the number of distinct paths coming out of question q_1 is

$$1 < |e_{q_i}| < a;$$

the number of paths coming out of a question z_i having q_i for descendant also satisfies the strict inequalities

$$a^{k-i} < |e_{z_i}| < a^{k-i+1},$$

which establishes relation (10) when $\beta > 0$.

Conversley, arborescences with maximal rank $\beta > 0$ and possessing Property (10) are such that the terminals are all of rank k+1 or k.

Theorem 3 A necessary and sufficient condition for a homogeneous arborescence to be balanced, is that the number of paths coming out of any question z_i or entering into any answer z_i of rank i is bounded by

$$a^{k-i} \leq |e_{z_i}| \leq a^{k-i+1}. \tag{14}$$

This arborescence is the support of an equiprobable questionnaire of optimal routing if $\beta \neq 0$ or, when $\beta = 0$, if the question with basis $\beta + 1$ is of rank k.

Any other equiprobable polychotomic questionnaire has a greater routing length.

Corollary 4 The polychotomic equiprobable questionnaires ($\beta \geq 0$) with optimal routing are characterized by the property: the probabilities of the vertices of any rank i lie between two fractions in which the numerators are two consecutive powers of a and the denominators are equal to the number of answers.

$$\frac{a^{k-i}}{N} \leq p(z_i) \leq \frac{a^{k-i+1}}{N} \tag{15}$$

6.6.2   Equiprobable Dichotomic Questionnaires

6.6.2.1   Optimal questionnaires

The dichotomic case has many applications in data-processing and, since β is always zero, it merits a special treatment.

Some of the properties of Chapters 4 and 5 are simplified.

Whatever the value of N, an optimal dichotomic questionnaire Q always has arborescent support; the number of questions is equal to N-1 and the number of arcs to 2(N-1); the number of vertices of any positive rank is even.

There are $F_k = 2^k - 1$ questions with rank less than the least rank of an answer; the binary notation for the vertices can be used in such a way that any number between 0 and $2^k - 1$ is the index s of a vertex $[r,s]$;

The homogeneous arborescence with N equiprobable answers and optimal routing has routing length

$$L_H = k + \frac{2\alpha}{N} \, . \qquad (16)$$

The dichotomic questionnaire with N equiprobable answers has routing length at least equal to

$$L_{max} = \frac{N-1}{N} (1 + \frac{N}{2})$$

L_{max} increases considerably more quickly than L_o.

For N = 1024
$$L_H = 10 \quad \text{and} \quad L_{max} \# 513$$

For N = 10^6
$$L_H \# 20 \quad \text{and} \quad L_{max} \# 5 \times 10^5 \, .$$

6.6.2.2 Routing in a dichotomic questionnaire which is not optimal

Let B be an arborescence which does not satisfy (10).

B has N terminal extremities,

$$N = 2^k + \alpha \quad (\alpha > 0).$$

I. The maximal rank of a terminal extremity of B is k+1.

I.1 B possesses a terminal vertex with rank i<k and N-1 terminal vertices with rank k or k+1.

We must have
$$2^{k-i+1} - 1 \leqslant 2^k - \alpha \, .$$

There exists an arborescence B having

1 terminal vertex with rank i,

$2^k \alpha - (2^{k-i+1} - 1)$ terminal vertices with rank k,

$2\alpha + (2^{k-i+1} - 2)$ terminal vertices with rank $k+1$.

The routing is

$$L_B = \frac{1}{N}\{1 \cdot i + [2^k - \alpha - (2^{k-i+1} - 1)]k + [2\alpha + 2^{k-i+1} - 2](k+1)\},$$

$$L_B = \frac{1}{N}\{(2^k + \alpha)k + 2\alpha + 2^{k-i+1} - k + i - 2\}$$

$$L_B = k + \frac{2\alpha}{N} + \frac{2^{k-i+1} - (k-i+2)}{N}$$

But $k-i \geq 1 \Rightarrow 2^{k-i+1} - (k-i+2) \geq 1$ and $2^{k-i+1} - (k-i+2)$ increases with $k-i$ thus,

$$L_B \geq k + \frac{2\alpha}{N} + \frac{1}{N}, \qquad (17)$$

where we have equality if $i = k-1$.

I.2 B possesses s terminal vertices having the ranks i_s:

$$i_1 \leq i_2 \leq i_3 \leq \ldots i_s \leq k,$$

with $\qquad 1 \leq i_r < k \quad (\forall_r)$

and

$$i_h < i_{h+1} = i_{h+m} < i_{h+m+1} \Rightarrow m < 2^{i_h+1},$$

and $N-s$ terminal vertices with rank k or $k+1$.

The routing is

$$L = \frac{1}{N}\left\{\sum_{r:=1}^{s} i_r + \left[2^k - \alpha - (\sum_{r:=1}^{s} 2^{k-i_r+1} - s)\right]k\right.$$

$$\left. + \left[2\alpha + (\sum_{r:=1}^{s} 2^{k-i_r+1} - s) - s\right](k+1)\right\}$$

$$L = \frac{1}{N}\left\{(2^k + \alpha)k + 2\alpha + \sum_{r:=1}^{s} 2^{k-i_r+1} + \sum_{r:=1}^{s} i_r - ks - 2s\right\}$$

$$L = k + \frac{2\alpha}{N} + \frac{1}{N}\left\{\sum_{r:=1}^{s}(2^{k-i_r+1}) + \sum_{r:=1}^{s}(i_r - k - 2)\right\},$$

$$L = k + \frac{2\alpha}{N} + \frac{1}{N}\left\{\sum_{r:=1}^{s} 2^{k-i_r+1} - \sum_{r:=1}^{s}(k - i_r + 2)\right\}$$

and since

$$k - i_r \geq 1 \Rightarrow 2^{k-i_r+1} - (k - i_r + 2) \geq 1,$$

$$L_B \geq k + \frac{2\alpha}{N} + \frac{s}{N}. \qquad (18)$$

There is equality if

$$i_r = k - 1. \quad (\forall r)$$

II. The maximal rank of a terminal of B is equal to $k+2$.

II.1 B possesses a terminal vertex with rank $i<k$ and $N-1$ terminal vertices with rank $k+1$ or $k+2$.

The arborescences studied in case I.1 above would correspond to $2^{k-i+1}-1 \leq 2^k - \alpha$.

Here we must have

$$2^{k-i+1} - 1 > 2^k - \alpha.$$

Put

$$t = 2^{k-i+1} - 1 - (2^k - \alpha)$$

[t is positive].

There exists an arborescence B having:

    1 terminal vertex with rank i,

    $N-1-2t$ terminal vertices with rank $k+1$,

    $2t$ "   "   "   " $k+2$.

The routing is

$$L = \frac{1}{N}\{i + (N - 1 - 2t)(k + 1) + 2t(k + 2)\}$$

$$= \frac{1}{N}\{Nk + N + 2t - (k - i + 1)\}$$

$$= k + \frac{2\alpha}{N} + \frac{1}{N}\{2^k - \alpha + 2t - (k - i + 1)\}$$

$$L = k + \frac{2\alpha}{N} + \frac{1}{N}\{2^{k-i+1} + t - (k - i + 2)\}$$

and consequently

$$L \geq k + \frac{2\alpha}{N} + \frac{1+t}{N}. \qquad (19)$$

There is equality if $k-i=1$.

The case $t=0$ corresponds to the limit of case I.1.

II.2 B possesses s terminal vertices having ranks i_s:

$$i_1 \leq i_2 \leq i_3 \leq \ldots \leq i_s < k,$$

with
$$1 \leq i_r < k \quad (\forall r)$$

and
$$i_h < i_{h+1} = i_{h+m} < i_{h+m+1} \Rightarrow m < 2^{i_{h+1}}$$

and N-s terminal vertices with rank k+1 or k+2 (t supposed positive).

Put
$$\sum_{r:=1}^{s} 2^{k-i_r+1} - s = 2^k - \alpha + t.$$

B possesses

    s terminal vertices with rank $1 \leq i_r < k$,

    N-s-2t "   "   "   "   k+1

    2t   "   "   "   "   k+2.

The routing is

$$L = \frac{1}{N} \left\{ \sum_{r:=1}^{s} i_r + (N - s - 2t)(k + 1) + 2t(k + 2) \right\}$$

$$= k + \frac{2\alpha}{N} + \frac{1}{N} \left\{ 2^k - \alpha + 2t + \sum_{r:=1}^{s} i_r - s(k + 1) \right\}$$

$$= k + \frac{2\alpha}{N} + \frac{1}{N} \left\{ \sum_{r:=1}^{s} 2^{k-i_r+1} - s + t + \sum_{r:=1}^{s} i_r - s(k + 1) \right\}$$

$$= k + \frac{2\alpha}{N} + \frac{1}{N} \left\{ \sum_{r:=1}^{s} 2^{k-i_r+1} - \sum_{r:=1}^{s} (k - i_r + 2) + t \right\} ;$$

$$L \geq k + \frac{2\alpha}{N} + \frac{s+t}{N}. \tag{20}$$

EXERCISES

1. Consider the algorithm, said Namffuh algorithm, deduced from Huffman's algorithm by making the selection of the a_1 greatest probabilities of answers and in inserting their sum in the ordered file of the vertices' probabilities; then keeping the recursive process downstream to upstream and using the

maximum number of vertices with the greatest probability at each step, show that

(1) the heterogeneous arborescence built by this algorithm is of maximum height, for fixed $\mathcal{A}$;

(2) the questionnaire thus constructed is the heterogeneous arborescent questionnaire of maximal routing length for given $(\mathcal{A}, \mathcal{P})$.

2. Evaluate the routing length of a polychotomic questionnaire in the strict sense and which is equiprobable ($p_e = \frac{1}{ak+\alpha(a-1)}$, $e \in E$) when there exists an answer with rank k-1, the other answers being of rank k or k+1.

3. Same exercise when the questionnaire is polychotomic in the broad sense ($\beta > 0$).

4. Consider anew the second exercise by supposing that one answer is of rank k-i, the others being of rank k and k+1.

Chapter Seven

INFORMATIONAL STUDY OF QUESTIONNAIRES

7.1 Introduction to Information

7.1.1 Hartley and Shannon's forms

We are familiar with the idea of information in everyday live (journals; newspapers; football results; zero, partial or complete knowledge of an economic, physical or physiological situation; etc.). However we need here to give a mathematical definition which should depend neither on the form (sound, letter, message) nor on the content (the exact meaning of the message) nor on the length (number of characters) of the set of signals received.

If I draw a letter from an alphabet of 26 letters, I receive one unit of information. If I read a series of p letters, typed effectively at random, then I receive p times this unit of information. There are $N=26^p$ possible "words" that could have been typed. The information $I(N)$ may be considered as quantified by $I(N)=\log_a N$ so that $I(N)=p$ when $a=26$ as in the present example.

The use of the decimal system would lead to using logarithms to the base $a=10$. The nature of data-processing machines (for computing and processing information), where the electric signals or elementary magnetic signs have only two possible values, leads to the most usual choice of a to be 2.

Actually the base of the logarithms used to measure a choice from among for example 26^p possibilities is of secondary importance. For if I close my eyes and type out r new letters, the result will be one of the $N'=26^r$ possible "words". Reading this I would say that I have received information $I(N')=r$, and I observe that the ratio of the informations

$$I(N')/I(N) = \log_{26} 26^r / \log_{26} 26^p = r/p$$

is the same with respect to any base because

$$\log_a 26^r / \log_a 26^p = r/p \quad \text{(for all a)}.$$

Thus the ratio of the informations is independent of the number of symbols referred to in the base. However the so-called measure of $I(N)$ may still come into the discussion.

Following Hartley if we take $a=2$, we have

$$I(N) = \log_2 N , \qquad (1)$$

where $I(N)$ is the information obtained by knowing what event

175

has been realized among the N elements of a given set E.

We will show that the logarithmic function with base 2 is the only function satisfying the three axioms of information in the sense of Hartley:

(H1) $I(N.M) = I(N) + I(M)$;   additivity

(H2) $I(N) \leq I(N+1)$;   inclusion

(H3) $I(2) = 1$;   normality

Proof Let N be an integer greater than 2. Then for any power N^r of N there is an integer $s(r)$ such that

$$2^{s(r)} \leq N^r < 2^{s(r)+1}$$

Hence

$$\frac{s(r)}{r} \leq \log_2 N < \frac{s(r)+1}{r}$$

and

$$\log_2 N = \lim_{r \to \infty} s(r)/r .$$

Axiom (H2) gives $N < M \Rightarrow I(N) \leq I(M)$ and also

$$I(2^{s(r)}) \leq I(N^r) < I(2^{s(r)+1}) ;$$

(H1) implies

$$I(N^r) = rI(N)$$

(H3) fixes $I(2)=1$ and consequently

$$s(r) \leq rI(N) < s(r) + 1$$

therefore

$$I(N) = \lim_{r \to \infty} \frac{s(r)}{r} ,$$

and consequently $I(N)=\log_2 N$ for $N \geq 2$.

We see that if $2^k < N \leq 2^{k+1}$, it will be necessary to receive k+1 units of information because it is not always possible to define submultiples if, for example, the comprehension (or detection) threshold is the binary digit 1.

Further, even in typing "at random" on the keyboard of my typewriter, I will have a marked tendency to miss out A, Z, Q, W on the left, and P, M on the right as well as B and N which are near the long space bar; the probabilities of hitting the different keys will not be the same. However, the information in the sense of Hartley does not take into account any asymmetry, that is, any preference for some of the 26^n events considered. We must therefore go further than this definition, to find an axiom system stronger than the one formed by (H1), (H2) and (H3).

To this end we will now give an intuitive presentation of Shannon's information in the case when the probabilities are rational and leave the axiomatic formulation till later (§2).

Let there be N equiprobable events grouped into n sets $E_1, E_2, \ldots, E_n$ all disjoint and such that:

$$\bigcup_{i:=1}^{n} E_i = E, \quad |E_i| = N_i \quad \text{and} \quad |E| = N,$$

Then $p_i = N_i/N$ is the rational probability that an event e belongs to E_i; $\sum_i p_i = 1$.

To characterize an element e amounts to determining:

1. The set E_i to which e belongs.

The information to be acquired here will be denoted by

$$I(1) = I(p_1, p_2, \ldots, p_n).$$

2. Which element e is considered from among the N_i possible events of the set E_i.

The information to be acquired in this second phase is denoted by $I(2)$ and is determined from Hartley's information.

Finally the characterization of an element e is the object of the information $I(3) = I(1) + I(2)$, the sum of the preceding informations.

But

$$I(3) = \log_2 N$$

in the sense of Hartley (and, as we will see, also in the sense of Shannon) because there are N events.

What we will call information (in the sense of Shannon), is the unknown quantity $I(1)$ depending on the probabilities of n events of type E_i (more general than the elementary events e) and we will compute $I(1) = I(3) - I(2)$.

Let $I(2_i)$ be the information obtained in the particular case when e belongs to the set E_i.

There are N_i events, and therefore

$$I(2_i) = \log_2 N_i.$$

But the probability that we are in such a case is $p_i = N_i/N$. We will write $I(2)$ for the mean value of $I(2_i)$ over the set of n possible values, that is

$$I(2) = \sum_{i:=1}^{n} p_i I(2_i)$$

Consequently

$$I(1) = \log_2 N - \sum_{i:=1}^{n} p_i \log_2 N_i$$

$$I(1) = \sum_{i:=1}^{n} p_i \log_2 N - \sum_{i:=1}^{n} p_i \log_2 N_i$$

and finally

$$I(p_1, p_2, \ldots, p_n) = \sum_{i:=1}^{n} p_i \log_2 \frac{1}{p_i} . \qquad (2)$$

This information characterizes the set $E_i \subset E$: this is Shannon's form. Note that the elements $e \in E_i$ have been introduced merely with a view to making possible, through the probabilities, the quantification $|E_i|$. Shannon's form then takes into account the probability of finding an event among the elements of the set E_i contained in E. Thus the objection raised against Hartley's form with regard to the indifference of the information to the repartition of the events disappears with Shannon's form.

In the case when $(\forall i) p_i = 1/n$ we have:

$$I = \sum_{i:=1}^{n} \frac{1}{N} \log_2 n = \log_2 n .$$

That is, if all the sets E_i have the same order, then the information in the sense of Shannon takes the same value as the information in the sense of Hartley. Hartley's logarithmic law has merely been weighted by probabilities in order to produce Shannon's form.

Example

Let $N=10$, $n=2$, $N_1=2$, $N_2=8$.

We may consider that the N events are represented by the picking up of ten distinct balls. The balls of the set E_1 are red and carry the numbers 0 and 5; the others are white and carry the numbers 1 to 4 and 6 to 9.

Then

$$I(3) = \log_2 10 = 3.32$$

whereas

$$I(2_1) = \log_2 N_1 = 1$$
$$I(2_2) = \log_2 N_2 = 3$$

$$I(2) = \tfrac{1}{5} \times 1 + \tfrac{4}{5} \times 3 = \tfrac{13}{5} = 2.6 .$$

It follows that

$$I(1) = 3.32 - 2.6 = 0.72 .$$

Formula (2) gives the same result if we take:

$$p_1 = \frac{1}{5} \text{ and } p_2 = \frac{4}{5}$$

that is

$$I\left(\frac{1}{5}, \frac{4}{5}\right) = I(1) = 0.72$$

We can also say that $I(1)$ is the information relative to colour whereas $I(2)$ is the information relative to number when the colour is determined; $I(2_2)$ is the information relative to number when we know that the ball is red.

7.1.2 Questionnaires in the sense of Shannon

In his initial memoir on the theory of information, Shannon studied the function (2) which represents the degree of indeterminacy in communication problems. He has established the link between choice and indeterminacy.

A communication system consists, according to Shannon, of five elements (Fig. 7.1).

1. A *source of information* which produces a message or a sequence of messages to be transmitted. The message can be of various types: a sequence of letters, a function of time and other variables or various combinations of functions.

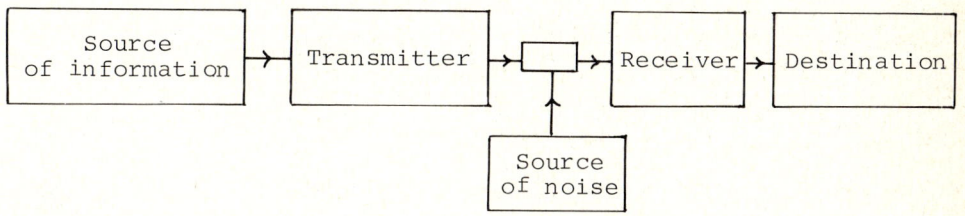

Figure 7.1

2. A *transmitter* which operates on the message in such a way as to produce a *signal* which can be *transmitted* along a channel.

3. The *transmission channel* is the medium effectively used to transmit the signal from the transmitter to the receiver. During the transmission or at one of the extremities of the channel the signal may be perturbed by its own incompatibility (inefficiency) or by an external noise source acting on the emitted signal to produce the received signal.

4. A *receiver*, which carries out the inverse operation to that of the transmitter and reconstitutes the *received message*

from the received signal.

5. A *destination*, which is the element to be acquainted with the message which has been emitted for it.

It is possible to consider a questionnaire as an element in a communication system. We will make use of the terms which are proper to data-processing to give a better picture but without restricting the scope of the theory.

Suppose that within a problem or a program operating on a computer there is a sub-program consisting in the selection of one answer from N. Then the questionnaire is the transmission element of the selection sub-program considered as a system of communication (Fig. 7.2).

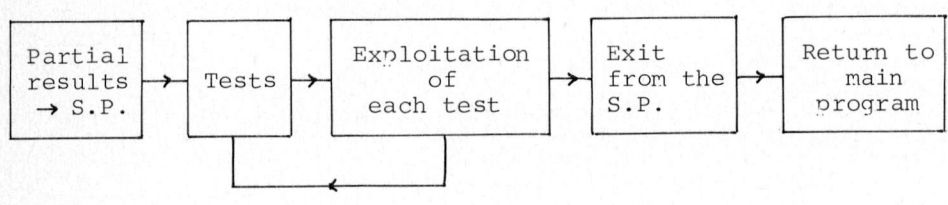

Figure 7.2

1. The source of information is the set of partial results acquired at the input of the selection sub-program (SP).

2. The transmitter is the part of the sub-program which is charged with verifying the acquired results (questions) the emission of a question corresponds to a signal.

3. The transmission channel is the part of the program which is charged with exploiting the answers to the questions with a view to leading to the receiver or causing the emission of a new signal to constitute a message.

4. The receiver is formed by the set of outcomes (answers) of the questionnaire. The message entering the receiver is coded as [r,1] according to the notation of the questionnaire (§5.4.1).

5. The destination is the rest of the principal program coming after the selection sub-program and working on the results of this program.

The loop, or more precisely the iteration circuit on the second and third elements of this communication system, must be open in the case of a message consisting of only one signal, that is the case of a unique question. In the case of a message composed of many questions and answers, each answer having the possibility of influencing the following question, the circuit used k times may be replaced by a sequence of k "second" elements each followed by a "third" element.

The questionnaire is formed by the second and third elements of the sub-program and the arc forming the circuit, and it is within the questionnaire that the origin of an inefficiency and the noise source lies. The questionnaire could be called a "data-processing channel".

The information received at the end of the selection sub-program is the difference between the indeterminacy before and after the sub-program. The measure of the information emitted by the source in the first element of the sub-program considered as a communication system is greater than or equal to the information at the exit of the sub-program with equality if and only if there is no noise or inefficiency.

Corresponding to each passage through the iteration circuit, and thus to each new question, there is a new value of the information.

The information acquired at the exit of the selection sub-program will be the information received at the exit of the data processing channel whatever its structure.

The average number of questions necessary to acquire the information transmitted to the exit of the selection channel in all the circumstances represents the information emitted by source A. In this manner we can consider that the routing length is a first form of information whose minimal value is achieved for L-optimal questionnaires. Information theory tells us that the information at the source will be closer to the information received by the receiver if there is no redundancy, that is, if any acquired result is not subject to a new question rejoining an earlier question. This result corresponds in the theory of graphs to having no cycles. It also signifies that there is no question whose answer is known a priori, otherwise the conditional information would be as weak at the outcome of this question as immediately before; only one arc would come out of the question. Axiom (q_3) of questionnaires and the result of graph theory show that any question of an optimal homogeneous questionnaire, when N-1 is not divisible by a-1, has at least $\beta+1 \geqslant 2$ outcomes, and it therefore translates exactly the requirements of the information theory. Thus graph theory has enabled us to elaborate the theory of questionnaires and information theory will make it possible to interpret its results.

The questionnaire considered as the transmission channel in the communication system formed by the selection sub-program of an answer is subject to noises produced by human or material causes, or to its own noise and to an inefficiency.

a. The human or material causes are programming errors and errors in the computing unit and in the access or transfer units of the computer used. In the case where the selection would have to be done by hand, these errors due to the use and not to the structure of the questionnaire can also be

considered as material errors. In all our discussion we will
suppose that the causes of errors of transmission, that is,
noise of the category (a) do not exist.

b. The inefficiency of the questionnaire is the consequence
of the divergence between the routing length which is always
rational if the $p(y_i)$ are rational, and an information which
takes the form of a transcendental function. We will say that
the inefficiency is due to the rationalization of the routing:
for the lengths of all the paths linking the root of the
arborescence to the set of answers are integers and the combination according to rational probabilities leads necessarily to
a length of routing which is rational, whereas I, always
transcendental, has rational values only for discreet sets of
values of N and of the $p(y_i)$. If I is rational the L-optimal
"realizable" questionnaire Q will be identical to the "ideal"
questionnaire without inefficiency, only under certain conditions which will be the subject of §8.1. A questionnaire
which is not L-optimal and which has a routing length greater
than that of Q processes the noise emitted by a bad choice of
questions.

7.2 Axiomatics of Information

7.2.1 Faddeev's axioms

The axioms for information may be written in the following
form (A) which was given by Faddeev for a finite set of
elementary random events.

(A0) Let $y_i \in E$ (i: = 1,2,...,N) be the possible results of
the search for a finite number of mutually exclusive random
events which are given the probabilities

$$p(y_i) = p_i, \quad \sum_{i=1}^{N} p_i = 1$$

in the probability space $(E, \mathcal{P}(E))$. We will sometimes replace
p_{N-1} and p_N by $p_{N-1}t$ and $p_{N-1}(1-t)$, where $0 < t < 1$.

The information which may be acquired by the search for one
of the eventualities $y_1, \ldots, y_N$ will be denoted by

$$I_N(p_1, p_2, \ldots, p_N) .$$

(A1) The unit of information is fixed by

$$I_2(\tfrac{1}{2}, \tfrac{1}{2}) = 1 .$$

(A2) I_N is a symmetric function, of the p_i:

$$I_N(p_1, p_2, \ldots, p_N) = I_N(p_{k(1)}, p_{k(2)}, \ldots, p_{k(N)})$$

where

$$\begin{pmatrix} 1, & 2, & \ldots, & N \\ k(1), & k(2), & \ldots, & k(N) \end{pmatrix} \text{ is a permutation }.$$

(A3) $\quad I_N(p_1, p_2, \ldots, p_{N-1}t, p_{N-1}(1-t))$
$= I_{N-1}(p_1, \ldots, p_{N-1}) + p_{N-1} I_2(t, 1-t)$.

(A4) $p \to I_2(p, 1-p)$ is continuous for $0 \leq p \leq 1$.

Note that the choice of the base of logarithms is related to axiom (A1) whereas (A3) imposes both additivity and weighting of information by probabilities.

The additivity is relative to two classes of events:

1. A class formed of N-2 elementary events or eventualities and of a non-elementary event;

2. A class formed of this event, which is itself formed of the last two eventualities.

The weighting is relative to the second class because all the eventualities are concerned once and only once in the N-1 events of the first class.

It is interesting to compare the operation carried out when regrouping two answers of a dichotomic questionnaire according to Huffman's algorithm and the method of calculating $I_N(p_1, p_2, \ldots, p_{N-1}t, p_{N-1}(1-t))$ according to (A3). Since there is a regrouping of two elementary events among N, we must determine the information for a set of N-1 elementary events among N, or a dichotomic L-optimal questionnaire with N-1 answers. Both these processes are therefore recursive and lead to the computation of the function $I_2(p, 1-p)$ of two variables or to the construction of an elementary questionnaire with two answers of probabilities p and 1-p.

This parallelism between Huffman's algorithm for dichotomic questionnaires and the information determined by the system of axioms (A) can be preserved for polychotomic questionnaires with basis a>2 (resp. for heterogeneous questionnaires) but it will then be necessary to regroup a (resp. a_k, $k \in K$) elementary events and we will be led to develop new systems of axioms. We will consider that the axiomatic (A) is that of information for dichotomic questionnaires. The system (A) has enabled Faddeev and Renyi (cf. for example the book by Renyi of 1966) to establish with the aid of a lemma by Erdos the fundamental property:

Theorem 1 The solution of the system of axioms (A) is Shannon's information function:

$$I_N(p_1, p_2, \ldots, p_N) = \sum_{i:=1}^{N} p_i \log_2 \frac{1}{p_i} .$$

We will not reproduce the proof one of the less technical aspects of which consists in proving (1) for Hartley's system of axioms (H).

For N=2, we then find (Fig. 7.3 and tables at the end of the volume) that

$$I_2(p, 1-p) = p \log_2 \frac{1}{p} + (1-p) \log_2 \frac{1}{1-p} , \qquad (3)$$

which implies that

$$I_2(p, 1-p) = 0 \Rightarrow p.(1-p) = 0$$

and also

$$(p_1 + p_2) I_2 \left(\frac{p_1}{p_1 + p_2}, \frac{p_2}{p_1 + p_2} \right) = p_1 \log_2 \frac{1}{p_1}$$
$$+ p_2 \log_2 \frac{1}{p_2} - (p_1 + p_2) \log_2 \frac{1}{p_1 + p_2} . \qquad (4)$$

In particular, for r>0;

$$I_2 \left(\frac{r}{r+1}, \frac{1}{r+1} \right) = \log_2(r+1) - \frac{r}{r+1} \log_2 r . \qquad (5)$$

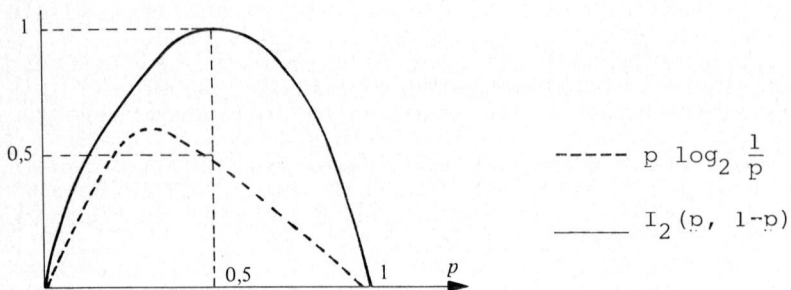

Figure 7.3

When the probability distribution is uniform

$$p_i = \frac{1}{N} \quad (\forall i) ,$$

and Shannon's information takes on the value

$$I_N \left(\frac{1}{N}, \frac{1}{N}, \ldots, \frac{1}{N} \right) = \log_2 N . \qquad (6)$$

Shannon's function is often called *entropy* because of its analogy with the quantity introduced by Boltzmann in the kinetic theory of gases: entropy characterizes indeterminacy, that is, the uncertainty a priori of an event which is going to happen. We remark that we could approach the discussion from this point of view instead of looking for what information we can receive after an event belonging to a certain set E has happened.

7.2.2 Some axiom systems

The axiom system given originally by Shannon and later by Khintchin are equivalent to Faddeev's system and require five axioms together with A_0 which is preserved. Khintchin's system includes the axioms concerning the properties known by the names:

normality (A1), *symmetry* (A2), *strong additivity* (7), *expansibility* (8) and *non-negativity* (9) with:

$$I_{NM}(p_1 t_{11}, \ldots, p_1 t_{1M}, p_2 t_{21}, \ldots, p_2 t_{2M}, \ldots, p_N t_{NM})$$
$$= I_N(p_1, \ldots, p_N) + \sum_{j:=1}^{N} p_j I_M(t_{j_1}, t_{j_2}, \ldots, t_{j_M}) , \qquad (7)$$

where
$$\sum_{k:=1}^{M} t_{j_k} = 1 ;$$

$$I_N(p_1, \ldots, p_{N-1}, 0) = I_{N-1}(p_1, \ldots, p_{N-1}) \qquad (8)$$

$$I_N(p_1, \ldots, p_N) \geq 0 . \qquad (9)$$

Among the other important properties related to the axiomatics enabling us to characterize Shannon's information, and catalogued by Aczel, we must point out the property of *nullity*:

$$I_N(0, 0, \ldots, 0, 1) = 0 , \qquad (10)$$

which is trivial for $I_2(0,1)$ and which may then be proved by the reader with the aid of (A3).

Daroczy and Pintacuda have reduced axiom (A3) to a weaker form by substituting for the second term of the right-hand side a function (not stated precisely) of two variables, that is, $L_2(p_{N-1}, t)$ instead of $p_{N-1} I_2(t, 1-t)$.

All these axiomatics lead to Shannon's information expressed with a unit of logarithms to base 2.

Axiom (A1) indicates that the information corresponding to an elementary equiprobable questionnaire with only one question with basis 2 has value $I_2(\frac{1}{2}, \frac{1}{2}) = 1$.

But we know that a polychotomic questionnaire Q_{a^k} having a^k equiprobable answers has routing length $L(Q_{a^k}) \geq k$.

If $a=2$, the information takes the value

$$I_{2^k}(2^{-k},2^{-k},\ldots,2^{-k}) = k$$

and consequently

$$I_{2^k}(2^{-k},2^{-k},\ldots,2^{-k}) \leq L(Q_{2^k})$$

If $a>2$, then

$$I_{a^k}(a^{-k},a^{-k},\ldots,a^{-k}) = k \log_2 a$$

that is

$$I_{a^k}(a^{-k},a^{-k},\ldots,a^{-k}) \leq \log_2 a \cdot L(\Omega_{a^k})$$

with equality if Q_{a^k} is L-optimal.

The factor $\log_2 a$ comes in because of the choice of the unit of information. We may therefore agree to adopt in the axiom system of information a choice of the unit which will make the information obtained by identifying the answers to an equiprobable elementary questionnaire having only one question with basis a equal to 1.

This leads to the axiom system (B) introduced at Besancon.

(B0) Identical to (A0) except for the restriction: $a-1 | N-1$; we will sometimes substitute $p_{N-a+1} t_j$ for p_{N-a+j} for $j:=1,\ldots,a$ with $\sum_{j:=1}^{a} t_j = 1$.

(B1) The unit of information is fixed by

$$I_a(\tfrac{1}{a}, \tfrac{1}{a}, \ldots, \tfrac{1}{a}) = 1.$$

(B2) Identical to (A2).

(B3) $I_N(p_1, p_2, \ldots, p_{N-a}, p_{N-a+1} t_1,$

$$p_{N-a+1} t_2, \ldots, p_{N-a+1} t_a)$$

$$= I_{N-a+1}(p_1, \ldots, p_{N-a}, p_{N-a+1})$$

$$+ p_{N-a+1} I_a(t_1, t_2, \ldots, t_a).$$

(B4) The function $p \to I_N(p_1, p_2, \ldots, p_{N-2}, p, t)$ is a continuous function of p, with

$$p + t = 1 - \sum_{i:=1}^{N-2} p_i.$$

This axiom system, equivalent to (A) for $a=2$, is valid only if N-1 is a multiple of a-1: This is the case for the number of answers of a polychotomic questionnaire in the strict sense.

Since axiom (B3) of additivity-weighting can be applied recursively for sets of N-k(a-1) answers, as long as N-k(a-1)>2a-1, the function solution of these axioms is the solution of system (A) except for a constant fixed by (B1). It is therefore

$$I_N(p_1, p_2, \ldots, p_N) = \sum_{i:=1}^{N} p_i \log_a \frac{1}{p_i} \qquad (11)$$

Note that it is possible to proceed to a choice in the order of arrangement of the elementary events y_i without modifying the value of I_N according to (B2).

We can reduce the computation of I_N to an exclusive computation of functions with a variables by recursive use of (B3). In particular if we order the events according to non-increasing probabilities, that is if:

$$p_1 \geq p_2 \geq \ldots \geq p_{N-a} \geq p_{N-a+1} \geq \ldots \geq p_N \ ,$$

then the second term of the right-hand side of (B3) will be weighted by the smallest possible factor p_{N-a+1}. Then the insertion of p_{N-a+1} in the list $p_1, p_2, \ldots, p_{N-a}$ between the two probabilities enclosing its value will enable us to apply (B3) with a factor $p_{N-2(a-1)}$ which will again be the smallest possible. Iterated use of this method in the course of the N-1/a-1 steps facilitates the evaluation of I_N according to an algorithm consisting partly of ordering the arrangement and partly of logarithmic calculation. The first part amounts exactly to Huffman's algorithm for the construction of L-optimal polychotomic questionnaires.

The extension of the proposed method of calculation, to the case of polychotomic questionnaires in the broad sense, makes it possible to have recourse once more to Huffman's algorithm, but the first use of (B3) must be made by taking only β+1 events, so that we will write this axiom in the following form, taking $0 < \beta \leq a-1$:

(B'3) $\quad I_N(p_1, p_2, \ldots, p_{N-\beta-2}, p_{N-\beta-1}, p_{N-\beta}t_1,$

$$p_{N-\beta}t_2, \ldots, p_{N-\beta}t_{\beta+1})$$

$$= I_{N-\beta}(p_1, p_2, \ldots, p_{N-\beta-1}, p_{N-\beta})$$

$$+ p_{N-\beta} I_{\beta+1}(t_1, t_2, \ldots, t_{\beta+1}) \ ,$$

where β is the smallest strictly positive integer such that $a-1 | N-\beta-1$ and where $\sum_{j=1}^{N} t_j = 1$.

We then substitute (B'3) for (B3) to constitute the axiom system (B') adapted to all polychotomic questionnaires.

However we are often required to make information computations bearing either on sets of N_1 events such that $a-1|N_1-1$ on $\beta+1$ eventualities. The axiom system (B) does not allow this last calculation, and we will write the coefficient of $P_{N-\beta}$ in the second term of the right-hand side of (B'3) in the form:

$$I_a(t_1, t_2, \ldots, t_{\beta+1}, 0, \ldots, 0) ;$$

taking account of (8) and (B), this amounts to evaluating

$$\log_a (\beta + 1) \, I'_{\beta+1}(t_1, t_2, \ldots, t_{\beta+1})$$

where I' is the information given by (B1), after the change of basis $a := \beta+1$. Thus the basis of one unique question leads to a change in the base of the logarithm and then to a homogeneity factor.

Finally the solution of system (B') remains as Shannon's information except for the logarithm base, given by (11).

The symmetry (B2) implies that the evaluation of $I_N(p_1, \ldots, p_N)$ is independent of the way in which the events are regrouped while using (B3), (events which are elementary for the first use of (B'3), according to a method already pointed out earlier, in the discussion of the parallelism between Huffman's algorithm and the process generated by the axiom (A3) of Faddeev).

Theorem 2 The information which it is possible to acquire by the search for one of the N answers of a polychotomic questionnaire depends only on the probability distribution $\mathcal{P}$ and is independent of the support, it is given by

$$I_N = \sum_{i:=1}^{N} p_i \log \frac{1}{p_i}$$

where the logarithmic base is 2 for the axioms (A) and a for the axioms (B').

More generally, consideration of heterogeneous questionnaires leads us to introduce an information function which is related not only to the set of probabilities of the eventualities but also to the different question bases.

A polychotomic questionnaire in the broad sense can be considered as a heterogeneous questionnaire having M-1 questions with basis a and one question with basis $\beta+1 (<a)$.

In (B') we have written

$$I_a(t_1, t_2, \ldots, t_{\beta+1}, 0, \ldots, 0)$$
$$= \log_a (\beta + 1) I_{\beta+1}(t_1, \ldots, t_{\beta+1}),$$

which amounts to using a as the base of the logarithms for all the M questions. It is clear that we can associate Shannon's information function with a heterogeneous questionnaire and extend Theorem 2 to it.

It is possible to say that a polychotomic questionnaire in the broad sense has the same basis for nearly all its questions but it is impossible to say the same thing about a heterogeneous arborescent questionnaire. To reduce the influence of a particular basis of questions we could consider a heterogeneous questionnaire and apply to it systematically a process analogous to that deduced from (B'3). It would then be justified to introduce for the base of the logarithms the basis of the corresponding question. But nothing can be said of the unicity of the function obtained when the support varies although the pair $(\mathscr{A}, \mathscr{P})$ is unchanged; we are thus led to axiom system (C), the solution function of which will be called *acquisition*.

(C0) Let $y_i \in E (i := 1, 2, \ldots, N)$ be the possible result of the search of a finite number of mutually exclusive random events with probabilities in the probability space $(E, \mathscr{P}(E))$:

$$p(y_i) = p_i; \quad \sum_{i:=1}^{N} p_i = 1.$$

(C1) The acquisition depends both on the probability distribution and on the bases $a_1, a_2, \ldots, a_M$ of the questions allowing identification of the eventualities. The probabilities of the eventualities and the bases are placed in non-increasing order:

$$p_1 \geqslant p_2 \geqslant \ldots \geqslant p_N \quad \text{and} \quad a_1 \geqslant a_2 \geqslant \ldots \geqslant a_M.$$

There are q_k questions with basis a_k.

The acquisition will be denoted by

$$A_N(p_1, p_2, \ldots, p_N; a_1, q_1, a_2, q_2, \ldots, a_M, q_M)$$

and if there is no ambiguity:

$$A_N(p_1, p_2, \ldots, p_N).$$

If we do not know the order of the probabilities we will write

$$\overline{A}_N(p_1, p_2, \ldots, p_N).$$

(C2) The unit of acquisition is:

$(\forall a) \quad A_a(\frac{1}{a}, \frac{1}{a}, \ldots, \frac{1}{a}; a, 1) = 1$.

(C3) $A_N(p_1, \ldots, p_{N-a_M}, p_{N-a_M+1} t_1, \ldots, p_{N-a_M+1} t_{a_M};$

$$a_1, q_1, \ldots, a_{M-1}, q_{M-1}, a_M, q_M)$$

$$= \bar{A}_{N-a_M+1}(p_1, \ldots, p_{N-a_M}, p_{N-a_M+1};$$

$$a_1, q_1, \ldots, a_{M-1}, q_{M-1}, a_M, q_M - 1)$$

$$+ p_{N-a_M+1} A_{a_M}(t_1, t_2, \ldots, t_{a_M}; a_M, 1)$$

with
$$\sum_{i:=1}^{a_M} t_j = 1 \quad ;$$

if $q_M - 1 = 0$, we will omit a_M from the list $\mathcal{A}$ and $q_M - 1$.

(C4) The function $p \to A_N(p_1, p_2, \ldots, p_{N-2}, p, t)$ is a continuous function of p for

$$p + t = 1 - \sum_{i:=1}^{N-2} p_i \quad .$$

In the case when all the questions (except possibly one) have the same basis, (C) reduces to system (B) (or (B')).

It is clear that the computation of

$$A_{a_M}(t_1, t_2, \ldots, t_{a_M}; a_M, 1)$$

leads to

$$A_{a_M} = \sum_{i:=1}^{a_M} t_i \log_{a_M} \frac{1}{t_i} \quad ;$$

By (C3) the computation of A_N requires the use of logarithms with non-decreasing bases. Moreover the reordering of the probability p_{N-a_M+1} which may be necessary in the course of the recursive process using (C3) is none other than that of Huffman's construction of an L-optimal questionnaire built on

$$(\mathcal{A}, \mathcal{P}) = (\{a_1, \ldots, a_M\}, \{p_1, \ldots, p_N\}) \quad .$$

Acquisition is therefore a function generalizing the information which has been associated essentially (but not exclusively) with the homogeneous case to the heterogeneous case.

7.3 Properties of Information

Here we establish some properties of information related essentially to the hypotheses of dependence or independence.

We will use the information function (11) defined by the axiom system (B'), and the logarithm will be expressed with base a if the associated questionnaire is polychotomic; in the case of dichotomic questionnaires it will then also be the function (2) defined by the system (A); when the associated questionnaire is heterogeneous, we will use a unique basis a which will be one of the bases a and we will then use axiom system (B') making use of the expansibility axiom of Khintchin (8) where necessary.

7.3.1 Convexity and Concavity

A function $x \to f(x)$ of a real variable x is said to be *convex on an interval* when any secant intersecting the curve $(x, f(x))$ in M_1 and M_2 is situated above this curve between the two points of intersection M_1 and M_2, that is, when

$$\lambda_1 f(x_1) + \lambda_2 f(x_2) \geq f(\lambda_1 x_1 + \lambda_2 x_2)$$

with

$$\lambda_1 + \lambda_2 = 1 \quad \text{and} \quad \lambda_1 \geq 0, \; \lambda_2 \geq 0 .$$

Property 1 Any convex function of a real variable satisfies *Jenssen's inequality*:

$$\sum_{k:=1}^{N} \lambda_k f(x_k) \geq f\left(\sum_{k:=1}^{N} \lambda_k x_k \right) , \qquad (12)$$

where

$$\lambda_k \geq 0 \; (\forall k) \quad \text{and} \quad \sum_{k:=1}^{N} \lambda_k = 1 .$$

This classical inequality will be applied to the convex function $f(x) = x \log x$ whatever the base of logarithms used:

$$\sum_{k:=1}^{N} \lambda_k x_k \log x_k \geq \left(\sum_{k:=1}^{N} \lambda_k x_k \right) \log \left(\sum_{k:=1}^{N} \lambda_k x_k \right) . \qquad (13)$$

An apparently neighbouring inequality has been obtained by Hardy, Littlewood and Polya when the λ_k are equal to 1 by means of a condition on the x_k:

Property 2 A sufficient condition for Polya's inequality

$$\sum_{k:=1}^{N} \varphi(x_k) \geq \varphi\left(\sum_{k:=1}^{N} x_k \right) \qquad (14)$$

to hold for all positive x_k is that $x^{-1} \varphi(x)$ is a decreasing function; the inequality is strict if $x^{-1} \varphi(x)$ decreases strictly and if $N>1$.

In the case when $\varphi(x)=x \log \frac{1}{x}$ we can directly establish the result:

if $x_k \geq 0$ ($\forall k$) and $\sum_{k=1}^{N} x_k \leq 1$, then

$$\sum_{k:=1}^{N} x_k \log \frac{1}{x_k} \geq (\sum_{k:=1}^{k=1} x_k) \log \frac{1}{\sum_{k:=1}^{N} x_k} \qquad (15)$$

using the concavity of $\varphi(x)=x \log \frac{1}{x}$ for $x>0$.

Inequality (15) is strict when there exist two indices j and ℓ such that $x_j>0$ and $x_\ell>0$.

For k=2, we find:

If $x_1 > 0$, $x_2 > 0$ and $x_1 + x_2 < 1$, then

$$x_1 \log \frac{1}{x_1} + x_2 \log \frac{1}{x_2} > (x_1 + x_2) \log \frac{1}{x_1 + x_2} \qquad (16)$$

which, taking into account (4), but not (9), implies the non-negativity of Shannon's information:

$$I_2 (\frac{x_1}{x_1 + x_2}, \frac{x_2}{x_1 + x_2}) \geq 0 ,$$

where the inequality is strict if x_1 and x_2 are strictly positive.

The function $\varphi(x)=x \log \frac{1}{x}$ is defined and positive for $x \in]0,1[$ and its concavity on this domain implies that its derivative is decreasing.

The slope of a secant intersecting the curve $(x, \varphi(x))$ at the points of abscissae x_1 and x_1+x_3 is then greater than the slope of a secant intersecting $(x, \varphi(x))$ at the points of abscissae x_1+x_2 and $x_1+x_2+x_3$ that is:

$$\frac{\varphi(x_1 + x_2 + x_3) - \varphi(x_1 + x_2)}{x_3} < \frac{\varphi(x_1 + x_3) - \varphi(x_1)}{x_3}$$

when $x_1 \geq 0$, $x_2 > 0$, $x_3 > 0$ and $x_1 + x_2 + x_3 \leq 1$.

Putting
$$x_o = x_1 + x_3 ,$$

this leads to:

$$(x_o + x_2) \log \frac{1}{x_o + x_2} - (x_1 + x_2) \log \frac{1}{x_1 + x_2}$$

$$< x_o \log \frac{1}{x_o} - x_1 \log \frac{1}{x_1}$$

and for $\quad x_2 > 0, \ x_o > x_1 \geq 0:$

$$x_o \log \frac{1}{x_o} + x_2 \log \frac{1}{x_2} - (x_o + x_2) \log \frac{1}{x_o + x_2}$$

$$> x_1 \log \frac{1}{x_1} + x_2 \log \frac{1}{x_2} - (x_1 + x_2) \log \frac{1}{x_1 + x_2} \qquad (17)$$

Putting

$$p_k = x_o, \ p_j = x_1, \ p_\ell = x_2$$

and remarking that, if $p_k + p_\ell \leq 1;$ then

$$p_k \log_2 \frac{1}{p_k} + p_\ell \log_2 \frac{1}{p_\ell} - (p_k + p_\ell) \log_2 \frac{1}{p_k + p_\ell}$$

$$= (p_k + p_\ell) I_2 (\frac{p_k}{p_k + p_\ell}, \frac{p_\ell}{p_k + p_\ell}) \qquad (18)$$

we obtain

Property 3 Shannon's information function satisfies the inequality

$$(p_k + p_\ell) I_2 (\frac{p_k}{p_k + p_\ell}, \frac{p_\ell}{p_k + p_\ell})$$

$$> (p_j + p_\ell) I_2 (\frac{p_j}{p_j + p_\ell}, \frac{p_\ell}{p_j + p_\ell}) \qquad (19)$$

when

$$p_k > p_j \geq 0, \ p_\ell > 0 \quad \text{and} \quad p_k + p_\ell \leq 1.$$

Consider two probability distributions

$$p = \{p_i\}, \ Q = \{q_i\}, \ \sum_{i:=1}^{N} p_i = \sum_{i:=1}^{N} q_i = 1.$$

Compare $\quad \sum_{i:=1}^{N} p_i \log \frac{1}{p_i} \quad$ and $\quad \sum_{i:=1}^{N} p_i \log \frac{1}{q_i}$

taking e as the base of logarithms; this is solely to simplify the comparison because all that matters is the unicity of the basis in these two expressions.

The logarithmic function $x \to \ln x$ defined for $x>0$ is concave, and therefore wholly represented by a curve below any tangent. At the point $x=1$, such that $\ln 1=0$, the slope of this tangent is 1. We can therefore write $\ln x \leq x-1$, equality holding only if $x=1$.

Taking $x = \dfrac{q_i}{p_i}$ we obtain successively:

$$\ln \frac{q_i}{p_i} \leq \frac{q_i}{p_i} - 1$$

$$p_i \ln \frac{q_i}{p_i} \leq q_i - p_i$$

and by summation

$$\sum_{i:=1}^{n} p_i \ln \frac{q_i}{p_i} \leq 0$$

and finally

Property 4 If $(p_1,\ldots,p_N)$ and $(q_1,\ldots,q_N)$ are complete probability distributions

$$(\sum_{i:=1}^{N} p_i = \sum_{i:=1}^{N} q_i = 1) ,$$

then

$$\sum_{i:=1}^{N} p_i \log \frac{1}{p_i} \leq \sum_{i:=1}^{N} p_i \log \frac{1}{q_i} ; \qquad (20)$$

with equality if and only if $p_i = q_i$ ($\forall i$).

Taking $q_i = \dfrac{1}{N}$ ($\forall i$) we have

$$\sum_{i:=1}^{N} p_i \log \frac{1}{p_i} \leq \log N .$$

Property 5 The information $I_N(p_1,\ldots,p_N)$ takes its maximal value

$$I_N(\tfrac{1}{N},\ldots,\tfrac{1}{N}) = \log N$$

when all the elementary events are equiprobable.

7.3.2 Independence and Dependence

We have seen that an experiment defined as a search for events belonging to a finite set having N answers can be associated with a questionnaire. If N is admissible as a basis, it can be an elementary questionnaire having the root as a unique question. Otherwise we will consider that it is an arborescent questionnaire when these answers are elementary events (the case discussed in §7.2).

Consider two experiments, A associated with a questionnaire Q_A and B associated with a questionnaire Q_B. Form the latticoid produce $Q_A Q_B$. Then the probabilities of the answers of $Q_A Q_B$ are the products of the probabilities of the answers of Q_A and Q_B and we will say that the questionnaires Ω_A and Q_B (resp. that the experiments A and B) are *independent*. Consider next a restricted product $Q_A \diamond Q_B$, letting the restriction of the set of answers of Q_A, that is of E, bear on the sub-set $E'=\{\xi_i\}$ consisting of only one answer of Q_A (or eventuality of A). We will say that the two experiments are independent and we will express the routing length by the sum $L(\Omega_A)+p(\xi_i)L(Q_B)$, according to (5-15) and the information by the use of the strong additivity (7), with the double restriction

(1) $\forall j, t_{ij} \neq 0$

(2) $\forall h \neq i, t_{hl} = 1$ and $\forall k \neq l, t_{hk} = 0$.

There is independence because the answers of B, asking only if ξ_i were the answer of A, do not depend on ξ_i.

But there will be a dependence if we have an experiment B_i, the nature of which is related to the outcome $\xi_i \in E$; this can be realized by effecting consecutively $|E|=N$ restricted products of questionnaires:

$$Q_A \diamond Q_{B_1} \diamond Q_{B_2} \diamond \cdots \diamond Q_{B_N}$$

where each restricted product bears on the one hand on a different answer $\xi_i \in E$ and on the other on a different questionnaire Ω_{B_i}.

The experiment B consisting of a questionnaire B_i related to ξ_i will depend on the experiment A (but it will not be the same for the answer of B).

The experiment A, itself realized by means of the questionnaire Q_A which can be conceived as the result of a sequence of restricted products (cf. 5-16) can be decomposed into experiments related each to the others: they are dependent but not their answers. The information is then given according to strong additivity.

It will not be the same with an experiment associated with a latticoid questionnaire: if two paths (at least) enter into some question i, the elementary experiment associated with question i will not be conditioned by the paths (μ_1 and μ_2 for example) if the flows are proportional (ex: compatible questionnaire of type B) but will be conditioned by the paths of any other compatible questionnaire C: the same experiment B_i will have outcomes whose probabilities

$$p((i,j|\mu_1)) \quad \text{and} \quad p((i,j|\mu_2))$$

are different. A latticoid questionnaire will therefore be

considered as the result of experiments which in general will be dependent.

For a latticoid questionnaire Ω can be obtained by a product $\Omega_A \Omega_B$ followed by contractions which can bear on the answers (of $\Omega_A \Omega_B$) corresponding to some one answer of Ω_B (Fig. 7.4). We will say then that the *experiment B depends on experiment A*.

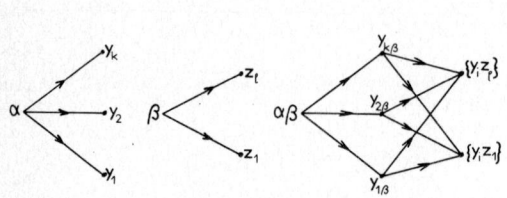

Figure 7.4

Consider therefore two sets of independent events A and B. The eventualities A_k and B_j corresponding to the different possible realizations of each of these sets of events are characterized by the probabilities

$$p(A_k) = p_k \quad \text{such that} \quad \sum_{A_k \in A} p_k = 1$$

$$p(B_j) = q_j \quad \text{such that} \quad \sum_{B_j \in B} q_j = 1 .$$

At the outcome of the event A (or B) nothing is fixed in what concerns B (or A).

The probability of the double event A B is

$$p(A_k B_j) = p(A_k) p(B_j | A_k) = p(A_k) p(B_j) = p_k q_j$$

where $p(B_j | A_k)$ is the conditional probability of eventuality B_j when the event A has happened in the form A_k; the independence, or absence of conditioning, means that $p(B_j | A_k) = p(B_j)$.

The information obtained at the outcome of the experiments A and B is:

$$I(AB) = \sum_k \sum_j p_k q_j \log \frac{1}{p_k q_j} ,$$

$$I(AB) = \sum_k \sum_j p_k q_j \left(\log \frac{1}{p_k} + \log \frac{1}{q_j} \right) ,$$

$$I(AB) = \sum_k p_k \log \frac{1}{p_k} \left(\sum_j q_j \right) + \sum_j q_j \log \frac{1}{q_j} \left(\sum_k p_k \right)$$

thus

$$I(AB) = \sum_k p_k \log \frac{1}{p_k} + \sum_j q_j \log \frac{1}{q_j} \tag{22}$$

that is
$$I(AB) = I(A) + I(B).$$

In the case when the events A and B are dependent, we consider the probabilities p_k, q_j a priori and the conditional probabilities (Fig. 7.5)

$$p_{jk} = p(A_k|B_j) \quad \text{and} \quad q_{k\ell} = p(B_\ell|A_k)$$

related by
$$\sum_k p_k = \sum_\ell q_\ell = 1 ;$$

$$\sum_k p_k q_{k\ell} = \sum_k p(A_k) p(B_\ell|A_k)$$

represents the probability of eventuality B_ℓ (whatever the outcome A_k of the event A) that is:

$$\sum_k p_k q_{k\ell} = q_\ell$$

and similarly

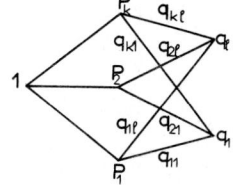

Figure 7.5

$$\sum_j q_j p_{jk} = p_k.$$

Moreover $\sum_\ell q_{k\ell}$ represents the set of possibilities of realization of the B_ℓ when A_k has happened, this event is therefore a certainty:

$$\sum_\ell q_{k\ell} = 1.$$

Consider the sequences ($\frac{\text{events}}{\text{probabilities}}$) relative to the experiments A and (A and B):

$$\begin{bmatrix} A_1 \cdots A_N \\ p_1 \cdots p_N \end{bmatrix} \quad \begin{bmatrix} A_1B_1 \cdots A_1B_\ell \cdots A_1B_M \cdots A_kB_\ell \cdots A_NB_M \\ p_1q_{11} \cdots p_1q_{1\ell} \cdots p_1q_{1M} \cdots p_kq_{k\ell} \cdots p_Nq_{NM} \end{bmatrix}$$

The probability of the event A_kB_ℓ is $p_k q_{k\ell}$. Then:

$$I(AB) = \sum_k \sum_\ell p_k q_{k\ell} (\log 1/p_k + \log 1/q_{k\ell})$$

$$= \sum_k p_k \log_2 1/p_k \left(\sum_\ell q_{k\ell}\right)$$

$$+ \sum_k p_k \left(\sum_\ell q_{k\ell} \log 1/q_{k\ell}\right).$$

The quantity

$$\sum_\ell q_{k\ell} \log 1/q_{k\ell} \qquad (23)$$

which we denote by $I(B|A_k)$ is the information relative to experiment B when experiment A has had outcome A_k; we will call this quantity *conditional information relative to outcome* A_k. We will then write:

$$I(AB) = I(A) + \sum_k p_k I(B|A_k) \; ;$$

whereas, in the independent case, the information was the sum of the informations of the two experiments taken in any order, here $I(AB)$ is the sum of $I(A)$ and of the weighted sum of the $I(B|A_k)$ which are already an information with local character (we could even say exact).

The quantity

$$I(B|A) = \sum_k p_k I(B|A_k) \qquad (24)$$

is called the *conditional information* of B (relative to the realization of the preliminary experiment A).

$I(B|A)$ is the mathematical expectation of $I(B|A_k)$ and we have

$$I(AB) = I(A) + I(B|A)$$

If there is independence, then $I(B|A) = I(B)$.

If there is dependence, the information on B a priori $I(B)$, must be greater than the information in the course of the experiment $I(B|A)$.

Applying Jenssen's inequality according to formula (13) in the case when $\lambda_k = p_k$ and $x_k = q_{k\ell}$, we find for any ℓ;

$$\sum_k p_k q_{k\ell} \log q_{k\ell} \geq (\sum_k p_k q_{k\ell}) \log (\sum_k p_k q_{k\ell}) = q_\ell \log q_\ell \; .$$

Summing over ℓ we find:

$$\sum_k p_k \sum_\ell q_{k\ell} \log_2 q_{k\ell} \geq \sum_\ell q_\ell \log q_\ell$$

or

$$I(B) = \sum_\ell q_\ell \log 1/q_\ell \geq \sum_k p_k \sum_\ell q_{k\ell} \log 1/q_{k\ell} = I(B|A)$$

and consequently

$$I(B) \geq I(B|A) \; .$$

Property 6 The information obtained at the outcome of a double experiment AB is related to the information of the simple experiments by

$$I(AB) = I(A) + I(B|A) \leq I(A) + I(B) \qquad (25)$$

with equality if and only if A and B are independent experiments.

We can write (25) in the form

$$I(AB) = I(B) + I(A|B)$$

taking for $I(A|B)$ a definition analogous to that of $I(B|A)$.

Note that then $I(A)-I(A|B) = I(B)-I(B|A) = I(A,B)$ and we will call $I(A,B)$ the *mutual information* of the two sets of events A and B: it is the quantity of information brought by each of the experiments to the other.

Taking again the numerical example from §7.1 of coloured balls carrying numbers we find:

$$I(AB) = 3.32, \; I(A) = 0.72, \; I(B) = 3.32, \; I(B|A) = 2.6$$

and $I(A|B) = 0$ (because the number determines the colour) where $I(A)$, $I(B|A)$ and $I(A,B)$ stand for what we earlier called $I(1)$, $I(2)$, $I(3)$ respectively. The mutual information is $I(A,B) = 0.72$.

In this example, A is the experiment determining the set E_i (one set among n), B is the experiment determining an element $e \in E_i$ (knowing the index i) and AB is the experiment determining the element $e \in E$; this experiment can be realized directly or with the aid of an auxiliary experiment C.

Let C be the experiment enabling us to form a sub-set $E' \subset E$ ($E' \neq \emptyset$ and $E' \neq E$) and to determine if $e \in E'$. Then experiment C will reduce the uncertainty a priori of the experiment AB.

Let $E'_i = E' \cap E_i$ and put $|E'|=N'$, $|E'_i|=N'_i$ and

$$(\forall i) \; r_i = \frac{N'_i}{N'} \quad \text{so that} \quad \sum_{i:=1}^{n} r_i = 1 \;.$$

Then the mutual information between AB and C is

$$I(AB,C) = \log \frac{N}{N'}$$

because, after the realization of C, there are but N' equiprobable eventualities.

Similarly the mutual information between C and an experiment B_i, realized knowing the result i of A, is

$$I(B_i, C) = \log \frac{N_i}{N'_i}.$$

But the outcome i has probability r_i, because of the fact that $E' \subset E$.

We will then write

$$I(B,C) = \sum_{i:=1}^{n} r_i \log \frac{N_i}{N'_i}.$$

The difference

$$I(AB, C) - I(B,C) = \sum_{i:=1}^{n} r_i \log \frac{r_i}{p_i}$$

represents the expectation of the information obtained by C.

We will call the *information gain* of the distribution $\mathcal{R} = (r_1, \ldots, r_n)$ relative to the distribution

$$\mathcal{P} = (p_1, \ldots, p_n)$$

the quantity which we will write as

$$I(\mathcal{R} || \mathcal{P}) = \sum_{i:=1}^{n} r_i \log \frac{r_i}{p_i} \qquad (26)$$

This quantity will be defined more generally for non-rational values of the probabilities.

For the numerical example already considered, we have

$$N_1 = 2, \quad N_2 = 8, \quad \mathcal{P} = (\tfrac{1}{5}, \tfrac{4}{5})$$

and taking

$$N'_1 = 1, \quad N'_2 = 5, \quad \mathcal{R} = (\tfrac{1}{6}, \tfrac{5}{6})$$

we obtain

$$I(\mathcal{R} || \mathcal{P}) = \tfrac{1}{6} \log_2 \tfrac{5}{6} + \tfrac{5}{6} \log_2 \tfrac{25}{24} = 0.005,$$

which we can compare with $I(A B)$, and

$$I(AB, C) = \log_2 \tfrac{5}{3} = 0.73.$$

For $N''_1 = 1$, $N''_2 = 6$, $R'' = (\tfrac{1}{4}, \tfrac{3}{4})$, we find $I(\mathcal{R}'', \mathcal{P}) = 0{,}011$.

The concept of information gain has enabled Aggarwal to form random questionnaires in which it is possible to detect the membership of answers to sub-sets $E_i \in E$ but without being able to separate the answers $e \in E_i$. These questionnaires constitute a generalisation of the questionnaires K (cf. Definition 4.14).

7.4 Processed Information and Transmitted Information

Definition 1 The *information transmitted* by a questionnaire Q is the function

$$I(Q) = I_N(p(y_1), p(y_2), \ldots, p(y_N))$$

$$= \sum_{y \in E} p(y) \log \frac{1}{p(y)} \qquad (27)$$

which depends only on the probability distribution of the answers; the base of the logarithm is a when the questionnaire is homogeneous, or a_h ($a_h \in \mathcal{A}$) when the questionnaire is heterogeneous. This same base of logarithms will be used for the questions of a latticoid questionnaire Q and those of the compatible arborescent questionnaires.

Definition 2 The *information processed* by a question i is the sum

$$J(i) = \sum_{j \in \Gamma i} \frac{p(i,j)}{p(i)} \log \frac{p(i)}{p(i,j)} . \qquad (28)$$

$J(i)$ is equal to the information $I(Q_i)$ transmitted by the elementary questionnaire consisting of the unique question i the outcomes of which have the conditional probabilities

$$p(j|i) = \frac{p(i,j)}{p(i)} \quad \text{such that} \quad \sum_{j \in \Gamma i} p(j|i) = 1 .$$

The application of (28) to an answer $y \in E$ would lead to $J(y) = 0$ because $\Gamma y = \emptyset$.

Definition 3 The *information processed* by a questionnaire Q is the sum

$$J(Q) = \sum_{i \in F} p(i) J(i) . \qquad (29)$$

We can also write

$$J(Q) = \sum_{i \in F} \sum_{j \in \Gamma i} p(i,j) \log \frac{p(i)}{p(i,j)} . \qquad (30)$$

We will specify the base of the logarithms used by writing $I_a(Q)$, $J_a(i)$, $J_a(Q)$.

We know (Property 4.19) that all the arborescent questionnaires compatible with Q have the same routing length but the probabilities of a question i^s are not the same in all these questionnaires when the image of a question i of Q consists of many vertices; in the same way the probabilities of the answers are not the same in all the questionnaires compatible with Q if Q is not an arborescence and has fewer arcs than the arborescence of its paths.

Let $C = (E_C \cup F_C, \Gamma_C, P_{\Gamma_C})$ be an arborescent questionnaire compatible with Q. Then the information processed by C is

$$J(C) = \sum_{i^s \in F_C} p_C(i^s) \sum_{u \in \Gamma_C i^s} \frac{p_C(i^s, j_u^s)}{p_C(i^s)} \log \frac{p_C(i^s)}{p_C(i^s, j_u^s)} \quad (31)$$

using the notation of §4.3.2.

It is possible to regroup the questions i^s belonging to the image of the same vertex i of Λ. Taking into account the definition of $\lambda_u^s(i)$ (Formula 4.18), this leads to

$$J(C) = \sum_{i \in F} \sum_{s \in \mu(i)} p_C(i^s) \sum_{u \in \Gamma i} \lambda_u^s(i) \frac{p(i, j_u)}{p(i)} \log \frac{p(i)}{\lambda_u^s(i) p(i, j_u)} \quad (32)$$

where $\mu(i)$ is the set of paths linking the root to i.

Property 4 enables us to give an upper bound

$$J(C) \leq \sum_{i \in F} \sum_{s \in \mu(i)} p_C(i^s) \sum_{u \in \Gamma i} \lambda_u^s(i) \frac{p(i, j_u)}{p(i)} \log \frac{p(i)}{p(i, j_u)} \quad (33)$$

because

$$\sum_{u \in \Gamma i} \frac{p(i, j_u)}{p(i)} = \sum_{u \in \Gamma i} \lambda_u^s(i) \frac{p(i, j_u)}{p(i)} = 1.$$

There is equality in (33) if and only if $\lambda_u^s(i) = 1$ ($\forall i, s, u$) that is only for the compatible questionnaire B having proportionality of the flow.

Property 7 The maximal value of the information processed by an arborescent questionnaire compatible with the latticoid questionnaire Q is obtained when the compatible questionnaire satisfies the hypothesis of proportionality of the flow.

Now reversing the order of summation:

$$J(C) \leq \sum_{i \in F} \sum_{u \in \Gamma i} \frac{p(i, j_u)}{p(i)} \log \frac{p(i)}{p(i, j_u)} \sum_{s \in \mu(i)} \lambda_u^s(i) p_C(i^s)$$

and since $\sum_{s \in \mu(i)} \lambda_u^s p_C(i^s) = p(i)$, we obtain, by (30):

$$J(C) \leq J(Q). \quad (34)$$

There is equality in (34) if and only if $\lambda_u^s(i) = 1$, ($\forall i, s, u$) and consequently

$$J(B) = J(Q). \quad (35)$$

INFORMATIONAL STUDY OF QUESTIONNAIRES

Theorem 3 The information processed by a latticoid questionnaire Q is equal to the maximal value of the information processed by a compatible arborescent questionnaire, that is, equal to the information processed by the questionnaire satisfying the hypothesis of proportionality of the flow.

Let us find a new expression for the processed information which will enable us to distinguish the root and the answers from questions of a latticoid questionnaire.

Let h, i, j be vertices such that $h\Gamma i$ and $i\Gamma j$ and

$$p(i) = \sum_{h \in \Gamma^{-1}i} p(h,i) = \sum_{j \in \Gamma i} p(i,j) . \qquad (36)$$

We may write (30) in the equivalent form

$$J(Q) = \sum_{h \in F} \sum_{i \in \Gamma h} p(h,i) \left(\log \frac{1}{p(h,i)} + \log p(h) \right)$$

The root α does not make any contribution through the term $\log p(\alpha)$ which is zero, but only through the terms in $\log \frac{1}{p(\alpha,i)}$ for $i \in \Gamma\alpha$; an answer y contributes only through the term in $\log \frac{1}{p(h,y)}$ where $h \in \Gamma^{-1}y$.

However a vertex i having both at least one ancestor and two proper descendants contributes through both the terms in $\log p(i)$ and $\log \frac{1}{p(h,i)}$.

We therefore write

$$J(Q) = \sum_{h \in F} \sum_{i \in \Gamma h} p(h,i) \log \frac{1}{p(h,i)}$$

$$+ \sum_{i \in \overline{F}} p(i) \log p(i) .$$

The first term of the right-hand side may be rewritten as

$$\sum_{i \in \overline{F}} \sum_{h \in \Gamma^{-1}i} p(h,i) \log \frac{1}{p(h,i)}$$

$$+ \sum_{i \in E} \sum_{h \in \Gamma^{-1}i} p(h,i) \log \frac{1}{p(h,i)}$$

Hence

$$J(Q) = \sum_{i \in \bar{F}} \sum_{h \in \Gamma^{-1}i} p(h,i) \log \frac{p(i)}{p(h,i)}$$

$$+ \sum_{i \in E} \sum_{h \in \Gamma^{-1}i} p(h,i) \log \frac{1}{p(h,i)} \quad . \quad (37)$$

The contribution of an internal question i to J(Q) lies in the first term.

Definition 4 The *information consumed* by an internal question or an answer is the sum

$$e(i) = \sum_{h \in \Gamma^{-1}i} p(h,i) \log \frac{p(i)}{p(h,i)} \quad . \quad (38)$$

The information consumed by an internal question or an answer is:

- zero, if $|\Gamma^{-1}i|=1$, that is, if there is only one arc entering into i,

- strictly positive otherwise, because then

$$p(i) > p(h,i) \quad (\forall h) \quad .$$

If $e(i)=0$ for all $i \in \bar{F}$, then

$$J(Q) = \sum_{i \in E} \sum_{h \in \Gamma^{-1}i} p(h,i) \log \frac{1}{p(h,i)} \quad .$$

The sub-latticoid generated by the set of questions is then an arborescence. A questionnaire Q possessing this last property is such that at any answer $i \in E$, by Polya's inequality (Property 2),

$$\sum_{h \in \Gamma^{-1}i} p(h,i) \log \frac{1}{p(h,i)} \geq$$

$$\left(\sum_{h \in \Gamma^{-1}i} p(h,i) \right) \log \frac{1}{\sum_{h \in \Gamma^{-1}i} p(h,i)} \quad (39)$$

with equality only if there is only one arc entering into y.

Summing over $i \in E$, we find

$$J(\Omega) \geq I(\Omega) \quad , \quad (40)$$

with equality only if the answers are, like the internal questions, terminal extremities of only one arc:

Theorem 4 The *processed information* is at least equal to the transmitted information, with equality if and only if the questionnaire has an arborescent support.

It is not possible to relate the processed and transmitted information by means of the consumed information; in fact, by (27), (37) and (38) we have

$$J(Q) = I(Q) + \sum_{i \in E \cup \overline{F}} e(i) \qquad (41)$$

Theorem 5 The difference between the information processed and the information transmitted by a questionnaire is equal to the sum of the information consumed at the vertices with more than one entering arc.

The sum $\sum_{i \in E \cup \overline{F}} e(i)$ which is equal to the decay undergone by the information processed by a latticoid questionnaire whose answers transmit only $I(Q)$, may be considered as a kind of information *processed in the reverse direction* according to the analogy between Definitions 2 and 4.

An *arborescent* questionnaire Q_A can be associated with a sequence of *independent experiments*: the information processed by each question is additive and the questionnaire Q_A transmits the direct sum $I(Q_A) = \sum_{i \in F_A} p(i)J(i)$; this generalizes Property 6 (the case of equality).

A latticoid questionnaire must be associated with a sequence of dependent experiments: the transmitted information is less than the sum of the information processed by each question: $I(Q) \leq \sum_{i \in F} p(i)J(i)$, and the difference $J(Q)-I(Q)$ which is the direct sum of the information consumed at vertices of indegree greater than 1 can be specified locally, which was not the case for Property 6 (the case of equality). The case of equality corresponds to the compatible questionnaire satisfying the hypothesis of proportionality of flow, whereas the inequality corresponds to a compatible questionnaire C.

Consider therefore a latticoid questionnaire Q having V distinct paths linking α to E, d arcs entering into the N vertices of E such that $V \geq d \geq N$.

Then with the questionnaire Q we can associate:

- the questionnaire B (compatible, with proportional flow);

- a compatible questionnaire C (C≠B);

- an arborescent questionnaire D having d answers with the same probabilities as the probabilities $p_C(h,y)$ of the d arcs entering into the set of answers of C;

- an arborescent questionnaire G with the same answers as Q.

Whatever the choices of the supports of D and G, these questionnaires are such that $J(D)=I(D)$ and $J(G)=I(G)$ because they are arborescent, the information transmitted by G is $I(G)=I(Q)$.

Moreover $I(G) \leq I(D) \leq I(C)$ according to axiom (A3) used while grouping the answers of C according to the paths ending by the same arcs of Q to form D, then, grouping the answers of D to form G.

We have proved:

Property 8 The information transmitted and the information processed by a latticoid questionnaire Q and by the arborescent associated questionnaires of type B, C, D, G are related by the inequalities:

$$J(Q) = J(B) = I(B) \geq J(C) = I(C) \geq I(D) \geq I(G) = I(Q) ;$$

moreover the inequalities are strict, except when

- there is only one arc entering into each answer of Q and then
$$I(D) = I(G) ,$$

- there is only one arc entering into each question of Q and then there exists a C such that $I(C)=I(D)$.

- Q is arborescent and then $J(Q)=I(Q)$.

Now consider the questionnaires Q and D in the case when Q has at least one question and answer with at least two entering arcs:
$$J(Q) > J(D) = I(D) > I(Q) ;$$

D is then an arborescent questionnaire built from a latticoid questionnaire Q such that D transmits more and processes less information than Q.

The difference $J(Q)-I(Q)$ expresses the fact that the existence of cycles in a latticoid graph leads to processing an information greater than that which is transmitted. This is related directly to the fact that the number of arcs of Q is greater than that of G. Q processes a greater information than G, but Q and G have the same answers (or exits) transmitting the same information. From the point of view of exploitation, the output of questionnaire G is as interesting as that of Q; G transmits less information than B because B can be considered as the best possible questionnaire able to process the same information as Q.

Remarks

1. If $e(i)>0$, then $e(i)$ can be greater than $J(i)$, in other words, the question i reduces the transmitted information. In the limit if all the answers coincide (quasi-strongly connected graph) it is clear that no information is transmitted and

certain vertices will only have consumed information.

2. If i is an internal vertex with one leaving arc with probability p(i,j) and other arcs with probability zero, then J(i)=0. The contraction of the set {i} ∪ Γi does not alter the information processed by such a graph which does not satisfy axiom (q₃) and is a quasi-questionnaire (cf. Definition 8.6).

7.5 Other Definitions of Information

From among all the possible axioms or definitions giving a measure of information, we will retain the ones due to Rényi and to Kampé de Fériet and Forte.

7.5.1 Information for Incomplete Distributions

Whereas the axiom systems developed in §7.2 were related to a complete distribution of eventualities (axiom A0), we may also be interested in incomplete distributions.

$$Q = \{q_1, \ldots, q_N\} \text{ where } 0 < q_i < 1$$

such that

$$\sum_{i:=1}^{N} q_i < 1$$

the treatment of which can be brought back to that of complete distributions after normalization

$$r_i = \frac{q_i}{\sum_{i:=1}^{N} q_i}.$$

An incomplete distribution is such that $\sum_{k=1}^{N} q_i$ can represent the probability of a composite event $E' \subset E$, E being the set on which the probability space is built. The r_i, are then the conditional probabilities knowing the set E'.

It is on the basis of incomplete distributions that Aczel has catalogued the properties of the information gain functions which we introduced at the end of §7.3.

The information can then be defined for a distribution bearing on one variable in such a way that the unit will no longer be fixed by (A1) but by

(R1) $\qquad\qquad I_1^*\left(\frac{1}{2}\right) = 1$

where the star recalls that the distribution may be incomplete.

The other axioms enabling us to determine the information function are:

(R2) there exists a function g continuous and increasing in R^+

such that

$$I_1^*(p_1,\ldots,p_n) = g^{-1}\left(\frac{\sum_{i:=1}^{N} p_i\, g(I_1^*(p_i))}{\sum_{i:=1}^{N} p_i}\right)$$

(R3) $I_{MN}^*(p_1 t_1,\ldots,p_1 t_N, p_2 t_1,\ldots,p_M t_1,\ldots,p_M t_N)$

$$= I_M^*(p_1,\ldots,p_M) + I_N^*(t_1,\ldots,t_N)$$

which is an axiom of *additivity* a particular case of (7) obtained for $t_{ji}=t_i\,(\forall j)$.

(R4) $I_1^*(p)$ is a continuous function of p.

Daroczy has resolved Rényi's axiom system (R) in the following form:

Theorem 6 The solution of system (R) for $\alpha>0$ and $\alpha\neq 1$ is given exclusively by Rényi's information:

$$I_N^\alpha(p_1,p_2,\ldots,p_N) = \frac{1}{1-\alpha}\log_2 \frac{\sum_{i:=1}^{N} p_i^\alpha}{\sum_{i:=1}^{N} p_i}$$

and for $\alpha=1$, by Shannon's information:

$$I_n^1(p_1,p_2,\ldots,p_N) = \frac{\sum_{i:=1}^{N} p_i \log_2 \frac{1}{p_i}}{\sum_{i:=1}^{N} p_i} \quad.$$

When the distributions are complete we obtain on the one hand

$$I_N(p_1,\ldots,p_N) = I_N^1(p_1,\ldots,p_N)$$

and on the other

$$I_N^\alpha(p_1,p_2,\ldots,p_N) = \frac{1}{1-\alpha}\log_2 \sum_{i:=1}^{N} p_i^\alpha, \qquad (42)$$

which will be used again later (§9.4).

Renyi has also established an axiomatic for information gain whose solution is given for incomplete distributions and for α>0 by

$$I_N((r_1,r_2,\ldots,r_N)\,||\,(p_1,p_2,\ldots,p_N))$$

$$= \frac{1}{\alpha-1}\log_2 \frac{\sum_{i:=1}^{N} r_i^{\alpha} p_i^{1-\alpha}}{\sum_{i:=1}^{N} r_i} \qquad (\alpha \neq 1)$$

$$I_N^1((r_1,r_2,\ldots,r_N)\,||\,(p_1,p_2,\ldots,p_N))$$

$$= \frac{\sum_{i:=1}^{N} r_i \log_2 \frac{r_i}{p_i}}{\sum_{i:=1}^{N} r_i} \qquad (p_i \neq 0).$$

7.5.2 Measure, Probability and Information

7.5.2.1 Probability and Information

Although our information study has been effected for events with given probabilities, it is important to recall that the effort was originally made to define information as a first concept from which probability would be derived (the approach suggested by Kolmogorov) or from which would proceed certain measures without involving the concept of probability.

One of the objections sometimes raised against information of Shannon or Rényi type for example is as follows:

Consider two random variables X and Y taking the values $x_1, x_2, \ldots, x_N$ or $y_1, y_2, \ldots, y_N$ with probabilities $p_1, p_2, \ldots, p_N$ in both cases.

The information (of Shannon or Rényi) brought by the realization of the experiment E_X with outcome X, that is, the information obtained by knowing if the answer is $x_1, x_2, \ldots$ or x_N is the same as the information obtained by the realization of E_Y with outcome Y. But the two values taken by X, x_1 and x_2 may be very close to each other while the values taken by Y with the same probabilities p_1 and p_2, that is y_1 and y_2 may be extreme values of Y. That is why certain authors have defined an information function for metric spaces (such as Aggarwal, Cannonge and Rigal).

Another approach suggested by Watanabe consists in weighting the answers, but still using Shannon's information "locally" that is,

$$I_N(p_1,\ldots,p_N) = \sum_{i:=1}^{N} k_i \log \frac{1}{p_i}$$

and we will find again a particular Watanabe's information in §9.2, adapted to the discussion of certain questionnaires; we can also think of acquisition as a Watanabe information bearing on the set of outcomes of all the questions of a Huffman heterogeneous questionnaire.

However Urbanik and Ingarden, and later B. Forte and J. Kampe de Feriet have introduced for the first time an information involving exclusively set theoretical axiomatic concepts from which probability can be completely eliminated.

7.5.2.2 Information for Measure Spaces

Let $(E, \mathcal{T})$ be a measurable space where E is a set finite denumerable or not and $\mathcal{T}$ is a σ-algebra (σ-field) of E. A positive measure h of the information furnished by the realisation of an event A can be defined from the properties:

(h1) $h(E) = 0$ and $h(\emptyset) = \infty$.

(h2) if A and B are non-empty sets of $\mathcal{T}$, then

$$A \subset B \Rightarrow h(B) \leq h(A) .$$

(h3) if $\{\mathcal{T}_i, i \in I\}$ is a family of independent σ-algebras in the sense of set theory (M-independence, of Banach), then,

a. independence

$$h(A_j \cap A_k) = h(A_j) + h(A_k) , \quad \forall A_j \in \mathcal{T}_j, A_k \in \mathcal{T}_k ;$$

b. σ-independence

$$h(\bigcap_{i:=1}^{\infty} A_i) = \sum_{i:=1}^{\infty} h(A_i), \quad \forall A_i \in \mathcal{T}_i \ (i \in I) .$$

(h4) Existence of a composition law for information:

$$h(A \cup B) = h(A) \star h(B) \quad \text{for} \quad A \cap B = \emptyset ,$$

$$h(\bigcup_{i:=1}^{\infty} A_i) = \mathop{\star}_{i:=1}^{\infty} h(A_i)$$

for A_i all disjoint,
where the operator $\star$ will be specified later.

Any three subsets of E, say A, B, C, for which A and B on the one hand, A and C on the other are independent, satisfy the distributive law of + relative to $\star$:

$$h(A) + h(B \cup C) = (h(A) + h(B \cap C)) \star$$
$$\star (h(A) + h(B \cap \overline{C})) \star (h(A) + h(\overline{B} \cap C))$$

which follows from (h3) and (h4) and the distributivity of $\cap$ relative to $\cup$.

Let μ be a finite measure on $(E, \mathcal{J})$; the measure μ such that $\mu(\emptyset)=0$ takes the value $\mu_0=\mu(E)$ and $0<\mu_0<\infty$.

For such a space, we will write the information measure of an event in the form $h(A)=\Theta(\mu(A))$ where $\Theta(x)$ is a continuous and strictly decreasing function defined for all $x \in \mu(A)$ and $A \in \mathcal{J}$.

We will write $\Theta(0)=h(\emptyset)=\infty$ and $\Theta(\mu_0)=h(E)=0$.

Consider a finite partition of an event A into mutually disjoint events (elementary or not)

$$\bigcup_{i:=1}^{N} A_i = A .$$

With this partition, we associate an experiment $\Pi_N(A)$ which will be complete if $A=E$, and incomplete if $A \neq E$ and we define a mapping:

$$\Phi_N(A) : \Pi_N(A) \to \overline{R}_+$$

which will be called Forte and Kampé de Fériet's information.

The axioms of $\Phi_N(A)$ are, according to Forte and Kampé de Fériet:

($\Phi 1$) $\Phi_N(A)$ is a symmetric function of the A_i such that

$$A = \bigcup_{i:=1}^{\infty} A_i$$

($\Phi 2$) $\Phi_{N+1}(A_1, \ldots, A_N, \emptyset) = \Phi_N(A_1, A_2, \ldots, A_N)$

($\Phi 3$) $\Phi_N(A_1, \ldots, A_N) - \Phi_{N-1}(A_1, \ldots, A_{N-2}, (A_{N-1} \cup A_N))$

$$= \Psi(\bigcup_{j:=1}^{N-2} A_j , A_{N-1}, A_N) .$$

where Ψ defines a gain of information.

Using this last axiom of branching, which generalizes (A3), n-1 times, we obtain:

$$\Phi_N(A_1, \ldots, A_N) - \Phi_1(A)$$

$$= \sum_{j:=0}^{N-2} \Psi(\bigcup_{k:=0}^{j} A_k, A_{j+1}, \bigcup_{k:=j+2}^{N} A_k) \qquad (43)$$

where $A_0 = \emptyset$.

But $\Phi_1(A)$ is the information of an event, that is $\Phi_1(A) = h(A)$.

The function
$$J_N(A_1,\ldots,A_N) = \Phi_N(A_1,\ldots,A_N) - h(A) \qquad (44)$$

represents the information of the partition $\Pi_N(A)$ conditioned by event A.

Let γ be a positive valued function satisfying:

$$\gamma(\mu,\mu_1,\mu_2) = \gamma(\mu,\mu_2,\mu_1) > 0$$

$$\gamma(\mu,\mu_1,0) = 0$$

$$\gamma(\mu,\mu_1,\mu_2) + \gamma(\mu,\mu_1 + \mu_2,\mu_3)$$
$$= \gamma(\mu,\mu_1,\mu_3) + \gamma(\mu,\mu_1 + \mu_3,\mu_2)$$

$$\gamma(\mu(A), \mu(A_2), \mu(A_3))$$
$$= \sum_{k:=1}^{p} \gamma(\mu(A \cap B), \mu(A_2 \cap B_k), \mu(A_3 \cap B_k)),$$

with
$$\mu > \mu_1 \qquad \mu > \mu_2,$$
$$A = A_1 \cup A_2 \cup A_3, \quad B = \bigcup_{k:=1}^{p} B_k$$
$$A_i \cap A_j = \emptyset, \quad B_k \cap B_\ell = \emptyset$$

Finally we take as information function on $(E, \mathcal{J}, \mu)$ the solution of the axioms (Φ), taking into account (h4) and (44) that is:
$$\Phi_N(A) = h(A) + J_N(A);$$

this information possesses Forte and Pintacuda's property

$$\Psi(A, B, C) = \gamma(\mu(A) + \mu(B) + \mu(C), \mu(B), \mu(C)) \qquad (45)$$

where A, B, C are mutually disjoint.

Forte and Pintacuda have furthermore established:

Theorem 7 $J_N(A)$ is an information on a measure space if and only if it is of the form

$$J_N(A) = \sum_{j:=1}^{N-1} \gamma(\mu(A), \mu(A_j), \sum_{k:=j+1}^{N} \mu(A_k))$$

7.5.2.3 Non-probabilitistic questionnaires

If in the axioms (q_4) and (q_5) the mapping

$$M_\Gamma : \Gamma \to \overline{R}_+ \qquad (46)$$

is substituted for the mapping $P_\Gamma : \Gamma \to]0,1[$ then we will no longer be able to allocate a probability of event $p(E_i)$ to a question i (see 4.2.3) but only a measure of event $\mu(E_i)$. We will then speak of questionnaires M generalising the probabilistic questionnaires Q or K, studied up to now.

Forte and Kampé de Fériet's information has enabled Schneider to study questionnaires M after having given the following definitions for a questionnaire M admitting N answers $e_1,\ldots,e_N$, a root α with measure $\mu(E_\alpha)=\mu(E)=\mu_o$. Let $i \in F$ be a question with measure $\mu(E_i)$ having b_i predecessors $x_1,\ldots,x_b$ and a_i successors $y_1,\ldots,y_a$:

Information transmitted by M:

$$I_\mu(M) = \Phi(M) = \sum_{\ell:=1}^{N-1} \gamma(\mu_o, \mu(e_\ell), \sum_{k:=\ell+1}^{N} \mu(e_k)) ; \qquad (47)$$

Information processed by i:

$$J_\mu(i) = \sum_{j:=1}^{a-1} \gamma(\mu(E_i), \mu(E_j), \sum_{k:=j+1}^{a} \mu(E_k)) \qquad (48)$$

with

$$\mu(E_i) = \sum_{j:=1}^{a} \mu(E_j) ;$$

Information processed by M:

$$J_\mu(M) = \sum_{i \in F} J(i) ; \qquad (49)$$

Information consumed by i:

$$e(i) = \sum_{h:=1}^{b-1} \gamma(\mu_o, \mu(E_h), \sum_{k:=h+1}^{b} \mu(E_k)) . \qquad (50)$$

Schneider has generalised Theorems 4 and 5 concerning processed information and transmitted information to arborescent questionnaires or questionnaires satisfying a hypothesis of proportionality of the flow.

Examples

1. M_Γ coincides with the mapping P_Γ in $]0,1[$.

Then the measure $\mu(E_i)$ reduces to the probability $p(E_i)$; $h(A) = \log\frac{1}{P(A)}$; $\Phi_N(A_1,\ldots,A_N)$ takes on the value of Shannon's information.

We find in particular:

$$J_\mu(i) = \sum_{j \in \Gamma i} \frac{p(i,j)}{p(i)} \log \frac{p(i)}{p(i,j)}$$

if the questionnaire is arborescent and

$$J_a(E_i) = J_\mu(i) = J(i) \quad \text{with} \quad |\Gamma i| = a$$

so that

$$\Phi_a(i) = J(i) - h(i),$$

$$\Phi_a(i) = \sum_{j \in \Gamma i} \frac{p(i,j)}{p(i)} \log \frac{p(i)}{p(i,j)} - \log \frac{1}{p(i)}$$

$$\Phi_a(i) = \frac{1}{p(i)} \sum_{j \in \Gamma i} p(i,j) \log \frac{1}{p(i,j)} ;$$

$\Phi_a(i)$ is the information provided by question i considered as an incomplete experiment (cf. Theorem 6).

2.
$$h(A) = \frac{1}{|A|}$$

and

$$\gamma(\mu_0,\mu_1,\mu_2) = \frac{1}{\mu_1} + \frac{1}{\mu_2} - \frac{1}{\mu_1 + \mu_2}$$

that is

$$J_\mu(i) = \frac{1}{\alpha_1} + \frac{1}{\alpha_2} - \frac{1}{\alpha_1 + \alpha_2}$$

when the question i is dichotomic and α_1 and α_2 are the numbers of elements corresponding to the two outcomes of i.

We see that, for $\alpha_1+\alpha_2=m$, fixed, $J_\mu(i)$ is maximal when $\alpha_1=1$ (or $\alpha_2=1$) that is when one of the elements is determined by one of the outcomes, whereas the $(m-1)$ other elements of E_i remain grouped on the other outcome.

Hence in this case it is the first question of an arborescence of maximal height which maximises the information processed by a question $J_\mu(i)$, which is the opposite of what happens with a balanced arborescence for Shannon's information and equiprobable answers.

EXERCISES

1. Calculate, for the values

$$\lambda_A = 0.5 \; ; \; \lambda_B = 1 \; ; \; \lambda_C = 1.5 \; ,$$

the informations
$$I(A), \; I(B), \; I(C), \; I(Q), \; J(Q) \; ,$$

of the questionnaires which were the subject of Exercise 2 of Chapter 4.

2. Let
$$\mathcal{P} = (p_1, \ldots, p_m, \ldots, p_N)$$

be a complete distribution and let $\sum_{i=1}^{m} p_i = 1 - \eta$. Using axiom A3 of Faddeev show that

$$I_N(p_1, \ldots, p_m, \ldots, p_N) \leq I_2(\eta, 1 - \eta)$$

$$+ (1 - \eta) \log m + \eta \log (N - m);$$

for $\eta < \varepsilon < \frac{1}{2}$; and deduce the strict inequality

$$I_N(p_1, \ldots, p_m, \ldots, p_N) < I_2(\varepsilon, 1 - \varepsilon)$$

$$+ \log m + \varepsilon \log N \; .$$

3. Evaluate the information gained by the reading of an arbitrary (French) registration number (for example 2506DL75) taking 2 as the base of logarithms. Make successively the two hypotheses:

(h1) All the numerals have probability 1/10 and the letters 1/25 (the letter I does not occur).

(h2) The ending 75 has probability 1/16, while the other 94 endings allowed in France all have probability 15/(16×94), and the remaining five possible endings cannot occur on a registration number. Further the letters (except I) all have probability 1/25, the three numerals preceding the letters (here 506) are equiprobable and the first numeral x (here 2) has a probability conditioned by the ending $P\{x=y|75\}=1/10$ for $y:=0,1,\ldots,9$ and $P\{x=0|\text{not } 75\}=1$.

4. Evaluate the information which the telegraph transmits in the message of twenty six characters:

    CATHERINE   PASSED   EXAM   GOOD

assuming that the alphabet has twenty six letters and a space, each with probability 1/27.

Supposing that the opening of the telegram corresponds to bringing in an answer among the twelve answers of a questionnaire (for example: Catherine passed Very good/Good/Good enough/passable. Catherine has failed. Philip is admissible at High School. Joseph is admitted/refused at university. Sacha arrives at the Gare du Nord at 3 p.m./at the Gare de l'Est at 8 a.m.; cousin Amedee will be buried tomorrow/the day after tomorrow) give a least upper bound of the information, with base 2, which is effectively acquired by the addressee; compare it to that transmitted (base 27, base 2) by the telegraph.

5. Let Ω be an arborescent questionnaire having a question i. Let $I(E(i))$ be the information transmitted by the experiment consisting in the sub-questionnaire having i as root and $E(i)$ as set of answers.

Show that the information processed by question i can be expressed by

$$J(i) = I(E(i)) - \sum_{k \in \Gamma i} \frac{p(k)}{p(i)} I(E(k)) , \qquad (51)$$

that is, by a *mutual information* which can also be written:

$$J(i) = I(B) - I(B|A) ;$$

interpret experiment B conditioned by A.

Chapter Eight

INFORMATION AND ROUTING LENGTH

Arborescent questionnaires are the only ones to possess the property of equality between the transmitted and processed information. Since they are equally solutions of Problem 2 of L-optimality, they represent an important class from both these points of view, and the main object of this chapter is therefore to study the connection between information and routing in arborescent questionnaires.

8.1 Information and Routing in Questionnaires

8.1.1 Inefficiency and noise

Let Q be a polychotomic questionnaire in the strict sense in which the transmitted information and the routing length are determined by

$$L(Q) = \sum_{i \in F} p(i) \quad \text{and} \quad I(Q) = \sum_{i \in F} p(i) J(i)$$

where $J(i)$ is the information processed by question i;

$$J(i) = \sum_{j \in \Gamma i} \frac{p(i,j)}{p(i)} \log \frac{p(i)}{p(i,j)} \leq 1 \qquad (1)$$

This relation is an equality if and only if

$$\frac{p(i,j)}{p(i)} = \frac{1}{a} \quad (\forall i,j), \qquad (2)$$

with $|\Gamma i| = a \, (\forall i)$ and it implies that

$$L(Q) \geq I(Q)$$

Definition 1 The *information contributed* by a question i of a polychotomic questionnaire Q is the expression:

$$I(i) = p(i) J(i) . \qquad (4)$$

We will then write:

$$I(Q) = \sum_{i \in F} I(i) ,$$

and we will call the function $I(Q) = J(Q)$ simply the *information of the questionnaire* Q.

A consequence of Theorem 7.2 is:

Property 1 All polychotomic questionnaires with the same basis a defined on the same probability distribution $\mathcal{P}$ transmit the same information.

We will exhibit the probability distributions compatible with the equation (2).

Let us write the probability at any vertex $[r,s]$ of a polychotomic questionnaire Q in the strict sense or not in the form

$$p(r,s) = a^{-r} + \varepsilon_{r,s} \ . \qquad (5)$$

The information processed by question $[r,s]$ is:

$$J(r,s) = \sum_{j:=0}^{a-1} \frac{a^{-r-1} + \varepsilon_{r+1,as+j}}{a^{-r} + \varepsilon_{r,s}} \log \frac{a^{-r} + \varepsilon_{r,s}}{a^{-r-1} + \varepsilon_{r+1,as+j}} \leq 1$$

with equality if and only if

$$\frac{a^{-r-1} + \varepsilon_{r+1,as+j}}{a^{-r} + \varepsilon_{r,s}} = \frac{1}{a} \quad (\forall j)$$

(or $\frac{1}{\beta+1}$ in the case when simultaneously N-1 is not divisible by a-1, r=h-1 and $[r+1,as+j]$ (j=0,1,...β) is an answer coming from the question of basis less than a, the rank of which is one less than the height of Q.

(1) N-1 is divisible by a-1.

$$J(r,s) = 1 \text{ if } a^{-r-1} + \varepsilon_{r+1,as+j} = \frac{a^{-r} + \varepsilon_{r,s}}{a} \quad (\forall j)$$

that is

$$\varepsilon_{r+1,as+j} = \frac{\varepsilon_{r,s}}{a} \qquad (6)$$

The a outcomes of the question $[r,s]$ must be equiprobable (these outcomes could be questions of rank r+1 or answers), $J(r,s)$ is therefore equal to 1, for all $r \geq 1$ and for all s, only if the a successors of $[r,s]$ are equiprobable.

Suppose there exists a vertex $[\xi,\eta]$ such that $\varepsilon_{\xi,\eta} \neq 0$.

Then $J(\xi,\eta)$ will be equal to 1 if the successors of $[\xi,\eta]$ have probability $a^{-r-1} + \frac{\varepsilon_{\xi,\eta}}{a}$ For the predecessor of $[\xi,\eta]$,

that is h, to process a unit information, all the vertices $k \in \Gamma h$ must have the same probability as $[\xi,\eta]$.

Rising step by step, it becomes necessary that all the vertices with rank 1 have the same probability,

$$a^{-1} + \varepsilon_{i,j}.$$

But $\sum_{j:=0}^{a-1} p(i,j) = 1$ whatever the structure of the polychotomic arborescence and since

$$\sum_{j:=0}^{a-1} p(1,j) = \sum_{j:=0}^{a-1} (a^{-1} + \varepsilon_{i,j}),$$

it is necessary that $\varepsilon_{i,j}$ be zero whatever j in order that $I(r,s)=1$.

Consequently $\varepsilon_{r,s}=0$ whatever r and s.

(2) N-1 is not divisible by a-1.

The vertices with maximal rank coming out of the question $[r',s']$, with basis less than a, are such that $J(r',s')$ is maximal if:

$$\frac{a^{-r'-1} + \varepsilon_{r'+1,as+j}}{a^{-r'} + \varepsilon_{r',s}} = \frac{1}{\beta+1}.$$

The study of the case when N-1 is divisible by a-1 implies here that

$$\varepsilon_{r',s} = 0,$$

and it follows that

$$a^{-r'-1} + \varepsilon_{r'+1,as+j} = \frac{a^{-r'}}{\beta+1} > a^{-h}$$

In other words the probability of all the answers coming out of $[r',s']$ is

$$p(h,s'') = \frac{a^{-h+1}}{\beta+1}$$

and the probability of all the other vertices (including $[r',s']$) is

$$p(r,s) = a^{-1} \quad (\forall r).$$

In this case, $J(r',s')$ takes the value

$$J(r',s') = \sum_{i:=0}^{\beta} \frac{1}{\beta+1} \log(\beta+1) = \log(\beta+1).$$

$\log(\beta+1) < 1$, because the base of logarithms used is $a > \beta+1$.

Example

Consider the L-optimal polychotomic questionnaire in the broad sense (Fig. 8.1) given by

$$a = 4 \; ; \mathcal{P} = (\frac{1}{4}, \frac{1}{4}, \frac{1}{4}, \frac{1}{16}, \frac{1}{16}, \frac{1}{16}, \frac{1}{48}, \frac{1}{48}, \frac{1}{48})$$

We find

$$L = 1 + \frac{1}{4} + \frac{1}{16} = 1,3125$$

$$I = 1 + \frac{1}{4} + \frac{1}{16} \log_4 3 = 1,2995 \ .$$

In conclusion there exists in a polychotomic questionnaire in the broad sense at least one question i which does not satisfy (2) and for which $I(i)<1$; there is then at least one vertex $[r,s] \in \Gamma i$ with probability different from a^{-r}. Accordingly:

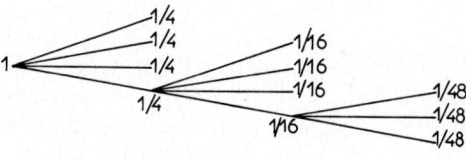

Figure 8.1

Theorem 1 The routing length of a polychotomic questionnaire is greater than or equal to the information of the questionnaire and it is equal if and only if the probability of any vertex with rank $r (0 \leqslant r \leqslant h)$ is $p(r,s)=a^{-r}$.

We can now define the noise and the inefficiency of a questionnaire (cf. §7.1.2).

Definition 2 The *inefficiency* of a polychotomic L-optimal questionnaire is the difference $L_H - I$ between its routing length and its information.

Definition 3 The *proper noise* of a homogeneous questionnaire Q is the difference $L(Q) - L_H$ between the routing lengths of this questionnaire and of the L-optimal questionnaire built with the same basis and the same probability distribution of the answers. The *noise* is the sum of the inefficiency and the proper noise.

We will denote the proper noise by

$$B_P = L - L_H \qquad (7)$$

and the inefficiency by

$$B_H = L_H - I(Q) \qquad (8)$$

L_H being the routing length of the L-optimal questionnaire whose set of questions is F_H.

It is also possible to write the inefficiency as

$$B_H = \sum_{i \in F_H} p(i)(1 - J(i))$$

and the noise $B = B_H + B_p$ as

$$B = \sum_{i \in F} p(i)(1 - J(i)). \qquad (9)$$

The noise can be reduced by an improvement of the routing whereas the inefficiency is determined intrinsically by a and $\mathcal{P}$; that is why we can define the noise on a homogeneous latticoid questionnaire.

The solution of the problem concerning the distribution $\mathcal{P}$ is then:

Property 2 A necessary and sufficient condition for a polychotomic L-optimal questionnaire to be without inefficiency is that the probability distribution has the form:

$$\mathcal{P} = (a^{-r_1}, a^{-r_2}, \ldots, a^{-r_N}) \text{ with } \sum_{i:=1}^{N} a^{-r_i} = 1.$$

in particular a-1 is a divisor of N-1 in such a questionnaire.

Definition 4 An *absolutely optimal questionnaire* is any polychotomic L-optimal questionnaire with zero inefficiency.

Remark Let Λ be a homogeneous latticoid questionnaire and G a polychotomic questionnaire whose answers have the same probability distribution $\mathcal{P}$ as those of Λ; let H be the L-optimal questionnaire built on $\mathcal{P}$, then

$$L(\Lambda) \geq L(G) \geq L(H) \geq I(G) = I(\Lambda)$$

in such a way that
$$B_p(\Lambda) = L(\Lambda) - L(H)$$
and
$$B_H(\Lambda) = L(H) - I(\Lambda).$$

8.1.2   L-Optimal Questionnaires ($L_H \geq I$)

We have seen that the noise of a questionnaire is equal to the sum of the "local inefficiencies" due to those questions which do not have a equiprobable outcomes (9).

The inefficiency can be evaluated in a particularly significant way in the case of balanced polychotomic questionnaires in which all the answers are equiprobable (noiseless questionnaires).

Equiprobable polychotomic questionnaires

1. If $N=a^k$, the L-optimal questionnaire has zero inefficiency; $L_H = I = k$.

2. Let
$$N = a^k + \alpha(a-1).$$

Then putting
$$N = ta^{k+1} \quad (\tfrac{1}{a} < t < 1) \tag{10}$$

we have
$$\alpha = \frac{at-1}{a-1} a^k.$$

The information is
$$I = k + 1 + \log_a t.$$

The routing length is
$$L_H = k + \frac{a^{k+1}}{N} \frac{at-1}{a-1}.$$

that is
$$L_H = k + \frac{at-1}{(a-1)t} \tag{11}$$

and consequently the inefficiency is
$$B_H = \log_a \frac{1}{t} - \frac{1-t}{(a-1)t}. \tag{12}$$

The inefficiency which is zero when $t=\frac{1}{a}$ or 1, therefore depends uniquely on t and not on k, that is, on the relative position of N with respect to two consecutive powers of a.

The value of the maximum inefficiency is obtained by differentiation:
$$\frac{dB_H}{dt} = \frac{1}{(a-1)t^2} - \frac{1}{t \ln a}$$

and $\frac{dB_H}{dt}$ vanishes for
$$t = \frac{\ln a}{a-1}, \tag{13}$$

Then B_H takes the value
$$B_{max} = \frac{1}{a-1} - \frac{1}{\ln a} - \log_a \frac{\ln a}{a-1}. \tag{14}$$

The value of B_{max} for $a=2$ was first pointed out by W. H. Burge.

The maximal values of B_H and the corresponding values of t are given as functions of a in the following table (Fig. 8.2)

a	t	B_{max}
2	0.693	0.086
3	0.549	0.135
4	0.462	0.169
5	0.402	0.194
6	0.358	0.215
7	0.324	0.231
8	0.297	0.246
9	0.274	0.258
10	0.256	0.269
11	0.240	0.278
12	0.226	0.287
13	0.214	0.295
14	0.203	0.302
15	0.193	0.309
16	0.185	0.315
17	0.177	0.321
18	0.170	0.326
19	0.164	0.331
20	0.158	0.335
50	0.080	0.411
100	0.047	0.459

The mean value $\bar{B}$ of the inefficiency is

$$\bar{B} = \int_{\frac{1}{a}}^{1} B_H dt \Big/ \int_{\frac{1}{a}}^{1} dt$$

that is

$$\bar{B} = \frac{1}{\ln a} - \frac{a \ln a}{(a-1)^2} . \qquad (15)$$

For $a=2$ and $a=3$ we find respectively $\bar{B}=0.0564$ and $\bar{B}=0.086$.

3. Let

$$N = a^k + \alpha(a-1) + \beta \quad (1 \leq \beta < a-1) .$$

and write

$$N = a^k + \alpha'(a-1) = ta^{k+1} .$$

with

$$\alpha' = \alpha + \frac{\beta}{a-1} \quad (\alpha \text{ and } \beta \text{ are integers})$$

If the routing of the questionnaire had length

$$k + \frac{a\alpha'}{N} ,$$

the inefficiency B_H would belong to the curve $B(t)$ (Fig. 8.2)

But the routing has length

$$L_H = k + \frac{a\alpha+\beta+1}{N}$$

and consequently the inefficiency decomposes into two parts: on the one hand

$$B_1 = \log_a \frac{1}{t} - \frac{1-t}{(a-1)t} \qquad (16)$$

which is the difference between the function

$$k + \frac{a\alpha'}{N},$$

defined and continuous for real α' varying from $\frac{1}{a-1}$ to a^k, and $\log_a N$, and on the other:

$$B_2 = \frac{a\alpha+\beta+1}{N} - \frac{a(\alpha+\frac{\beta}{a-1})}{N}$$

$$= \frac{1}{N}(1 - \frac{\beta}{a-1}); \qquad (17)$$

B_2 takes its maximal value for $\beta=1$:

$$B_2 \leq \frac{a-2}{(a-1)N},$$

then decreases to zero as β increases from 1 to $a-1$.

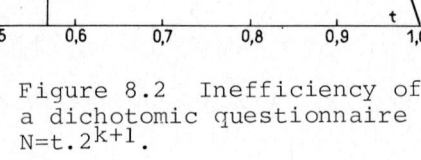

Figure 8.2 Inefficiency of a dichotomic questionnaire $N=t.2^{k+1}$.

In conclusion the routing length of an arborescence with N equiprobable answers is obtained by the superposition of three elements:

- the information of the questionnaire, I

- the inefficiency B_1 related to t that is to the ratio $\frac{N}{a^{k+1}}$;

- the inefficiency B_2 related to β that is to the remainder of the division of $N-a^k$ by $a-1$.

I is the logarithmic function: $I=\log_a N$.

B_1 is the function defined for integral values of α which can be extended to the closed interval $[\frac{1}{a},1]$:

$$N = ta^{k+1}$$

$$B_1 = \log_a \frac{1}{t} - \frac{1-t}{(a-1)t} \cdot$$

B_2 is a function defined for integral values of β which can be extended to the whole interval $[1, a-1[$:

$$B_2 = \frac{1}{N}\left(1 - \frac{\beta}{a-1}\right),$$

$$L = 1 \qquad \text{if } \alpha = \beta = 0,$$
$$L = I + B_1 \qquad \text{if } \alpha > 0, \beta = 0,$$
$$L = I + B_1 + B_2 \text{ if } \beta > 0,$$

B_1 is independent of N and of the power of a immediately less than N_1; B_{1_1} can be represented by a unique curve defined on the interval $\frac{1}{a} \leqslant t < 1$ (Fig. 8.1).

In contrast B_2 is inversely proportional to N; B_2 decreases for given α as β increases from 1 to a-1.

The value $L_H = k+1$ is attained as soon as

$$\alpha = a^k - 1 \quad \text{and} \quad \beta \geqslant 1,$$

because then we have

$$L_H = (k + 1 + \log_a t) + B_1 + B_2,$$

with

$$B_1 = \log_a \frac{1}{t} - \frac{1-t}{(a-1)t}, \quad t = 1 - \frac{a-1-\beta}{a^{k+1}}$$

and

$$B_2 = \frac{1}{N}\left(1 - \frac{\beta}{a-1}\right) = \frac{a-1-\beta}{(a-1)a^{k+1}t} = \frac{1-t}{(a-1)t},$$

that is

$$B_1 + B_2 = \log_a \frac{1}{t} \cdot$$

The evolution of L_H as a function of N is effected according to the superposition of two families of curves in festoon shape with relative steps a^k and a (Fig. 8.3)

Application: *Rational approximation of the logarithmic function*

The calculation of $I = \log_a x$ may be approximated without use of logarithms by the value

$$I + B_1 = k + \frac{a\alpha'}{x}$$

defined for $0 \leqslant \alpha' = \frac{x-a^k}{a-1} \leqslant a^k$ and for integer k with $a^k \leqslant x \leqslant a^{k+1}$.

$I+B_1$ is defined as a function of x by an arc of a hyperbola when k is given. The difference B_1 from the logarithmic function is always positive or zero and less than B_{max}. Thus this approximation of the logarithmic function enables us to deal with an error of known bound. Furthermore drawing the unique curve $B_H=B_1(t)$ defined on the segment $\frac{1}{a} \leqslant t \leqslant 1$ enables us to specify the error for any x (Fig. 8.2.) In practice, the preceding table shows that only the case a=2 leads to a small error.

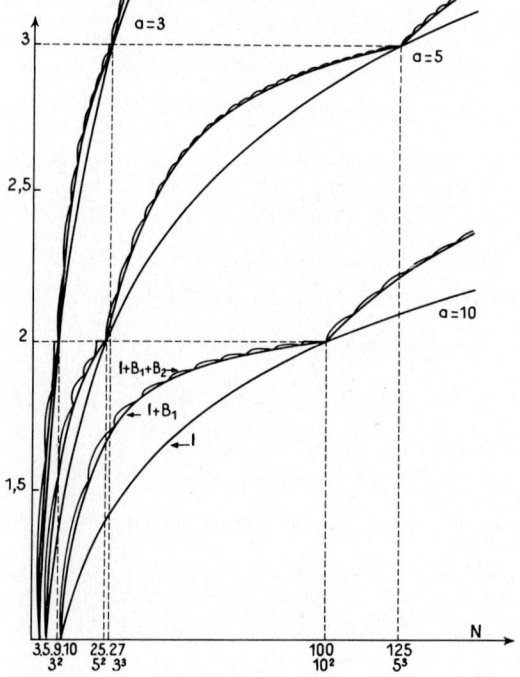

Figure 8.3 Inefficiency of Polychotomic Questionnaires

$I = \log_a N$

B_1, inefficiency related to t
B_2, inefficiency related to β
$(0<\beta<a-1)$

Examples of absolutely optimal questionnaires

- If $N=a^k+\alpha(a-1)$ and if there exist $N-a\alpha$ answers with probability a^{-k} and $a\alpha$ answers with probability a^{-k-1}, then the balanced arborescence having answers of the first class at rank k and the other a answers at rank k+1 has

$$L_H = I = k + \frac{a\alpha}{a^{k+1}}.$$

- Suppose that the answers of a given dichotomic questionnaire have probabilities

$$\frac{1}{2}, \frac{1}{4}, \frac{1}{8}, \ldots, 2^{-N+2}, 2^{-N+1}, 2^{-N+1}$$

and that its support is an arborescence of maximal height. Then the answers have rank r and probability

$$2^{-r} \quad (r := 1, \ldots, N-1) .$$

The questions have rank r-1 and probability $2^{-(r-1)}$ in such a way that

$$I = L_H = \sum_{r=1}^{N-1} 2^{-(r-1)} = 2 - 2^{-(N-2)}$$

8.1.3 Questionnaire product of two polychotomic questionnaires

The informations of equiprobable questionnaires are related only to the basis and the number of answers and since

$$\log_a N_1 N_2 = \log_a N_1 + \log_a N_2 ,$$

we directly obtain

$$I_{1 \times 2} = I_1 + I_2 ,$$

where the questionnaire $Q_{1 \times 2}$ is the product of the equiprobable polychotomic questionnaires Q_1 and Q_2 with basis a.

More generally, if Q_1 and Q_2 are two polychotomic questionnaires which are not equiprobable but which have the same basis then

$$I_{1 \times 2} = \sum_{\xi_j \eta_\ell \in E_{1 \times 2}} p_1(\xi_j) p_2(\eta_\ell) \log \frac{1}{p_1(\xi_j) p_2(\eta_\ell)} ,$$

$$I_{1 \times 2} = \sum_{\xi_j \in E_1} p_1(\xi_j) \sum_{\eta_\ell \in E_2} p_2(\eta_\ell) (\log \frac{1}{p_1(\xi_j)} + \log \frac{1}{p_2(\eta_\ell)}) ,$$

$$I_{1 \times 2} = \sum_{\xi_j \in E_1} p_1(\xi_j) \log \frac{1}{p_1(\xi_j)} \sum_{\eta_\ell \in E_2} p_2(\eta_\ell)$$
$$+ \sum_{\eta_\ell \in E_2} p_2(\eta_\ell) \log \frac{1}{p_2(\eta_\ell)} \sum_{\xi_j \in E_1} p_1(\xi_j)$$

and since

$$\sum_{\eta_\ell \in E_2} p_2(\eta_\ell) = \sum_{\xi_j \in E_1} p_1(\xi_j) = 1$$

$$I_{1 \times 2} = I_1 + I_2 , \qquad (18)$$

in other words, there is *independence* between the questionnaires Q_1 and Q_2.

The noise being the difference between the routing length and the information (see 8.7) we find immediately:

$$B_{1 \times 2} = (L_1 - I_1) + (L_2 - I_2). \qquad (19)$$

Theorem 2 The noise of a product questionnaire is the sum of the noises of the two factor questionnaires. If the factor questionnaires have neither inefficiency nor proper noise the same is true for the product questionnaire.

If A_1 and A_2 are optimal questionnaires and if $A_{1 \times 2}$ is an optimal questionnaire, then the noise of $A_{1 \times 2}$ which is equal to $B_1 + B_2$ is minimal and

$$B_1 + B_2 = 0 \Rightarrow B_{1 \times 2} = B_{2 \times 1} = 0.$$

Denoting the optimal questionnaire with $N_1 N_2$ answers, its inefficiency, its routing length and the information transmitted by it by A_{12}, B_{12}, L_{12} and I_{12}, we obtain the following relations

$$I_{12} = I_{1 \times 2}$$

and $L_{12} \leq L_1 + L_2$ (L_1 and L_2 being minimal) and consequently

$$B_{12} \leq B_{1 \times 2}. \qquad (20)$$

Property 3 The noise of the product questionnaire of two optimal questionnaires with N_1 and N_2 answers is at least equal to the inefficiency of the optimal questionnaire with $N_1 N_2$ answers.

Particular case: equiprobable questionnaires

Suppose that we have two optimal questionnaires:

A_1 characterized by

$$N_1 = t_1 a^{k_1 + 1} \quad (\tfrac{1}{a} \leq t_1 < 1)$$

and

A_2 characterized by

$$N_2 = t_2 a^{k_2 + 1} \quad (\tfrac{1}{a} \leq t_2 < 1),$$

then the product

$A_{1 \times 2}$ is characterized by

$$N_1 N_2 = t_1 t_2 a^{k_1+k_2+2} \quad (\tfrac{1}{a^2} \leq t_1 t_2 < 1) .$$

If the ranks of the answers of $A_{1 \times 2}$ are k_1+k_2, k_1+k_2+1 and k_1+k_2+2, then the routing of $A_{1 \times 2}$ will not be optimal (Property 6.3)

1.
$$t_1 t_2 = \tfrac{1}{a^2}$$

then
$$N_1 = a^{k_1}, \quad N_2 = a^{k_2}, \quad N_1 N_2 = a^{k_1+k_2} .$$

The questionnaires A_1, A_2 and $A_{1 \times 2}$ have all their answers of the same rank (k_1, k_2 and k_1+k_2) and therefore all three are optimal. The noise of questionnaire $A_{1 \times 2}$ is zero and consequently:
$$L_{12} = L_{1 \times 2} = I_1 + I_2 ,$$

in accordance with the preceding theorem.

Property 4 If the number of answers of a questionnaire is equal to the power k of the basis of polychotomy, then it is possible to consider this questionnaire as the product of two questionnaires with a^{k_1} and a^{k_2} answers ($k_1+k_2=k$). The decomposition of this questionnaire into two questionnaires of smaller order, and then the product operation, generates no noise.

Example: ($a=3$, $N_1 N_2 = 243$) leads to two trichotomic questionnaires of respective heights 2 and 3:
$$N_1 = 3^2 \quad \text{and} \quad N_2 = 3^3 ,$$

with
$$p_1(\xi_i) = \tfrac{1}{3^2}, \quad p_2(\eta_\ell) = \tfrac{1}{3^3}$$

and
$$p(\xi_i \eta_\ell) = \tfrac{1}{3^2} \tfrac{1}{3^3} = \tfrac{1}{243} .$$

2.
$$t_1 - \tfrac{1}{a}, \quad t_2 > \tfrac{1}{a} ,$$

then
$$N_1 = a^{k_1}, \quad N_2 = t_2 a^{k_2+1} \quad \text{and} \quad N_1 N_2 = t_2 a^{k_1+k_2+1} .$$

If $N_2 = a^{k_2} + \alpha_2 (a - 1)$, then
$$N_1 N_2 = a^{k_1+k_2} + \alpha_2 a^{k_1} (a - 1) \quad \text{and} \quad 0 < \alpha_2 a^{k_1} < a^{k_1+k_2} .$$

The routing lengths are

$$L_1 = k_1,$$
$$L_2 = k_2 + a\frac{\alpha_2}{N_2},$$
$$L_{1\times 2} = k_1 + (k_2 + a\frac{\alpha_2}{N_2})$$

and
$$L_{12} = (k_1 + k_2) + a\frac{\alpha_2 a^{k_1}}{N_1 N_2}.$$

Therefore
$$L_{1\times 2} = L_{12} \quad \text{and} \quad B_{1\times 2} = B_{12}. \qquad (21)$$

But if $N_2 = a^{k_2} + \alpha_2(a-1) + \beta_2$ ($\beta_2 > 0$), we write

$$N_1 N_2 = a^{k_1+k_2} + \alpha_2 a^{k_1}(a-1) + \beta_2 a^{k_1}$$
$$= a^{k_1+k_2} + (\alpha_2 a^{k_1} + \gamma)(a-1) + \delta$$

with
$$0 < \alpha_2 a^{k_1} < a^{k_1+k_2}, \quad a^{k_1} \leq \beta_2 a^{k_1} < (a-1)a^{k_1},$$

$$\beta_2 a^{k_1} = \gamma(a-1) + \delta, \quad \gamma + 1 < a^{k_1} \quad \text{and} \quad 0 < \delta < a-1.$$

The optimal questionnaire with $N_1 N_2$ answers has routing length

$$L_{12} = (k_1 + k_2) + \frac{a(\alpha_2 a^{k_1} + \gamma) + \delta + 1}{a^{k_1} N_2}$$

whereas the questionnaires appearing in the product have routing lengths

$$L_1 = k_1 \quad \text{and} \quad L_2 = k_2 + \frac{a\alpha_2 + \beta_2 + 1}{N_2}$$

In such a way that

$$L_{1\times 2} = k_1 + k_2 + \frac{a\alpha_2 + \beta_2}{N_2} + \frac{1}{N_2}$$

and therefore

$$L_{1\times 2} = L_{12} + \frac{a^{k_1} - (\gamma + 1)}{a^{k_1} N_2}$$

that is
$$B_{1\times 2} > B_{12}. \qquad (22)$$

3. $t_1 > \frac{1}{a}$ and $t_2 > \frac{1}{a}$ and $\beta_1 = \beta_2 = 0$.

We cannot have $t_1 t_2 = \frac{1}{a}$ otherwise

$$\frac{N_1}{a^{k_1+1}} \frac{N_2}{a^{k_2+1}} = \frac{1}{a}, \text{ then } N_1 N_2 = a^{k_1+k_2+1}.$$

But N_1 and N_2 being integers we should have $N_1 = a^\mu$, $N_2 = a^\nu$ contrary to the hypothesis $t_1 > \frac{1}{a}$ and $t_2 > \frac{1}{a}$.

Put
$$N_1 N_2 = t_1 t_2 a^{k_1+k_2+2}.$$

First case

$$\frac{1}{a^2} < t_1 t_2 < \frac{1}{a}.$$

$$N_1 N_2 = (at_1 t_2) a^{k_1+k_2+1}.$$

then, by equation (11)

$$L_{12} = (k_1 + k_2) + \frac{a^2 t_1 t_2 - 1}{(a-1) a t_1 t_2}$$

whereas

$$L_{1\times 2} = k_1 + \frac{a t_1 - 1}{(a-1) t_1} + k_2 + \frac{a t_2 - 1}{(a-1) t_2},$$

that is

$$L_{1\times 2} = k_1 + k_2 + \frac{(a t_1 - 1) a t_2 + (a t_2 - 1) a t_1}{(a-1) a t_1 t_2}$$

that is

$$L_{1\times 2} = k_1 + k_2 + \frac{a^2 t_1 t_2 - 1}{(a-1) a t_1 t_2}$$
$$+ \frac{a^2 t_1 t_2 - 1 - a(t_1 + t_2)}{(a-1) a t_1 t_2}$$

and therefore

$$L_{1\times 2} = L_{12} + \frac{(a t_1 - 1)(a t_2 - 1)}{(a-1) a t_1 t_2}. \qquad (23)$$

Second case

$$\frac{1}{a} < t_1 t_2 < 1$$

$$N_1 N_2 = t_1 t_2 a^{k_1+k_2+2}$$

and

$$L_{12} = k_1 + k_2 + 1 + \frac{at_1 t_2 - 1}{(a-1)t_1 t_2},$$

whereas

$$L_{1 \times 2} = k_1 + k_2 + \frac{(at_1 - 1)t_2 + (at_2 - 1)t_1}{(a-1)t_1 t_2}$$

that is

$$L_{1 \times 2} = k_1 + k_2 + \frac{(at_1 t_2 - t_1 t_2) + (at_1 t_2 - 1) + (t_1 t_2 + 1 - t_1 - t_2)}{(a-1)t_1 t_2}$$

and therefore

$$L_{1 \times 2} = L_{12} + \frac{(1 - t_1)(1 - t_2)}{(a-1)t_1 t_2} \qquad (24)$$

and

$$B_{1 \times 2} > B_{12}.$$

4. If β_1 is different from zero, there exists in $A_{1 \times 2}$ a question z_1 having fewer than a outcomes, the rank of which is not maximal because the sub-arborescences are issued from the $\beta+1$ descendants of q_1. $A_{1 \times 2}$ is then not an optimal questionnaire (Property 5.2).

If β_2 is different from zero, the routing length $L_{1 \times 2}$ being equal to $L_{2 \times 1}$ and $A_{2 \times 1}$ not being an optimal questionnaire for the same reason, it follows that

$$L_{12} < L_{1 \times 2} \text{ if } \beta_1 \text{ or } \beta_2 \text{ is different from zero.}$$

In conclusion, setting

$$N_1 = t_1 a^{k_1+1} = a^{k_1} + \alpha_1(a - 1) + \beta_1,$$

$$N_2 = t_2 a^{k_2+1} = a^{k_1} + \alpha_2(a - 1) + \beta_2;$$

$$\beta_1 = \beta_2 = \alpha_1 \alpha_2 = 0 \Rightarrow L_{1 \times 2} = L_{12};$$

$$\left. \begin{array}{l} \alpha_1 = \beta_1 = 0 \\ \beta_2 a^{k_1} > 0 \end{array} \right\} \Rightarrow L_{1 \times 2} = L_{12} + \frac{N_1 - (\gamma + 1)}{N_1 N_2};$$

$$\left.\begin{array}{l}\beta_1 = \beta_2 = 0 \\ \alpha_1 \alpha_2 \neq 0\end{array}\right\} \Rightarrow L_{1\times2} = L_{12} + \frac{(at_1 - 1)(at_2 - 1)}{(a - 1)at_1 t_2}$$

$$\text{if } t_1 t_2 < \frac{1}{a}$$

or
$$L_{1\times2} = L_{12} + \frac{(1 - t_1)(1 - t_2)}{(a - 1)t_1 t_2}$$

$$\text{if } t_1 t_2 > \frac{1}{a} \ ;$$

$$\left.\begin{array}{l}\beta_1 \neq 0 \\ \text{or } \beta_2 \neq 0 \\ \text{or } \beta_1 \neq 0 \text{ and } \beta_2 \neq 0\end{array}\right\} \Rightarrow L_{1\times2} > L_{12} \ .$$

Figure 8.4 represents the evolution of the noise $B = L_{1\times2} - L_{12}$, as a function of t_1, for different values of t_2 when $a=2$ and accordingly $\beta_1=\beta_2=0$. We have also drawn in the diagram the edge envelope of the curves corresponding to (23) and (24) that is:

$$(L_{1\times2} - L_{12})_{max} = \frac{(1 - t_1)(at_1 - 1)}{(a - 1)t_1}$$

a value which is never attained as it does not correspond to integral values of N_1 and N_2.

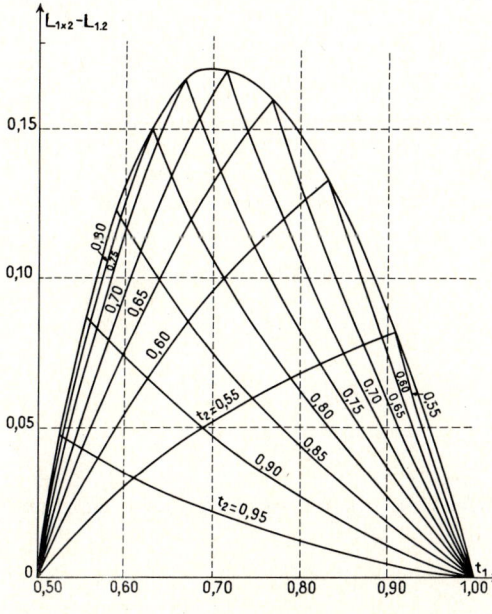

Figure 8.4 Noise of the product of dichotomic questionnaires.

Case of the restricted product

Suppose it desired to realize a polychotomic questionnaire with basis a possessing $N=N_1+N_2$ answers partitioned into two classes C_1 such that $\sum_{i=1}^{N_1} p(y_i) = 1-\eta$

and C_2 such that $\sum_{i=N_1+1}^{N_1+N_2} p(y_i) = \eta$,

η being considered as arbitrarily small, for example

$$\eta < p(y_i) \quad (\forall i < N_1) .$$

It is possible to form a polychotomic questionnaire Q with N_1+1 answers on the distribution $\mathcal{P} = (p_1,\ldots,p_{N_1},\eta)$ where $p(y_i)=p_i$ and a second polychotomic questionnaire R having N_2 answers with probabilities $(q_1,\ldots,q_{N_2})$ such that

$$q_j = \frac{p(y_j)}{\eta}, \quad \sum_{j:=1}^{N_2} q_j = 1 ,$$

and then the restricted product $Q \diamond R$, where the set $E' \subset E(Q)$ consists of only one vertex, say Y, the probability of which is η.

In view of axiom (A3) or (B'3), if $a-1|N_1-1$, we can write

$$I(Q \diamond R) = I(Q) + \eta I(R) . \qquad (25)$$

Similarly by (5.15), we can write

$$L(Q \diamond R) = L(Q) + \eta L(R) . \qquad (26)$$

If Q and R are L-optimal, then $Q \diamond R$ is also L-optimal because application of Huffman's algorithm to the construction of the questionnaire having N_1+N_2 answers will lead to forming a question with probability η that is, to forming a subquestionnaire in which the routing possesses the same length as R.

We may then say that questionnaire Q enables us to attain one of the N_1+N_2 answers with probability $1-\eta$ and, with probability η it will be necessary to have recourse to the questionnaire R.

The routing length $L(Q \diamond R)$ is close to $L(Q)$ if $\eta L(R)$ is small. Set $\varepsilon = \eta L(R)$.

Then $L(Q \diamond R) \approx L(Q)$, where the symbol $\approx$ signifies: within an ε-neighbourhood. It is then the same for the information: $I(Q \diamond R) \approx I(Q)$.

That is, with a probability $1-\eta$ of not having to ask a "subsidiary" question - or with a risk η of not making any error - we can approximate questionnaire $\Omega_\diamond R$ by questionnaire Q which will transmit, to within ε the same information for the same routing length, to within ε.

Certain applications will require truncating the set of answers in this way when η is small and N_2 is large compared with N_1. The loss of information will be negligable (to within ε) and the economy in routing length also. But if the risk η is accepted, it will be possible to avoid constructing and memorizing questionnaire R.

This can be generalized either by partitioning the answers of R into even more classes if N_2 is very large (and in the limit, infinite) or by making more decompositions of the same type requiring as many restricted products $Q \diamond R_1 \diamond R_2 \diamond R_3 \ldots \diamond R_s$.

Each of the restricted products will correspond to asking the sub-questionnaire R_k if and only if an answer $y_k (k<N_1)$ is obtained by the questionnaire Q.

8.1.4 Heterogeneous questionnaires

In the case of homogeneous questionnaires, the unit of information was a logarithm expressed with the basis of the questionnaire. A homogeneous questionnaire consisting of a unique question and of a answers which are not equiprobable transmits a quantity of information which is less than unity and the length of its routing, measure of the mean number of questions, is 1.

For heterogeneous questionnaires, the unit of routing length is still the elementary arc linking two consecutive vertices of the same path: that is, the difference of ranks of 1. But what unit of information must we take?

In the homogeneous case we have written inequality (3) because

$$J_a(i) = \sum_{j \in \Gamma i} \frac{p(i,j)}{p(i)} \log_a \frac{p(i)}{p(i,j)} \leq 1 \qquad (27)$$

with $|\Gamma i| \leq a$

and equality can hold only in the polychotomic case in the strict sense (the index a recalls the base of the logarithm).

But a change of base for the logarithm will lead to computing

$$J_b(i) = \log_b a \cdot \sum_{j \in \Gamma i} \frac{p(i,j)}{p(i)} \log_a \frac{p(i)}{p(i,j)}$$

which can be strictly greater than 1 when $b<a$; then the maximal basis a_M will imply systematically $J_{a_M}(\Omega) < L(\Omega)$.

Example

Consider two questionnaires Q and R built on $(\mathcal{A}, \mathcal{P})$, where

$$\mathcal{A} = (4, 3, 2), \quad \mathcal{P} = \{\tfrac{1}{4}, \tfrac{1}{4}, \tfrac{1}{8}, \tfrac{1}{8}, \tfrac{1}{12}, \tfrac{1}{12}, \tfrac{1}{12}\}$$

and where

Q is an L-optimal questionnaire

R is a questionnaire of height 3 in which the basis of the root is 2, the answer of rank 1 has probability $\tfrac{1}{12}$, there are two answers of rank 2 of respective probabilities $\tfrac{1}{12}$ and $\tfrac{1}{8}$ and the four other answers are of rank 3.

We find $L(Q) = 1.5$ and $L(R) = 2.625$.

Then, writing I_a for the information of the questionnaires Q and R, expressed with a as base for the logarithms we obtain

$$I_2 = 2.65, \quad I_3 = 1.67 \text{ and } I_4 = 1.32,$$

so that

$$I_4 < L(Q) < I_3 < L(R) < I_2 .$$

Following F. Dubail we introduce an informational concept related directly to a given questionnaire $Q = (X, \Gamma, P_\Gamma)$:

Definition 5 The *heterogeneous information processed by a given questionnaire* Q is the quantity

$$H(Q) = \sum_{i \in F} \sum_{j \in \Gamma i} p(i,j) \log_{a_i} \frac{p(i)}{p(i,j)}. \tag{28}$$

It is clear that the term

$$J_{a_i}(i) = \sum_{j \in \Gamma i} \frac{p(i,j)}{p(i)} \log_{a_i} \frac{p(i)}{p(i,j)}$$

can always be interpreted as an information (with basis a_i) processed by the question i; $p(i) J_{a_i}(i)$ will be called the *heterogeneous information contributed* by the question i. Furthermore $H(Q) = I_{a_i}(Q)$ if Q is homogeneous with basis a_i. Suppose Q is arborescent: even in this case $H(Q)$ is related to Q and does not depend only on the probability distribution of the answers, but depends on the support (X, Γ) and also on the labelling, that is, on the allocation of the probabilities to the answers. This may be seen through the following example.

Example Taking the pair $(\mathcal{A}, \mathcal{P})$ given by

$$\mathcal{A} = (3, 2), \quad \mathcal{P} = \left\{\tfrac{1}{3}, \tfrac{1}{3}, \tfrac{1}{6}, \tfrac{1}{6}\right\} ;$$

let Q be an L-optimal questionnaire and R a questionnaire deduced from Q by permutation of two trivial sub-arborescences whose vertices have probability $\frac{1}{3}$ and $\frac{1}{6}$ (Fig. 8.5)

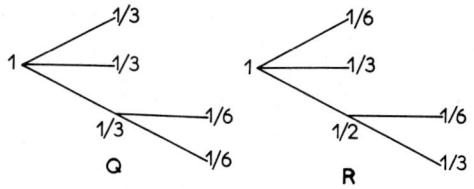

Figure 8.5

We calculate:

$$H(Q) = (\frac{1}{3} \log_3 3 + \frac{1}{3} \log_3 3 + \frac{1}{3} \log_3 3)$$
$$+ \frac{1}{3}(\frac{1}{2} \log_2 2 + \frac{1}{2} \log_2 2) = 1.33$$

$$H(R) = (\frac{1}{2} \log_3 2 + \frac{1}{3} \log_3 3 + \frac{1}{6} \log_3 6)$$
$$+ \frac{1}{2}(\frac{2}{3} \log_2 \frac{3}{2} + \frac{1}{3} \log_2 3) = 1.38 ,$$

and remark that $L(Q)=H(Q)<H(R)<L(R)$.

Thus heterogeneous information does not represent only a function of the answers' probability distribution.

However it is possible for the heterogeneous information to be equal to the routing length.

For if $J_{a_i}(i)=1 (\forall i \in F)$, then $H(\Omega) = \sum_{i \in F} p(i)$.

The following property gives a direct generalization of Theorem 1.

Property 5 The heterogeneous information processed by an arborescent questionnaire Q is equal to its routing length if and only if the probability of any vertex i can be written in the form

$$p(i) = \prod_{j:=0}^{r-1} \frac{1}{a_{x(j)}}$$

where the path linking the root to i consists of r arcs whose origins are

$$\alpha = x(0), x(1), \ldots, x(r-1) ;$$

Otherwise $H(Q) \leq L(Q)$.

Note that a questionnaire Q may be non-optimal but still process a heterogeneous information equal to its routing length (Fig. 8.6).

The problem arises of determining an information function which depends only on the pair $(\mathcal{A}, \mathcal{P})$ and, for that, the value of the heterogeneous information for a particular questionnaire must be determined.

A solution can be offered by an L-optimal questionnaire:

Property 6 The heterogeneous information processed by a Huffman questionnaire is equal to the acquisition function $A(\Omega)$ of any questionnaire built on the same pair $(\mathcal{A}, \mathcal{P})$ and moreover

$$A(\Omega) \leq L(Q) ;$$

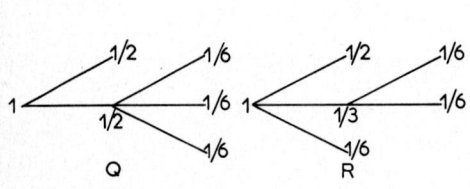

Figure 8.6

with equality if and only if the probabilities of any question satisfy the relation of Property 5.

This property follows directly from the fact that the acquisition $A(Q)$ is defined, according to axiom (C3) (§7.2.2) precisely to be $H(Q)$ when Q is built according to Huffman's algorithm (§6.4) applied to the pair $(\mathcal{A}, \mathcal{P})$.

We can then decide to characterize the quantity of information processed by a heterogeneous questionnaire Q by means of the acquisition $A(Q)$ which depends only on $(\mathcal{A}, \mathcal{P})$ as implied by axiom (C1).

The first drawback to this choice appears on making the following observation:

- $I(Q)$ can be computed by means of logarithms without constructing any questionnaire.

- $L(Q)$ can be computed without the aid of logarithms but with the construction of Q.

- $A(Q)$ can be computed by means of logarithms and requires the construction of the Huffman questionnaire Q_H associated with Q.

In the homogeneous case the computation of $I(\Omega)$ enabled us to avoid the construction of Q; in the heterogeneous case, there is no way of escaping the construction of a questionnaire Q to calculate $H(Q)$, Q_H to compute $A(Q)$.

The second drawback of the acquisition function will appear in §8.2.4 where we will show that the questionnaire Q_m built on ($\mathcal{A}, \mathcal{P}$) for which $H(Q_m)$ presents the minimal value of the heterogeneous information may be distinct from the Huffman questionnaire.

A particular case

The product of two polychotomic questionnaires in the strict sense ($\beta=0$) with different bases is a heterogeneous questionnaire.

Consider a questionnaire A_1 with basis a_1 and N_1 answers and a questionnaire A_2 with basis a_2 and N_2 answers such that

$$a_1 > a_2 .$$

Then the number of questions, M_1 and M_2 are related to N_1 and N_2 by

$$N_1 = (a_1 - 1)M_1 + 1$$

$$N_2 = (a_2 - 1)M_2 + 1$$

and the number of answers of the product is $N_1 N_2$.

The product questionnaire $A_1 A_2$ possesses M_1 questions with basis a_1 and $N_1 M_2$ questions with basis a_2, whereas the product questionnaire $A_2 A_1$ possesses $N_2 M_1$ questions with basis a_1 and M_2 questions with basis a_2.

For example let A_1 have

$$a_1 = 4, M_1 = 3, N_1 = 10$$

and A_2

$$a_2 = 2, M_2 = 2, N_2 = 3 .$$

Then we deduce from these that $N_1 N_2 = 30$ and $A_1 A_2$ possesses 3 questions with basis 4 and 20 questions with basis 2 whereas $A_2 A_1$ possesses 9 questions with basis 4 and 2 questions with basis 2.

Moreover in $A_2 A_1$, the questions with smallest basis (a_2) have a rank generally smaller (and at least once) than the questions with basis a_1 so that $A_2 A_1$ certainly does not have an optimal routing. On the contrary, in $A_1 A_2$ the questions with greatest basis (a_1) have a rank generally smaller than the questions with basis a_2 and hence $A_1 A_2$ can have an optimal routing.

If A_1 and A_2 are L-optimal questionnaires which are

equiprobable and if $N_1 = a_1^{k_1}$, then $A_1 A_2$ possesses a minimal routing length $L_{1 \times 2}$. However the routing length $L_{2 \times 1} = L_{1 \times 2}$ can be reduced for the questionnaire $A_2 A_1$ by application of rule RO (cf. §6.2.1).

Thus the product of two homogeneous questionnaires with different bases can lead after certain transfers of sub-arborescences to a difference of routing between the questionnaires $(A_1 A_2)^T$ and $(A_2 A_1)^T$ deduced from $A_1 A_2$ and $A_2 A_1$. The optimal routing of the questionnaire with $N_1 N_2$ answers is that of the questionnaire having the greatest number of questions with basis a_1 that is $(A_2 A_1)^T$.

By (28), the heterogeneous information of the product questionnaire is

$$H(A_1 A_2) = \sum_{ib \in F_1\{b\}} \sum_{j \in \Gamma i} p_\Lambda(ib, jb) \log_{a_1} \frac{p_\Lambda(ib)}{p_\Lambda(ib, jb)}$$

$$+ \sum_{\xi k \in E_1 F_2} \sum_{\ell \in \Gamma k} p_\Lambda(\xi k, \xi \ell) \log_{a_2} \frac{p_\Lambda(\xi k)}{p_\Lambda(\xi k, \xi \ell)}$$

where $E_1 = \{\xi\}$, $E_2 = \{\eta\}$ are the sets of answers of A_1 and A_2, F_1 and F_2 are the sets of questions of A_1 and A_2, a, b, are the roots of A_1 and A_2, and p_Γ, p_Δ, p_Λ are the mappings (probabilities of the arcs) valuating A_1, A_2 and $A_1 A_2$.

We find:

$$H(A_1 A_2) = \sum_{i \in F_1} \sum_{j \in \Gamma i} p_\Delta(b) p_\Gamma(i, j) \log_{a_1} \frac{p_\Gamma(i)}{p_\Gamma(i, j)}$$

$$+ \sum_{k \in F_2} \sum_{\xi \in E_1} p_\Gamma(\xi) \sum_{\ell \in \Gamma k} p_\Delta(k, \ell) \log_{a_2} \frac{p_\Delta(k)}{p_\Delta(k, \ell)}$$

that is $H(A_1 A_2) = H(A_1) + H(A_2)$

or furthermore

$$H(A_1 A_2) = I_{a_1}(A_1) + I_{a_2}(A_2) . \qquad (29)$$

We can define a class of arborescent questionnaires satisfying a supplementary axiom:

Axiom SH: A questionnaire is said to be *semi-homogeneous* if at every rank the questions all have the same basis.

Such questionnaires can be introduced when the number of answers is of the form

$$N_j = \sum_{i:=1}^{h_j} a_i$$

and when the number of questions with given basis a_i is no longer fixed or when it is imposed to formulate questions with the same basis at the same searching phase: the height is h_j.

A semi-homogeneous arborescent questionnaire can be formed by a product of elementary questionnaires whose bases may be distinct.

We can then generalize (29):

Property 7 The product of s semi-homogeneous arborescent questionnaires $A_1, A_2, \ldots, A_s$ having $N_j = \prod_{i:=1}^{h_j} a_i$ answers (j=1,...,s) is a semi-homogeneous questionnaire $A_1 A_2 \ldots A_s$ with $N = \prod_{i=1}^{s} N_j$ answers the heterogeneous information of which is the sum of the heterogeneous informations of the A_j. The routing length of $A_1 A_2 \ldots A_s$ is independent of the order in which the bases are allocated to the ranks if in each arborescent questionnaire A_j the answers have a rank equal to the height.

The answers then have rank $h = \sum_{j=1}^{s} h_j$ and the routing length is therefore L=h, whatever the order of the bases.

8.2 Contribution of Information

Here we consider how to determine the information contributed by a set of questions F' strictly contained in F. To that end the discussion will be placed successively within the context of algorithms of Shannon-Fano and Huffman types by considering that F' consists of all the questions of a sub-questionnaire with the same root as the questionnaire (a process acting upstream to downstream) or that F' is constituted according to the order of formation of questions given by Huffman's algorithm (downstream to upstream).

8.2.1 Maximisation of the processed information and Shannon-Fano's algorithm

We have seen in 6.5 that the researches on coding had a tendency to set up as a principle the partition by equalization of probabilities of given rank. We have shown how this often very fruitful method can fail if it contradicts rule R3 and we have shown that the efficient strategy for formulating a questionnaire is to construct it according to Huffman's algorithm. The interpretation of this strategy can be made by means of information theory.

To partition the successors of a question of rank r in such a way that their probabilities are as near to each other as possible amounts to maximising the function

$$J(r,s) = \sum_{i:=1}^{a-1} \frac{p(r+1, as+i)}{p(r,s)} \log_a \frac{p(r,s)}{p(r+1, as+i)}$$

and for this it is necessary to determine among the answers descending from [r,s] those which descend from [r+1, as+i], for various values of i. This process acts upstream to downstream with application at any rank and for any sub-arborescence of a principle of maximization of processed information J(r,s).

But Huffman's algorithm which proceeds downstream to upstream shows that this process does not lead to the optimal strategy.

On the contrary after having applied Huffman's algorithm it is possible to modify the probabilities of the questions by transfers of sub-questionnaires without increasing the routing length if the rules R1 and R2 are always respected: the difference between the probabilities of questions of the same rank can thus be reduced. Proceeding always downstream to upstream it can happen that we find a questionnaire which maximizes J(r,s) with the same routing length as the one from which we started, but this depends entirely on the distribution $\mathcal{P}$.

Thus a questionnaire without inefficiency will not generally be obtained by maximization of the processed information J(r,s). In particular the root does not always satisfy:

$$\text{Max } J(\alpha)$$

where the maximum is taken on the set of possible probability distributions of the vertices of rank 1 and the maximal reduction of inefficiency at the first question does not always give the optimal strategy.

This remark is illustrated by the arborescences I and II of Figure 6.5 where J(1,0) takes the values 1 and

$$0.44 \log \frac{1}{0.44} + 0.56 \log \frac{1}{0.56} = 0.990$$

respectively and L takes the values 2.38 and 2.29 respectively.

8.2.2 Partitions in equiprobable dichotomic questionnaires and choice

To make a partition of the set of answers descending from a dichotomic question requires that the number of answers of each class satisfies the inequalities 6.15 of corollary 6.4 (6.6.1). If it is possible to realize this condition for all the questions, then a valid method can proceed upstream to downstream.

Let $N=2^k+\alpha$ be the number of answers descending from a question q; the partition will have to create two classes having N_1 and $N-N_1$ answers and such that

$$2^{k-1} \leq N_1 \leq 2^{k-1} + \alpha, \quad \text{if } 0 \leq \alpha \leq 2^{k-1}$$

or
$$\alpha \leq N_1 \leq 2^k, \quad \text{if } 2^{k-1} \leq \alpha < 2^k. \tag{30}$$

The number of realizable $(N_1, N-N_1)$ is

$$\alpha + 1, \quad \text{if } 0 \leq \alpha \leq 2^{k-1}$$

or
$$2^k - \alpha + 1, \quad \text{if } 2^{k-1} \leq \alpha < 2^k,$$

but the pairs $(N_1, N-N_1)$ and $(N-N_1, N_1)$ differ only by the class of the first successor of q (Γ_q^0 or Γ_q^1).

The number of distinct pairs $(N_1, N-N_1)$ where $N_1 \leq N-N_1$ is

$$\left. \begin{array}{c} \frac{\alpha}{2} + 1 \\ \\ \frac{\alpha+1}{2} \end{array} \right\} \quad \text{if } 0 \leq \alpha \leq 2^{k-1} \quad \text{and} \quad \left\{ \begin{array}{l} \alpha \text{ is even,} \\ \\ \alpha \text{ is odd,} \end{array} \right.$$

$$\left. \begin{array}{c} \frac{2^k - \alpha}{2} + 1 \\ \\ \frac{2^k - \alpha + 1}{2} \end{array} \right\} \quad \text{if } 2^{k-1} \leq \alpha < 2^k \quad \text{and} \quad \left\{ \begin{array}{l} \alpha \text{ is even,} \\ \\ \alpha \text{ is odd.} \end{array} \right.$$

Consequently, if $N=2^k-1$, 2^k or 2^k+1, the ratio of the probabilities of two successors of the root q of the given arborescence is fixed; but this ratio can vary between limits dependent on α and k when N gets away from powers of 2 and has maximal amplitude when $\alpha=2^{k-1}$, that is when N is the arithmetic mean between two consecutive powers. All the corresponding partitions enable us to obtain an optimal questionnaire if they are realized for every question of the arborescence. There therefore exists a choice in the values of the probabilities which the successors of q must have but this choice decreases progressively as the rank q increases. If it has lead to the creation of $2^{k'}-1, 2^{k'}$ or $2^{k'}+1$ paths coming out of one of the successors of a question q, then there is no other possible choice downstream.

The values $N=2^k-1$, 2^k or 2^k+1, correspond to minimal inefficiency. For the values near to $N=2^k+2^{k-1}$, that is $N=3/4.2^{k+1}$, for which the choice allows us to opt between $2^{k-2}+1$ distinct pairs, the inefficiency is on the contrary near to its maximal value $B_{max}=0.086$.

Thus a kind of equilibrium is established between the value of the inefficiency and the possibilities of choice. In principle when we proceed upstream to downstream to formulate a questionnaire by writing the conditions to be imposed on the questions, it is preferable to choose the values of N_1 and $N-N_1$ as near to each other as possible to preserve the maximum choice at the following rank.

However the criterion of the inequalities (30) must always be respected, if we want to avoid irreversible damage to the strategy of maximization of processed information upstream to downstream.

The following table gives the evolution of the number of distinct pairs for $16 \leqslant N \leqslant 32$.

N	extreme pairs		number of distinct pairs	N		extreme pairs	
16	8,8		1	32		16,16	
17	8,9		1	31		15,16	
18	8,10	9,9	2	30	14,16		15,15
19	8,11	9,10	2	29	13,16		14,15
20	8,12	10,10	3	28	12,16		14,14
21	8,13	10,11	3	27	11,16		13,14
22	8,14	11,11	4	26	10,16		13,13
23	8,15	11,12	4	25	9,16		12,13
24	8,16	12,12	5				

8.2.3   Minimization of the contributed information and Huffman's algorithm

8.2.3.1   Dichotomic questionnaires

The information $I(h)$ and $I(i)$ contributed by questions h and i whose outcomes have probabilities p_k, p_ℓ and p_j, p_ℓ respectively such that $p_k > p_j$ are related by Property 7.3, i.e. $I(h) > I(i)$. This leads to the following result:

Property 8 For a given probability distribution $\mathcal{P}$, the minimal information that a dichotomic question can contribute is $I(i)$, where i is a question whose two outcomes are the answers with probabilities equal to the two smallest elements of $\mathcal{P}$.

8.2.3.2   Polychotomic questionnaires in the strict sense

The basis a is such that $a-1 | N-1$.

Let us compare the information contributed by two questions differing only by two outcomes y_j and y_k ($p_k > p_j$). We will suppose that the a+1 distinct outcomes considered are all answers such that

$$\sum_{r:=1}^{a-1} p_r = p_\ell, \quad p_h = p_\ell + p_k, \quad p_i = p_\ell + p_j,$$

and furthermore that the answers y_r (for $r:=1,\ldots,a-1$) are distinct from the answers y_k and y_j.

$$I(h) = p_h I_a \left(\frac{p_1}{p_h}, \frac{p_2}{p_h}, \ldots, \frac{p_{a-1}}{p_h}, \frac{p_k}{p_h}\right)$$

$$I(i) = p_i I_a \left(\frac{p_1}{p_i}, \frac{p_2}{p_i}, \ldots, \frac{p_{a-1}}{p_i}, \frac{p_j}{p_i}\right) .$$

We then find that

$$I(h) - I(i) = \sum_{r:=1}^{a-1} p_r \left(\log \frac{p_h}{p_r} - \log \frac{p_i}{p_r}\right)$$

$$+ p_k \log \frac{p_h}{p_k} - p_j \log \frac{p_i}{p_j} .$$

and hence that

$$\log_2 a \ (I(h) - I(i)) = (p_k + p_\ell) I_2 \left(\frac{p_k}{p_k + p_\ell}, \frac{p_\ell}{p_k + p_\ell}\right)$$

$$- (p_j + p_\ell) I_2 \left(\frac{p_j}{p_j + p_\ell}, \frac{p_\ell}{p_j + p_\ell}\right) \quad (31)$$

which by Property 7.3, is strictly positive.

It is possible to reduce $I(i)$ if and only if the question with probability p_i does not have the a vertices with the smallest probabilities as its successors.

Property 9 Given a probability distribution $\mathcal{P}$ and a polychotomic questionnaire in the strict sense, then the minimal information which a question can contribute is obtained by the question formed by regrouping the a answers with smallest probabilities.

8.2.3.3 Heterogeneous questionnaires

We use the contribution of heterogeneous information and compare the contributions of information of two questions with bases a and a+1. Let

$$p_1, p_2, \ldots, p_a, p_{a+1} = p$$

be the a+1 smallest probabilities placed in non-decreasing order and $a = \sum_{j:=1}^{a} p_j$, with $q+p<1$.

Then we consider the sign of

$$\Delta = (q + p) I_{a+1} \left(\frac{p_1}{q+p}, \ldots, \frac{p_a}{q+b}, \frac{q}{q+b}\right) - q I_a \left(\frac{p_1}{q}, \ldots, \frac{p_a}{q}\right)$$

that is

$$\Delta = \sum_{j:=1}^{a+1} p_j \log_{a+1} \frac{q+p}{p_j} - \sum_{j:=1}^{a} p_j \log_a \frac{q}{p_j} .$$

$$\Delta = \sum_{j:=1}^{a} (\log_{a+1} a - 1) p_j \log_a \frac{1}{p_j} + \sum_{j:=1}^{a} p_j \log_{a+1} (q+p)$$

$$+ p \log_{a+1} \frac{q+p}{p} - \sum_{j:=1}^{a} p_j \log_a q ,$$

$$\Delta = p \log_{a+1} (1 + \frac{q}{p}) + q \log_{a+1} (1 + \frac{p}{q})$$

$$+ q(\log_{a+1} a - 1) \sum_{j:=1}^{a} \frac{p_j}{q} \log_a \frac{q}{p_j} .$$

But

$$\sum_{j:=1}^{a} \frac{p_j}{q} \log_a \frac{q}{p_j} \leq 1 \quad \text{and} \quad \log_{a+1} a - 1 < 0$$

and this enables us to find the lower bound:

$$\Delta \geq p \log_{a+1} (1 + \frac{q}{p}) + q \log_{a+1} (1 + \frac{p}{q})$$

$$+ q(\log_{a+1} a - 1) ,$$

$$\Delta \geq p \log_{a+1} (1 + \frac{q}{p}) + q \log_{a+1} \frac{a}{a+1} (1 + \frac{p}{q}) .$$

But $\frac{p}{q} \geq \frac{1}{a}$ and consequently $\frac{a}{a+1}(1+\frac{p}{q}) \geq 1$, which implies that

$$\Delta \geq p \log_{a+1} (1 + \frac{q}{p}) \tag{32}$$

and therefore
$$\Delta > 0 .$$

We can then compare the contributions of heterogeneous information of two questions with bases a and b > a. If the a answers with smallest probabilities are grouped as the outcomes of a question with basis a then the contribution of heterogeneous information is minimal.

Property 10 The minimum of the contribution of heterogeneous information which it is possible to realize by re-grouping a or b answers is obtained when the smallest probabilities are allocated to the answers to the question with minimal basis.

8.2.3.4 Polychotomic questionnaires in the broad sense

It is now possible to discuss the case of polychotomic questionnaires in the broad sense that is such that N-1 is not a multiple of a-1.

If β is the remainder of the division of N-1 by a-1, then the arborescent questionnaire has a question with basis β+1 and all the other questions have bases a>β+1.

Property 10 implies that

$$\sum_{j:=1}^{\beta+1} p_j \log_{\beta+1} \frac{\sum_{j:=1}^{\beta+1} p_j}{p_j} < \sum_{j:=1}^{a} p_j \log_a \frac{\sum_{j:=1}^{a} p_j}{p_j}$$

and since $\log_a x < \log_{\beta+1} x$, we find:

$$\sum_{j:=1}^{\beta+1} p_j \log_a \frac{\sum_{j:=1}^{\beta+1} p_j}{p_j} < \sum_{j:=1}^{a} p_j \log_a \frac{\sum_{j:=1}^{a} p_j}{p_j} \qquad (33)$$

which extends Propery 9 to polychotomic questionnaires in the broad sense.

Property 11 In a polychotomic questionnaire in the broad sense the minimal information that a question can contribute is obtained for the question with basis β+1<a formed by regrouping the answers with smallest probabilities.

8.2.3.5 Informational interpretation of Huffman's algorithm

We know that the questionnaire $Q_{N-\beta}$ formed on the N-(β+1) answers with smallest probabilities and the question with basis β+1 is homogeneous in the strict sense: we may then recursively apply Property 9 to it. (33) enables us to extend Property 10 to the contribution of information of a heterogeneous questionnaire (with a unique base for the logarithms).

Example

Consider three sub-questionnaires A, B, C formed by starting from a question with basis 2 or 3 on the answers of an incomplete system with probabilities (1/16, 1/8, 1/8, 1/4) (Fig. 8.7)

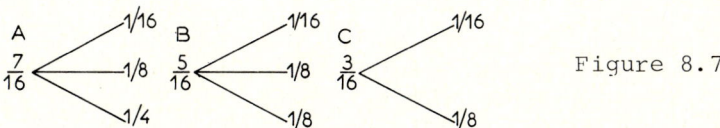

Figure 8.7

Denote by H(A), H(B), H(C) the contribution of heterogeneous information for A, B, C. Then

$$H(A) = \frac{7}{16} I_3 (\frac{1}{7}, \frac{2}{7}, \frac{4}{7}) \# \frac{5.52}{16}$$

$$H(B) = \frac{5}{16} I_3 (\frac{1}{5}, \frac{2}{5}, \frac{2}{5}) \# \frac{4.75}{16} < H(A)$$

$$H(C) = \frac{3}{16} I_2 (\frac{1}{3}, \frac{2}{3}) \qquad \# \frac{2.75}{16} < H(B) .$$

Moreover $H(B)-H(C) \# \frac{1}{8}$ which is effectively greater than the minimal value given by (32) which here becomes

$$[H(B) - H(C)] \geq \frac{1}{8} \log_3 (1 + \frac{3/16}{1/8}) \# \frac{0.83}{8} ;$$

Taking 3 for the base of the logarithms in the sub-questionnaires B and C we further obtain

$$I(B) - I(C) = H(B) - H(C) . \log_3 2 \# \frac{3.02}{16} .$$

The recursive aspect of Huffman's algorithm enables us in all the cases discussed to interprete each phase (or each question formation) as the regrouping of vertices minimizing the contributed information. We can compare this construction of the questionnaire with minimal routing length to the construction proposed by Fano which consists in maximizing the processed information from the root to the answers (§8.2.1). There is identity between the two techniques only if the process comprises only two questions or in particular cases such as that where the questions formed by Huffman's process have a probability less than the answers which do not descend from them (each question then has outcomes which are all answers except one).

We have thus proved:

Theorem 3 In the homogeneous case (dichotomic or polychotomic, whatever N and a) and in the heterogeneous case, each phase of Huffman's algorithm amounts to constructing the question minimizing the contributed information and the contributed heterogeneous information.

8.2.4 Heterogeneous information and acquisition

In view of the preceding theorem about Huffman's algorithm, we can say that the construction of an L-optimal questionnaire comes to a process with as many phases as there are questions, each phase corresponding to a minimization of the contributed information for given sets of probabilities and bases. The first phase is determined by the smallest basis and the answers with smallest probability, then each of the successive phases is determined by the smallest unused basis and by the vertices with smallest probability after insertion of the questions already formed in the list of vertices (questions and answers not taken before).

We then obtain for a given probability distribution and set of bases
$$\mathcal{A} = \{a_1, a_2, \ldots, a_M\}$$
with
$$p_1 \leq p_2 \leq \cdots \leq p_N \qquad \text{and} \qquad a_1 \leq a_2 \leq \cdots \leq a_M:$$
$$\min L(p_1, \ldots, p_N) = \sum_{i:=1}^{a_1} p_i + \min L(p_{a_1+1}, p_{a_1+2}, \ldots, p_N, q)$$
where $q = \sum_{i=1}^{a_1} p_i$,

and the same in the later phases following the process of dynamical programming.

As far as the information which does not depend on the arborescence supporting the questionnaire built on $(\mathcal{P}, \mathcal{A})$ and goes, there is no interesting global minimization.

However I did at one time make the following conjecture:

"Huffman's questionnaire presents the minimal value of heterogeneous information,
$$H(Q) = \sum_{x_i \in Q} \sum_{z \in \Gamma x_i} p(z) \log a_i \frac{p(x_i)}{P(z)}$$
among all the heterogeneous questionnaires built on the same pair ($\mathcal{P}, \mathcal{A}$), including the other L-optimal questionnaires."

In the case when for any k the probability of the question formed at the kth phase is less than all the probabilities of the answers not yet grouped, it is evident that the arborescence is of maximal height, each question becoming an outcome of the question formed in the following phase. There is then a constant accord between the Huffman and Fano algorithms and the maximization of information processed upstream to downstream or the minimization of the contributed information downstream to upstream are equivalent. In such a case the conjecture is true.

However the construction of Huffman's algorithm may give rise to various questions before taking them as outcomes of a new question and a forest is formed as a sub-graph before forming a sub-arborescence. For each connected component of this forest where the probability of a question is less than that of the answers which do not descend from it, the conjecture is certainly true. But if the conjecture is false it will fail in the case of a sub-graph which is a forest. We therefore try to construct a *heterogeneous forest* of minimal order with five terminals.

250   GRAPHS AND QUESTIONNAIRES

Let $\mathcal{P} = \{p_1, p_2\ r_1, r_2, s_3\}$ and $\mathcal{A} = \{2, 2, 3\}$

with $p_1 < p_2 < r_1 < r_2 < s_3$ and $p_1 + p_2 > r_2$.

and consider the questions with probability:

$$q_1 = p_1 + p_2 \qquad\qquad q'_1 = p_2 + r_1$$

$$q_2 = r_1 + r_2 \qquad\qquad q'_2 = p_1 + r_2$$

$$1 = q_1 + q_2 + s_3 \qquad\qquad 1 = q'_1 + q'_2 + s_3.$$

Then we form two questionnaires (Figure 8.8) one of which, Q_H, having questions with probabilities q_1, q_2, 1 is of Huffman type, and the other of which having questions with probabilities q'_1, q'_2, 1 generalizes that of Fano in the heterogeneous case.

We will compare the value of H for each of these two questionnaires denoting them respectively by H_H and H_F.

We will also put:

$$f_x = x \log_2 \tfrac{1}{x}, \quad \varphi_x = x \log_3 \tfrac{1}{x}$$

$$H_H = (\varphi_{q_1} + \varphi_{q_2} + \varphi_{s_3} - \varphi_{q_1+q_2+s_3})$$
$$+ (f_{p_1} + f_{p_2} - f_{q_1}) + (f_{r_1} + f_{r_2} - f_{q_2})$$

$$H_F = (\varphi_{q'_1} + \varphi_{q'_2} + \varphi_{s_3} - \varphi_{q'_1+q'_2+s_3})$$
$$+ (f_{p_2} + f_{r_1} - f_{q'_1}) + (f_{p_1} + f_{r_2} - f_{q'_2})$$

We find again trivially the fact that with the exclusive use of only one base of logarithms (i.e. $\varphi_x = f_x$) leading to equality of H_H and H_F, the information processed by an arborescent questionnaire depends only on the probabilities of the answers. However the probabilities of the questions affect the computation of H as soon as the questionnaire is heterogeneous.

Noting that $\varphi_{q_1+q_2+s_3} = 0$ we obtain

$$H_H - H_F = \varphi_{q_1} + \varphi_{q_2} - (f_{q_1} + f_{q_2})$$
$$- [(\varphi_{q'_1} + \varphi_{q'_2} - (f_{q'_1} + f_{q'_2})]$$

$$H_H - H_F = (\varphi_{q_1} + \varphi_{q_2} - \varphi_{q_1+q_2}) - (f_{q_1} + f_{q_2} - f_{q_1+q_2})$$
$$- [(\varphi_{q_1'} + \varphi_{q_2'} - \varphi_{q_1'+q_2'}) - (f_{q_1'} + f_{q_2'} - f_{q_1'+q_2'})]$$

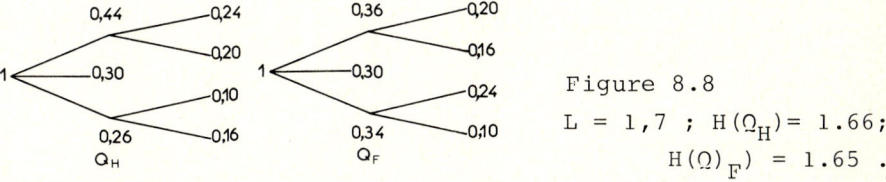

Figure 8.8
$L = 1,7 \; ; \; H(\Omega_H) = 1.66;$
$H(\Omega)_F = 1.65.$

$$H_H - H_F = (1 - \log_3 2)\,[\,(f_{q_1'} + f_{q_2'} - f_{q_1'+q_2'})$$
$$-(f_{q_1} + f_{q_2} - f_{q_1+q_2})\,].$$

Since
$$q_1 + q_2 = q_1' + q_2' = 1 - s_3$$

and
$$q_1 < q_1' \;,\; q_1' < q_2$$
$$q_1 < q_2' \;,\; q_2' < q_2$$

it follows that
$$f_{q_1'} + f_{q_2'} - f_{q_1'+q_2'} > f_{q_1} + f_{q_2} - f_{q_1+q_2} \,. \tag{35}$$

We can interpret this inequality as follows: since the questionnaire Q_F is more balanced that Q_H, the information processed by the root (with probability 1) is greater for Q_F than for Q_H.

So that $H_H - H_F > 0$.

In this example the minimal value taken on by H no longer corresponds to the Huffman's questionnaire and this disproves the conjecture.

Thus we see that the recursive procedure bearing on the minimum of the routing length of homogeneous or heterogeneous questionnaires cannot be applied to obtain the minimum of H if $H(\Omega)$ is not identical with $A(Q)$.

Massoumi has shown that in general the questionnaire minimizing H is L-optimal and can be deduced from Huffman's questionnaire by a succession of transfers tending to increase down stream to upstream the information processed by the different questions.

8.3 Quasi-Questionnaires

8.3.1 Quasi-answers and quasi-questions

Definition 6 A *quasi-questionnaire* is a lower quasi-strongly connected valuated graph

$$Q = (X, \Gamma, P_\Gamma)$$

in which the set of vertices admits a partition

$$X = E \cup \bar{E} \cup F \cup \Phi$$

where

E and $\bar{E}$ consist of terminal vertices,

F and Φ consist of non-terminal vertices,

and which satisfies the axioms (q_1), (q_2) and
(qq4) there exists a mapping P_Γ from the set of arcs

$$\Gamma = \{(i,j) \mid i \in F \cup \Phi\}$$

in $[0,1]$,

$$P_\Gamma : (i,j) \mapsto p(i,j)$$

such that

$$\sum_{j \in \Gamma i} p(i,j) = \sum_{h \in \Gamma^{-1} i} p(h,i) \neq 0$$

for all $i \in F \cup \Phi$ with both Γi and $\Gamma^{-1} i$ non-empty,

(qq5) this mapping P_Γ is such that

$$\sum_{i \in E \cup \bar{E}} \sum_{h \in \Gamma^{-1} i} p(h,i) = 1 \ .$$

Definition 7 The vertices of E are defined by $\forall e$, $\exists h$, $p(h,e) \neq 0$, and are called *answers* of Q; the vertices of $\bar{E}$ are defined by $\forall \bar{e}$, $\forall h \in \Gamma^{-1} \bar{e}$, $p(h,\bar{e}) = 0$ and are called *quasi-answers* of Q.

Definition 8 The vertices of F, defined by

$$\forall q \in F \quad \forall j \in \Gamma q, \ p(q,j) \neq 0$$

are called *questions of* Q; the vertices of Φ defined by

$$\forall q \in \Phi \quad \exists j \in \Gamma q, \ p(q,j) = 0$$

are called *quasi-questions* of Q.

It is possible to associate a flow with each vertex of a quasi-questionnaire Q and to define a probability for each arc, vertex and path of Q as has been done in Chapter 4 for questionnaires; the probability 0 corresponds to the empty event which belongs to $\mathfrak{P}(E \cup \overline{E})$.

Property 12 An arborescent quasi-questionnaire is a questionnaire if and only if all its terminals have a non-zero probability.

This condition is necessary. If it is satisfied ($\overline{E}=\emptyset$), then all the arcs with terminal extremity in E have non-zero probability and therefore the answers have no quasi-questions among their predecessors. Contraction at the set formed by a question and all its successors which are terminals creates an answer and no quasi-question. Repetition of this operation forms answers and never quasi-questions. Thus the condition is also sufficient.

Note that an arborescent questionnaire can be considered as the partial graph of a latticoid quasi-questionnaire. It suffices to adjoin an arc between two vertices x and y not already linked by an arc without forming any circuit, then to take
$$\Gamma' = \Gamma \cup \{x,y\} ,$$
$P_{\Gamma'}$: if $i\Gamma j$, then $(i,j) \mapsto p_{\Gamma'}(i,j) = p_{\Gamma}(i,j)$

and
$$p_{\Gamma'}(x,y) = 0 .$$

A valuated graph satisfying axioms (q_1), (q_2), (q_4) and (q_5) may be considered to be a quasi-questionnaire: for this F is partitioned into $F' \cup \Phi$ such that
$$i \in F' \Leftrightarrow \forall j \in \Gamma i \quad p(i,j) > 0$$
$$h \in \Phi \Leftrightarrow |\Gamma h| = 1;$$
and then (X,Γ,P_Γ) is transformed into $(X',\Gamma',P_{\Gamma'})$ by adjunction of a vertex k which does not belong to X, adjunction of an arc (i,k) where $i \in \Phi$ and by the valuation $p_{\Gamma'}(i,k)=0$, $p_{\Gamma'}(i,j)=p_\Gamma(i,j)$ if $i\Gamma j$. k will be a quasi-answer and h a quasi-question in the sense of Definitions 7 and 8. The range of the mapping $P_{\Gamma'}$ is $[0,1]$ in accordance with (qq4) and not $]0,1[$ as in (q4).

On the other hand such a graph can be considered as a questionnaire if axiom (q3) is omitted, as was already noted in the last section of §4.5 and in §5.3.1.

Graphs in which all the arcs coming out of an internal vertex would have valuation 0 have been ruled out as quasi-questionnaires. If such a vertex x arose in the course of construction, all the arcs coming out of x could be suppressed while preserving the valuation $\underline{p}(x)=0$. Any isolated vertices could also be suppressed as also those which belong to a sub-latticoid whose root had zero probability and where there were no entering arcs with non-zero probability. These suppressions could be performed recursively.

For the remainder of this section (§8.3) we consider only *polychotomic quasi-questionnaires* and they will be referred to simply as quasi-questionnaires.

Graphs in which there exists an internal vertex with zero valuation will be called *probabilized arborescences*.

The routing length of a quasi-questionnaire $\overline{K}$ can be defined by one or the other of the definitions

$$\overline{L} = \sum_{e \in E} p(e)r(e) \;,\; \overline{L} = \sum_{e \in E \cup \overline{E}} p(e)r(e)$$

or

$$\overline{L} = \sum_{q \in F \cup \Phi} p(q)$$

which are equivalent by Property 4.18.

With any polychotomic questionnaire Q built on $(a, \mathcal{P})$ may be associated an infinity of quasi-questionnaires obtained by substitutiton of π for $\mathcal{P}$: these are distributions of the same N non-zero probabilities and of a multiple of $(a-1)$ zero probabilities. But conversely corresponding to any quasi-questionnaire $\overline{K}$ there is only one pair $(a, \mathcal{P})$, where $\mathcal{P}$ is a distribution of N non-zero probabilities. Let $\overline{N}$ be the number of quasi-answers of $\overline{K}$ and $\overline{M}>1$ the number of quasi-questions.

We seek to make a transfer of terminal vertices (cf. §5.1.1) by permuting the probabilities of two vertices and answer e_i with rank $r(e_i)$ and a quasi-answer with rank r_h. If $r_h < r(e_i)$, then the new quasi-questionnaire has routing length

$$\overline{L} - p(e_i) [r(e_i) - r_h] . \tag{36}$$

If it is possible to repeat this type of transfer, we will do so.

If in the course of this process it happens that an arborescence has a vertex y which only has quasi-answers as descendants, then, in place of the sub-arborescence coming out of y, we substitute a single vertex y taken to be a quasi-answer. This reduces $\overline{N}$ by a multiple of $a-1$ and hence $\overline{M}$ will be decreased.

Let h be the greatest rank of answers. Then there exists no quasi-answer of rank $r_h > h$, for otherwise there would be a vertex of rank h none of whose descendants are answers.

If all the quasi-answers have rank h it is possible to proceed once more to a transfer between an answer and a quasi-answer; transfers of this type will not reduce the routing length but will enable us to arrange for a quasi-answers to succeed to the same vertex. We may then reduce the number of quasi-answers and quasi-questions.

$\overline{N}$ cannot be reduced by one of these transfers if and only if $\overline{N} < a-1$.

All the transfers made successively enable us to form a homogeneous questionnaire in the strict sense K' (if N'=0) or in the broad sense (if $0 < N' < a-1$) in which the answers e_i have rank $r'(e_i)$ such that $r'(e_i) \leqslant r(e_i)$ ($\forall i$).

Consequently this questionnaire K' has routing length

$$L' \leqslant \overline{L} . \qquad (37)$$

If moreover a quasi-answer had rank $r_h < h$, then the inequality (37) would become strict

$$L' < \overline{L} .$$

If all the quasi-answers have maximal rank then:

- if $\overline{K}$ has $\overline{N} < a-1$ quasi-answers, then $\overline{N}-1$ transfers at most will enable us to form a homogeneous questionnaire in the broad sense with routing length $L' = \overline{L}$.

- if $\overline{K}$ possesses $\overline{N} \geqslant a-1$ quasi-answers, then transfers will enable us to form at least one vertex which is not an endpoint and from which at most one answer comes out. If there is no answer coming out, this vertex will be replaced by a quasi-answer with rank less than r_M in such a way that a new transfer will enable us to reduce the routing length; otherwise we will be brought back to the above case:

- any quasi-question with rank r from which descends only one answer e_i can be replaced by this answer, whence a reduction of the routing length by $[r(e_i)-r] p(e_i)$.

We have thus established:

Property 13 Any quasi-questionnaire $\overline{K}$ built on $(a, \mathcal{P})$ with routing length $\overline{L}$, and having $\overline{N} > a-1$ quasi-answers, can be reduced to a questionnaire K' built on $(a, \mathcal{P})$ with routing length $L' < \overline{L}$.

We remark that the condition $\overline{N} > a-1$ is related to the fact that if $N = a^k + \alpha(a-1) + \beta$ then the same arborescence K can be considered either as a questionnaire or as a quasi-questionnaire.

8.3.2   <u>Instantaneous codes</u>

Relationships between coding and questionnaires have already been established in §6.5.

The pair $(a, \mathcal{P})$ represents the basis of questions and the probability distribution of the complete system of answers (questionnaire point of view) or the number of letters of the alphabet and the probabilities of the words of the code still forming a complete system of eventualities E (coding point of view). The words of an instantaneous code are among the terminals of a probabilized arborescence, those whose probability is not zero. The problem of instantaneous coding is to determine the ordered sequence of arcs of the path linking the root to the words of the code: this coding is employed also in the theory of questionnaires to represent the answers. But here we are also concerned with representing the questions or internal vertices. Since the characters used to code a question, that is to index its position in the arborescence, are prefixes of the characters of the terminal vertices, the decoding requires a knowledge of the ranks of the vertices (§5.4.1).

An instantaneous code is a quasi-questionnaire because it satisfies Definition 6.

Let A be a homogeneous arborescence in the strict sense. Then every vertex of A has either zero or a successors. Suppose that the set of terminals E is of order N, $E=\{e_i | i=1,2,\ldots,N\}$ say, and that the ranks are denoted by $r(e_i)$.

We know that it is possible to construct on A an absolutely optimal questionnaire K_0 by giving probability $p(e_i)=a^{-r(e_i)}$ to the answers with rank $r(e_i)$ (Property 2).

- Successive computation of the probabilities of the questions (from the answers to the root) shows that all the vertices of K_0 have a probability related to the rank by $p(x)=a^{-r(x)}$. Because the successors of a question x with rank r are vertices (questions or answers) with rank $r+1$ and probabilities $a^{-(r+1)}$ and thus $p(x)=\Sigma a^{-(r+1)}=a^{-r}$. In particular the root α is such that $p(\alpha)=a^0=1$. Since $p(\alpha) = \sum_{e_i \in E} a^{-r(e_i)}$, we have

$$\sum_{e_i \in E} a^{-r(e_i)} = 1 . \qquad (38)$$

This relation (38) is satisfied by any questionnaire built on A.

Consider therefore the partition $E=E' \cup \overline{E}$ of the terminals of A and let $q(\tilde{e})$ be the probabilities of an instantaneous code such that:

$$\sum_{e_i \in E'} q(e_i) = 1 \quad \text{and} \quad \begin{cases} \forall \overline{e} \in \overline{E}) \; q(\overline{e}) = 0 \\ \forall e_i \in E', q(e_i) \neq 0 \end{cases}$$

From (38) it follows that:

$$\sum_{e_i \in E'} a^{-r(e_i)} + \sum_{\overline{e} \in \overline{E}} a^{-r(\overline{e})} = 1 \qquad (39)$$

Hence
$$\sum_{e_i \in E'} a^{-r(e_i)} < 1 \qquad (40)$$

and however E' is a complete system of eventualities; $\bar{E}$ is the set of quasi-answers.

The quasi-questionnaire is a questionnaire if and only if $\bar{E} = \emptyset$.

We have proved:

Theorem 4 An instantaneous code built on the complete system of eventualities $E=\{e_i\}$ satisfies Kraft's inequality $\sum_{e_i \in E} a^{-r(e_i)} \leq 1$; it is a polychotomic questionnaire in the strict sense if and only if equality holds.

The existence of an instantaneous code for which the rank of any code-word e_i is the integer immediately greater than the logarithm (with base a) of the inverse of the probability of e_i was indicated by Shannon as early as 1948 and studied by Feinstein. It is based on the property converse to the preceding theorem; If Kraft's inequality $\sum_{i=1}^{N} a^{-r_i} \leq 1$ holds then it is possible to construct an instantaneous code with N words of lengths $r_1 \leq r_2 \ldots \leq r_N$.

For, start with a polychotomic arborescence with basis a having a^{r_N} terminals with rank r_N. Take a vertex with rank r_1, then a vertex with rank r_2, and so on until r_{N-1} in such a way that each is neither a descendant nor an ancestor (nor coincident) of the preceding one. Let us contract the sub-arborescences issued from these N-1 vertices.

This is possible because
$$\sum_{i:=1}^{N-1} a^{-r_i} \leq 1 - a^{-r_N},$$

$$\sum_{i:=1}^{N-1} a^{r_N\,r_1} \leq a^{r_N} - 1;$$

the left hand side indicates the number of terminals with rank r_N suppressed in the arborescence from which we started; we can then still place the last terminal at rank r_N.

Now consider the code words with length $r(e_i)$ such that
$$\log \frac{1}{p(e_i)} \leq r(e_i) < \log \frac{1}{p(e_i)} + 1. \qquad (41)$$

The left hand inequality leads to $a^{-r(e_i)} \leq p(e_i)$ and to

Kraft's inequality. We can then construct a *Feinstein code* satisfying (41) on the answers $\{e_i\}$ with probabilities $p(e_i)$; the word e_i is coded with $r(e_i)$ characters.

Multiplying (41) by $p(e_i)$ and summing over the set of $e_i \in E$, where E is the set of words of the code C, we obtain the double inequality

$$I_N(E) \leq \bar{L}(C) < I_N(E) + 1 , \qquad (42)$$

where $I_N(E)$ is Shannon's information of code C expressed with logarithms to the base a and $L(C)$ is the routing length of the quasi-questionnaire C.

This leads to the result known as the *noiseless coding theorem*.

Theorem 5 There exists an instantaneous code C such that

$$\log \frac{1}{p(e_i)} \leq r(e_i) < \log \frac{1}{p(e_i)} + 1$$

for any code word e_i; its routing length $\bar{L}(C)$ is related to the information by

$$I_N(E) \leq \bar{L}(C) < I_N(E) + 1 .$$

We can also write

$$r(e_i) = \lfloor \log \frac{1}{p(e_i)} \rfloor + 1 \qquad (43)$$

and

$$\bar{L}(C) = \sum_{e_i \in E} p(e_i) \left(\lfloor \log \frac{1}{p(e_i)} \rfloor + 1 \right) \qquad (44)$$

where $\lfloor x \rfloor$ is the greatest integer strictly less than x.

Now consider a Feinstein code C, in which the ranks of the words are determined by (41).

If $\log \frac{1}{p(e_i)}$ is an integer for every e_i, then

$$r(e_i) = \log \frac{1}{p(e_i)} \quad (\forall \; i)$$

and consequently $\sum_{e_i \in E} a^{-r(e_i)} = 1$ therefore C is a questionnaire.

If $\log \frac{1}{p(e_i)}$ is not an integer for at least one word e_i then by (41), $r(e_i) > \log \frac{1}{p(e_i)}$, that is:

$$p(e_i) > a^{-r(e_i)} . \qquad (45)$$

Since $p(e_i) \geq a^{-r(e_i)}$, for every word of C, if (45) is true for at least one word, then

$$\sum_{e_i \in E} a^{-r(e_i)} < \sum_{e_i \in E} p(e_i). \text{ Hence } \sum_{e_i \in E} a^{-r(e_i)} < 1;$$

and by Theorem 4, C is a quasi-questionnaire.

Theorem 6 A code C satisfying the inequalities (41) of the noiseless coding theorem is a polychotomic questionnaire in the strict sense if and only if $r(e_i) = \log \frac{1}{p(e_i)}$ ($\forall i$).

Homogeneous questionnaires in the broad sense have in fact a property similar to homogeneous questionnaires in the strict sense if the sub-questionnaire with one question of basis $\beta+1 < a$ has answers with the same probability of the form $a^{-r}_{\overline{\beta+1}}$. It is then necessary for all the other $N-(\beta+1)$ answers e_i forming a sub-set $E_1 \subset E$ to have a probability of the form $a^{-r(e_i)}$, and then the quasi-questionnaire is a homogeneous questionnaire in the broad sense:

Corollary 7 A code C satisfying inequalities (41) of the noiseless coding theorem is a polychotomic questionnaire in the broad sense if and only if the following conditions are satisfied:

$N-(\beta+1)$ is a multiple of $a-1$ and $\beta+1 < a$

$N-(\beta+1)$ of the answers have probabilities of the form $a^{-r(e_i)}$

$(\beta+1)$ of the answers have probabilities of the form $\frac{a^{-s}}{\beta+1}$

$s \geq \underset{e_i \in E_1}{\text{Sup}} r(e_i)$, where E_1 is the set of $N-(\beta+1)$ answers of the first type;

otherwise C is a quasi-questionnaire.

The last condition imposes the usual restrictions on polychotomic L-optimal questionnaires in the broad sense: the outcomes of the question with basis $\beta+1$ are the answers with the smallest probabilities.[1]

8.3.3 Upper Bounds for L-optimal questionnaires

L-optimal questionnaires can be constructed except for an equivalence by means of Huffman's algorithm. But as far as we

---

[1] These results may be compared with Renyi's posthumous article on the theory of information research in which the strict compatibility (relation 4.2) is rediscovered.

know there is no formula giving L_H directly without using this algorithm. Only E. Moore seems to have obtained a result in this direction.

We will study the ranks of the vertices of optimal questionnaires.

We now suppose that the ranks of the answers of the questionnaire K_H are all at least equal to k_0. For any rank $r<k_0$, there are exactly a^r vertices and the sum of the probabilities of the vertices of any rank ($0 \leq r \leq k_0$) is 1.

Let X_r be the set of vertices with rank r:

Case (1) $r \leq k_0$. Then we have

$$|X_r| = a^r \qquad (46)$$

and

$$\sum_{x \in X_r} p(x) = 1 . \qquad (47)$$

If K_H is an L-optimal questionnaire, then moreover:

$$x \in X_r \text{ and } y \in X_{r+1} \Rightarrow p(x) \geq p(y) \qquad (48)$$

and for the predecessor x_0 of y

$$(y \in \Gamma x_0) : p(x_0) > p(y) ; \qquad (49)$$

However these properties are not sufficient to characterize an L-optimal questionnaire.

If in K_H there was a vertex y with rank $r \leq k_0+1$ such that $p(y) \geq \frac{1}{a^{r-1}}$, then by (46), (48) and (49) we would have

$$\sum_{x \in X_{r-1}} p(x) > a^{r-1} p(y) \geq 1 ,$$

for the whole rank r-1 in contradiction with (47).

Consequently

$$r(x) \leq k_0 + 1 \Rightarrow p(x) < \frac{1}{a^{r(x)-1}} \qquad (50)$$

or $r(x) < \log \frac{1}{p(x)} + 1$,

which is the second inequality of (41) generalized to the questions of K.

Case (2) $r>k_0+1$. K_H being still an L-optimal questionnaire.

If for every vertex with rank r-1 we had

$$p(x_{r-1}) \leq a^{-\ell+1},$$

could there be a vertex with rank r+1 such that

$$p(x_{r+1}) \geq a^{-\ell} ?$$

In this case all the vertices with rank r would have probability $p(x_r) \geq a^{-\ell}$ and for at least one question with rank r: $p(q_r) > a^{-\ell}$. The questions of rank r-1 would have probability $p(q_{r-1}) \geq a^{-\ell+1}$, and for at least one question with rank r-1:

$$p(q_{r-1}) > a^{-\ell+1},$$

and this is a contradiction.

Thus we have:

Theorem 8 In an L-optimal questionnaire in which all the answers have rank at least equal to k_0, the probability of all the vertices x with rank $r(x) \leq k_0+1$ is such that $p(x) < \dfrac{1}{a^{r(x)-1}}$; if for $r > k_0+1$ any vertex with rank r-1 is such that $p(x_{r-1}) \leq a^{-\ell+1}$, then all the vertices with rank r+1 have probability $p(x_{r+1}) < a^{-\ell}$.

Theorem 8 allows a comparison of the ranks of the answers in the Feinstein and Huffman Codes.

Let $r_C(e)$ and $r_H(e)$ be the ranks of the same answer, with probability p(e) in these two codes and let: $\inf_{e \in E} r_H(e) = k_0$.

Then

$r_H(e) = k_0 \Rightarrow p(e) < a^{-k_0+1}$ and $r_C(e) \geq k_0$

$r_H(e) = k_0 + 1 \Rightarrow p(e) < a^{-k_0}$ and $r_C(e) \geq k_0 + 1$

$r_H(e) = k_0 + 2 \Rightarrow p(e) < a^{-k_0}$ and $r_C(e) \geq k_0 + 1$

$r_H(e) = k_0 + 3 \Rightarrow p(e) < a^{-k_0-1}$ and $r_C(e) \geq k_0 + 2$

and similarly

$r_H(e) = k_0 + 2h \Rightarrow p(e) < a^{-k_0-h+1}$

and
$$r_C(e) \geq k_0 + h$$

$r_H(e) = k_0 + 2h + 1 \Rightarrow p(e) < a^{-k_0-1}$

and
$$r_C(e) \geq k_0 + h + 1,$$

that is

$$r_H(e) = k_o + 2h \quad \Rightarrow r_c(e) \geq r_H(e) - h .$$

$$r_H(e) = k_o + 2h + 1 \quad \Rightarrow r_c(e) \geq r_H(e) - h .$$

Now let e_1 be the answer with greatest probability:

$$p(e_1) = \operatorname*{Sup}_{e_i \in E} p(e_i) ;$$

Then the rank of e_1 is k_o, that is $r_H(e) = k_o$ and moreover

$$r_c(e_1) = \left\lfloor \log \frac{1}{p(e_1)} \right\rfloor + 1 .$$

The above inequalities then lead to

$$k_o \leq \left\lfloor \log \frac{1}{p(e_1)} \right\rfloor + 1 .$$

Suppose further that

$$k_o \leq \left\lfloor \log \frac{1}{p(e_1)} \right\rfloor - 1$$

that is: $p(e_1) \leq a^{-k_o-1}$;

Since the questionnaire is L-optimal, the vertices with rank k_o+1 have a probability:

$$p(x_{k_o+1}) \leq p(e_1) \leq a^{-k_o-1}$$

and the questions with rank k_o:

$$p(q_{k_o}) \leq a^{-k_o}$$

in such a way that

$$\sum_{x \in X_{k_o}} p(x) \leq (a^{k_o} - 1)a^{-k_o} + a^{-k_o-1} < 1$$

which is impossible.

Thus

$$k_o = \left\lfloor \log \frac{1}{p(e_1)} \right\rfloor + \delta$$

where $\delta = 0$ or 1 and k_o has no other possible value (see Fig. 6.3). This allows us to state:

Theorem 9 In an L-optimal questionnaire K_H the answer with greatest probability e_1 has rank $r_H(e_1)=k_o$, where

$$k_o = \left\lfloor \log \frac{1}{p(e_1)} \right\rfloor \quad \text{or} \quad k_o = \left\lfloor \log \frac{1}{p(e_1)} \right\rfloor + 1 \;;$$

for $k \geq 0$, the ranks of the answers $r_H(e)$ are related to the ranks $r_C(e)$ of Feinstein's code by

$$\left. \begin{array}{l} r_H(e) = k_o + 2h \\ \text{or} \\ r_H(e) = k_o + 2h + 1 \end{array} \right\} \Rightarrow r_C(e) \geq r_H(e) - h \qquad (51)$$

The estimation of a bound for the routing length of the optimal questionnaire results from the fact that

$$L_H \leq L' \leq \overline{L}(C)$$

where L_H is the routing length of an L-optimal questionnaire

$\overline{L}(C)$ is the routing length of a Feinstein code

L' is the routing length of a questionnaire K' deduced from Feinstein's code.

all these three (quasi) questionnaires being built on the same pair $(a, \mathcal{P})$.

In addition we can enclose the preceding inequalities between informations

$$I_N(E) \leq L_H \leq L' \leq \overline{L}(C) < I_N(E) + 1 . \qquad (52)$$

Inequalities (44) and (52) then enable us to assert that

$$L_H < \sum_{e_i \in E} p(e_i) \left(\left\lfloor \log \frac{1}{p(e_i)} \right\rfloor + 1 \right) , \qquad (53)$$

whenever at least one answer has a probability which is not a power of $\frac{1}{d}$ (according to Property 13 and Theorem 6).

Now suppose that this condition is realized. Then the code C is a quasi-questionnaire and by Property 13, there exists indeed a questionnaire K' such that $L' < \overline{L}(C)$.

The evaluation of (43) needed by (53) requires the determination, for any answer e_i, of

$$c_i = \left\lfloor \log \frac{1}{p(e_i)} \right\rfloor + 1 .$$

Put $\sup_i c_i = c_N$ and

$$\sum_{e_i \in E} a^{-c_i} = 1 - \rho \qquad (54)$$

where

$$\rho = \sum_{\overline{e}_j \in \overline{E}} a^{-r(\overline{e}_j)}$$

will be called the *residue corresponding to the quasi-answers* of C.

ρ is the sum of $\overline{N}=|\overline{E}|$ powers (distinct or not) of $\frac{1}{a}$ corresponding to the quasi-answers of the quasi-questionnaire C. These powers can be obtained explicitly by means of (22): we use a number system with base a to express $\sum_{e_i \in E} a^{-c_i}$ and it will then be possible to write

$$\rho = \sum_{r_i := 1}^{N} t_i a^{-r_i}$$

with $0 \leqslant t_i < a$; $\overline{N}$ will then be the sum of the non-zero digits $\overline{N} = \sum_{i:=1}^{cN} t_i$.

Examples

1. If $\mathcal{P} = \{0.90: 0.10\}$, then $c_1=1$ and $c_2=4$;

$\sum_{e_i \in E} 2^{-c_i} \Rightarrow 0.1001$ in such a way that $\rho=0.0111$:

Thus $\overline{N}=3$ and the three quasi-answers have ranks 2, 3 and 4.

2. Similarly for $a=3$,

$$\mathcal{P} = \{\frac{32}{100}, \frac{25}{100}, \frac{20}{100}, \frac{15}{100}, \frac{5}{100}, \frac{2}{100}, \frac{1}{100}\} ,$$

(Fig. 8.9),

we find $c_i: = 2,2,2,2,3,4,5$.

Thus $\sum_{i=1}^{N} 3^{-c_i} \Rightarrow 0.11111 = 1-\rho$, because $4 \times 3^{-2} = 3^{-1} + 3^{-2}$

and $\rho = 0.11112$. Hence $\overline{N} = \sum_{i=1}^{cN} t_i = 6$.

We will denote the ranks of the quasi-answers by $\bar{r}_j$, where j runs from 1 to $\bar{N}$.

A quasi-questionnaire K_1 will arise from C by a succession of $\bar{N}$ permutations (at most) between answers and quasi-answers, each being followed possibly by a contraction in the case where a vertex which is not an endpoint would be formed with zero probability.

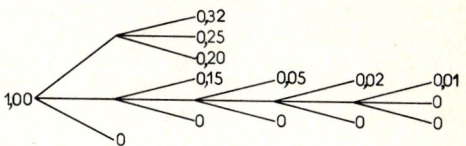

Figure 8.9

The maximal rank of an answer from C is

$$c_N = \bar{r}_N = 1 + \underset{e_i \in E}{\text{Max}} \left\lfloor \log \frac{1}{p(e_i)} \right\rfloor \qquad (55)$$

and corresponds to an answer, e_N, which has the smallest probability p_N.

If e_N is the only answer with rank c_N, then before any permutation, we can replace the quasi-question from which e_N comes out by an answer with probability p_N; e_N then has rank c_N-1 and the reduction in the routing length is p_N (recursive operation).

If the difference or rank between e_N and the answer with immediately greater probability e_{N-1} is c_N-c_N' we will then have reduced the routing length by $p_N(c_N-c_N')$ and reduced the number of quasi-answers by $(a-1)(c_N-c_N')$.

Now suppose that $c_N=c_N'$ and let c_N be the rank of e_N in C. If there exists a quasi-answer with rank $\bar{c}_j<c_N$, then the permutation of e_{N-1} with this quasi-answer gives a reduction of $(c_N-\bar{c}_j)p(e_{N-1})$. But we can increase this reduction by performing a permutation between the quasi-answer with minimal rank $\bar{c}_0$ and the answer with maximal probability among the answers with rank greater than $\bar{c}_0$; then in the same way for the other phases (j) after substitution of a new value for $\bar{c}_0$ (greater than or equal to the old one.

This algorithm can be formulated as follows:

- Let $\bar{c}_0(j)$ be the smallest rank of a quasi-answer in the course of phase j (for j starting from 1). Then the permutation of a quasi-answer with rank $\bar{c}_0(j)$ for the answer with the greatest probability having a rank c_i greater than $\bar{c}_0(j)$ causes a reduction of

$$\underset{c_i > \bar{c}_0(j)}{\text{Max}} \{p(e_i)\} \times (c_i - \bar{c}_0(j))$$

- At the following step (j+1) we substitute $\bar{c}_0(j+1)$ for

$\bar{c}_o(j)$, where $\bar{c}_o(j+1) \geq \bar{c}_o(j)$.

The algorithm will stop at phase t when $c_i \leq \bar{c}_o(t)$ for every answer.

By this means we obtain a quasi-questionnaire $\bar{K}_1$ possessing $\bar{N}_1$ quasi-answers and with routing length

$$L_1 = \bar{L}(C) - \sum_{j:=1}^{t} (c_i - \bar{c}_o(j)) \underset{c_i > \bar{c}_o(j)}{\text{Max}} \{p(e_i)\} \quad (56)$$

The answers of C and of K_1 have Property 6.2.

The answers ordered according to non-decreasing probabilities have non-decreasing ranks.

The questionnaire K' preserves this property; K' is deduced from K_1 according to Property 8.13.

Furthermore the routing length of K' is less than that of K: whenever (a-1) quasi-answers have been suppressed then an answer with probability $p_i > p_N$ can receive a rank less by 1.

Therefore

$$L' < L_1 - (\bar{N}_1 \cdot / \cdot (a - 1)) p_N \quad (57)$$

This formula could be improved either by writing $\bar{N}_1$ in polynomial form

$$\bar{N}_1 = \Sigma \gamma_\nu (a - 1)^\nu$$

and taking into account the successive remainders, or by substituting for p_N the $(\bar{N}_1 \cdot / \cdot (a-1))$ smallest probabilities among the $p(e_i)$.

From (56) and (57) we deduce:

Property 14 An upper bound for the routing length of an L-optimal polychotomic questionnaire is given by

$$L_H \leq \bar{L}(C) - \sum_{j:=1}^{t} (c_i - \bar{c}_o(j)) \underset{c_i > \bar{c}_o(j)}{\text{Max}} \{p(e_i)\}$$

$$- p_N (\bar{N}_1 \cdot / \cdot (a - 1)) \quad (58)$$

where $\bar{L}(C)$ is defined by (44).

$\bar{N}_1$ is the number of quasi-answers of the quasi-questionnaire $\bar{K}_1$ all of whose quasi-answers have maximal rank; p_N is the smallest probability of $\mathcal{P}$; the ranks $c_i, \bar{c}_o(j)$ and t have been explicitly stated in (56).

EXERCISES

1. Indicate the ranks of the code-words of the quasi-questionnaires built according to the algorithm of Feinstein's code for

$a = 2$ and $\mathcal{P} = \{0.425; 0.25; 0.08125$ (four times)$\}$

$a = 3$ and $\mathcal{P} = \{\frac{1}{3}, \frac{1}{9}, \frac{1}{9}, \frac{4}{135}$ (fifteen times)$\}$

2. Give an upper bound to the routing length of the L-optimal questionnaires built on the pairs

$a = 2$ and

$$\mathcal{P} = \{\frac{50}{100}, \frac{12}{100}, \frac{12}{100}, \frac{10}{100}, \frac{7}{100}, \frac{5}{100}, \frac{3}{100}, \frac{1}{100}\} \; ;$$

$a = 3$ and

$$\mathcal{P} = \{\frac{22}{45}, \frac{1}{5}, \frac{1}{15} \text{ (three times)}, \frac{1}{54} \text{ (six times)}\} \; .$$

3. Let Q be the dichotomic balanced questionnaire having for answers the vertices indexed by [r,s] and characterized by the probabilities

$p(2,0) = 0.40, \; p(2,1) = 0.11, \; p(2,2) = 0.36, \; p(2,3) = 0.09$.

Show that Q can be decomposed into a product of two independent elementary questionnaires A and B such that $I(Q)=I(A)+I(B)$. Determine A and B.

4. Let R be the balanced dichotomic questionnaires having for answers the vertices with probabilities:

$p(2,0) = 0.40, \; p(2,1) = 0.15, \; p(2,2) = 0.40, \; p(2,3) = 0.05$.

Show that R cannot be considered as the product of two questionnaires. Find three elementary questionnaires C, D, K such that $(K \diamond C) \diamond D = R$.

Deduce that the questions of rank 1 in R do not represent the same experiment. Then put

$$I(R) = I(A) + I(B/A)$$

and relate the experiments A with outcomes A_0 and A_1 and B/A_0, B/A_1 to the questionnaires K, C and D.

Chapter Nine

CONDITIONING OF THE QUESTIONS AND ANSWERS

9.1 Limitations and extensions

The general title of this chapter indicates the transition between the theory of L-optimal questionnaires and applications to concrete cases where it is not possible to realize such questionnaires and where there may exist limitations among the outcomes of a question i of a questionnaire. These limitations may be related to a kind of conditioning of an answer (formalized in paragraph 2 as a utility function) or a question (formalized in paragraph 3 as a cost). A binding between the cost associated with a question and the basis of the question will also be introduced (paragraph 3).

In all these cases we will define a length generalizing the routing length and it is this length which it will be interesting to minimize and eventually to compare with an information. Campbell questionnaires will allow an extension to Renyi's information of the properties established in the case of L-optimal questionnaires for Shannon's information (paragraph 4). Infinite questionnaires will afford the occasion to discuss problems related to graphs with circuits and to define flow charts with very general applications in data processing. Next we will define formable questions and realizable questionnaires by allowing only certain events, that is by restricting the set of events associated with the questions of an arborescent questionnaire to a proper subset of $\mathfrak{P}(E)$.

This approach will allow us to present a collection of tools adapted to the realization of questionnaires used in operational sciences (data processing, human sciences, applied statistics, operations research ...)

9.2 Utilities of the Answers

9.2.1 Useful length

Definition 1 A *utility* is a valuation of the answers $\mathfrak{U}: E \to R^+$ such that there exists an answer e for which $u(e) \neq 0$. The product $p(e)u(e)=v(e)$ defined for any answer of a questionnaire admitting a utility will be called the *preference of* e.

A questionnaire with utility is therefore a quadruple (X, $\Gamma, P_\Gamma, \mathfrak{U}$).

We will suppose throughout §9.2 that Q is an *arborescent questionnaire*. However the results may be generalized to latticoid questionnaires without any great difficulty.

In a questionnaire (with utility), the mathematical

expectation

$$L_u = \frac{\sum_{e \in E} u(e)p(e)r(e)}{\sum_{e \in E} u(e)p(e)} \qquad (1)$$

will be called the *useful length of the questionnaire*.

A question i of an arborescent questionnaire all of whose outcomes are answers has probability

$$p(i) = \sum_{j \in \Gamma i} p(j) \text{, where } \Gamma i \subset E.$$

We may therefore define the *preference* of such a question to be

$$v(i) = \sum_{j \in \Gamma i} v(j) = \sum_{j \in \Gamma i} p(j)u(j) \qquad (2)$$

and the utility of i to be

$$u(i) = \frac{v(i)}{p(i)}. \qquad (3)$$

In the same way as we determined $p(i)$ for any $i \in F$ in a questionnaire in which the answers were not valuated by utilities it is possible to determine $v(i)$ and $u(i)$ for any $i \in F$ by means of (2) and (3).

Note that if $u(j)$=constant, for $j \in \Gamma i$, then $u(i)=u(j)$.

The preference of the root is $v(\alpha) = \sum_{e \in E} u(e)p(e)$, that is the mean value of the utilities.

Normalize by setting

$$w(x) = \frac{v(x)}{v(\alpha)} \quad \text{for} \quad x \in X, \qquad (4)$$

that is for $i \in F$:

$$w(i) = \frac{\sum_{j \in \Gamma i} p(j)u(j)}{\sum_{e \in E} p(e)u(e)}$$

for $e \in E$

$$w(e) = \frac{p(e)u(e)}{\sum_{e \in E} p(e)u(e)}$$

in such a way that

$$w(i) = \sum_{e \in E(i)} w(e) \text{ and } \sum_{e \in E} w(e) = w(\alpha) = 1. \qquad (5)$$

Now consider the triple $Q_w(X,\Gamma,W)$, where (X,Γ) is the support of an arborescence and W a valuation $W: X \to [0,1]$. It is a questionnaire, a quasi-questionnaire or a probabilized arborescence according as u(e) possesses zero, one or many zeros, and the problem of determining the useful length of $Q=(X,\Gamma,P_\Gamma)$ admitting $\mathfrak{A}$ is brought back to that of the routing length of Q_W.

The useful length of Ω_W takes the two forms analogous to those of the routing length (formulae 4.32 and 4.36):

$$L_u = \sum_{e \in E} w(e) \, r(e) , \qquad (6)$$

and

$$L_u = \sum_{i \in F} w(i) , \qquad (7)$$

in such a way that the substitution of w(i) for p(i) enables us to deduce the algorithm for minimizing L_u from Huffman's algorithm:

Property 1 Huffman's algorithm enables us to construct the preferable, or L_u-optimal questionnaire in which L_u is minimal for fixed ($\mathcal{A}$, $\mathcal{P}$, $\mathfrak{A}$) (homogeneous or heterogeneous case).

The vertices with smallest rank are those for which w is maximal; we will say that they are the *preferable vertices*.

Example Consider the dichotomic questionnaire defined by

$$p(e) := \frac{3}{8} ; \frac{1}{4} ; \frac{1}{8} ; \frac{1}{8} ; \frac{1}{16} ; \frac{1}{16}$$

$$u(e) := 150 ; 50 ; 10 ; 200 ; 5 ; 100$$

then
$$v(e) := 56.25 ; 12.5 ; 1.25 ; 25 ; 0.3125 ; 6.25$$

and $\qquad v(\alpha) = 101.5625$.

The preferable questionnaire Q_P constructed by means of Huffman's algorithm (with the w) and shown in Fig. 9.1 with the vertices coded in the form $v(i)|p(i)$ has useful length L_u=1.738 and usual routing length L=2.5625.

Note that L is greater than the routing length of the L-optimal questionnaire Q_H constructed on the same distribution of answers.

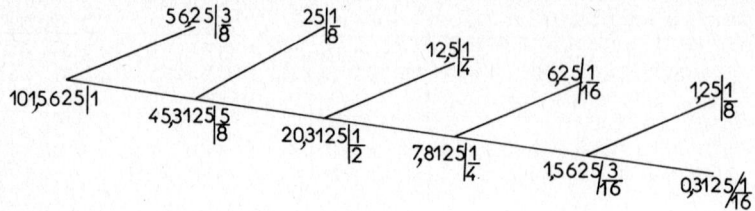

Figure 9.1

Among all the L-optimal questionnaires with supports not isomorphic to Q_P, consider the one with minimal height (Fig. 9.2) constructed by using Remark 4 of §6.4 relative to questions with probability equal to that of an answer. Its useful length is $L_u(Q_H) = 2.323$, and its routing length is $L(Q_H) = 2.375$.

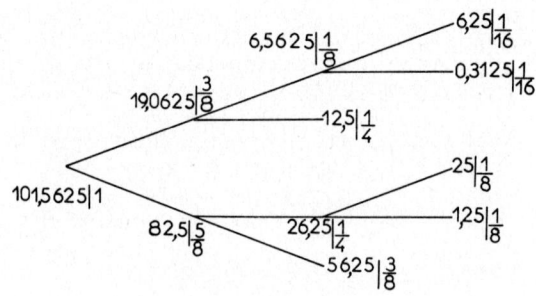

Figure 9.2

9.2.2 Useful information

We will proceed to the informational study of questionnaires with utility by introducing the "quantitative and qualitative" information of Belis and Guiasu.

Consider a set of N events admitting the complete probability distribution $\mathcal{P} = (p_1, \ldots, p_N)$ and the valuation $\mathcal{U} = (u_1, \ldots, u_N)$ and let $G(p,u)$ be the useful information of Belis-Guiasu obtained for an event characterized by the pair (p,u). The axioms are:

(G1) $G(p \cdot q, u) = G(p,u) + G(q,u)$;

(G2) $G(p, \lambda \cdot u) = \lambda \cdot G(p,u)$;

(G3) $G(p,u)$ is a continuous function of $p \in [0,1]$.

(G1) expresses the stochastic independence for constant utility whereas (G2) expresses the proportionality of G to u. Belis and Guiasu have shown that the information relative to an event is necessarily of the form

$$G(p,u) = k\, u\, \log \frac{1}{p}. \qquad (8)$$

For

for any $u \geq 0$ and $0 \leq p, q \leq 1$, (G2) implies

$$G(p,u) = u\, G(p,1),$$

and (G1) implies:

$$G(pq,1) = G(p,1) + G(q,1);$$

Putting $G(p,1) = F(\log p)$ we have

$$F(\log p + \log q) = F(\log p) + F(\log q),$$

a functional equation which has a unique continous solution $F(\log p) = k' \log p$, where k' is an arbitrary constant.

In the case of a complete system of events determined by $(\mathcal{P}, \mathfrak{U})$ we have

$$G_N(\mathcal{P}, \mathfrak{U}) = \sum_{j:=1}^{N} p_j G(p_j, u_j) = k \sum_{j:=1}^{N} p_j u_j \log \frac{1}{p_j}. \qquad (9)$$

If $\mathfrak{U}$ is a constant mapping ($u_j = u_0, \forall j$), then except for the factor $k u_0$, we find Shannon's information once again.

If $\mathcal{P}$ is a constant distribution, then

$$G_N(\{\tfrac{1}{N}, \ldots, \tfrac{1}{N}\}, \mathfrak{U}) = k\, v(\alpha) \log N.$$

The constant k can be fixed as usual by the choice of the base of the logarithm; we can then normalize by means of $v(\alpha)$. We will fix k by $k\,v(\alpha)=1$ and take the logarithm to the unique base, in the case of a polychotomic questionnaire.

Property 2 The useful information, satisfying the axioms (G1), (G2), (G3) and $k v(\alpha)=1$ is the function:

$$G_N(\mathcal{P}, \mathfrak{U}) = \sum_{e \in E} w(e) \log \frac{1}{p(e)} \qquad (10)$$

Belis and Guiaşu's information appears as a Watanabe type information (cf. §7.5.2.1).

It is now possible to associate Belis and Guiasu's information with any question of a questionnaire with utility on the answers by considering the elementary questionnaire formed by one question and its outcomes, that is:

$$G(i) = \sum_{j \in \Gamma i} \frac{v(j)}{v(i)} \log \frac{p(i)}{p(j)} \qquad (11)$$

The *useful information* of Q will then be written as

$$G(Q) = \sum_{i \in F} w(i) G(i) ,$$

that is

$$G(Q) = \sum_{i \in F} \sum_{j \in \Gamma i} w(j) \log \frac{p(i)}{p(j)} \qquad (12)$$

and we can show that (12) is equivalent to (10) if E is the set of answers of the arborescent questionnaire Q built on $(\mathcal{P}, \mathfrak{U})$.

In the case when $u(e) = u_0 = $ constant, we find:

$$G(Q) = J(Q) \quad \text{and} \quad G(i) = J(i)$$

and then
$$G(Q) \leq L_u(Q) .$$

However this inequality does not hold for every valuation $\mathfrak{U}$.

Consider for example a question i whose two outcomes j_1 and j_2 have different utilities. Property 7.4 does not enable us to compare

$$\sum_{j \in \Gamma i} \frac{v(j)}{v(i)} \log \frac{p(i)}{p(j)} \quad \text{and} \quad \sum_{j \in \Gamma i} \frac{p(j)}{p(i)} \log \frac{p(i)}{p(j)} ,$$

so that G(i) can be greater than 1. However an elementary questionnaire of which i is the root and the only question has useful length equal to 1.

The inequality $J(i) \leq 1$ which implies $J(Q) \leq L(Q)$ cannot be generalized because G(i) is not bounded above by 1.

Examples Let:

$$\mathcal{P} = \{3/4, 1/4\} , \quad \mathfrak{U} = \{8/9, 4/3\} , \quad \mathfrak{U}' = \{2/3, 9/5\}$$

and consider the elementary questionnaires Q built on $(\mathcal{P}, \mathfrak{U})$ and Q' built on $(\mathcal{P}, \mathfrak{U}')$.

Then $G(Q) = 0.943 < L_u(Q) = L_u(Q') = 1 < G(Q') = 1.166$.

S. Guiaşu and myself have studied the divergence between $G(Q)$ and $L_u(Q)$ for questionnaires with utility which are polychotomic in the strict sense and we have established:

Property 3 The lower bound of the useful length of a polychotomic questionnaire in the strict sense is

$$G(Q) + \log u(\alpha) - \sum_{e \in E} w(e) \log u(e) .$$

More precisely the lower bound is at most equal to G(Q) and is equal to it if and only if the utilities of the answers are constant ($u(e)=u(\alpha)$, $\forall\ e$); moreover the lower bound is attained if and only if the ranks of the answers are $r_i = a^{-w_i}$ a result which generalizes Theorem 8.1.

For the elementary questionnaires Q and Q' we find respectively
$$L_u(Q) \geqslant 0.918 \quad \text{and} \quad L_u(Q') \geqslant 0.998 \ .$$

Applications Application of the useful information G_N facilitates informational evaluation by taking into account the qualitative aspect brought by the utility function besides the purely stochastic aspect related to $\mathcal{P}$. This answers the objection raised in §7.5.2.1.

As soon as the distribution $\mathcal{P}$ is imprecise, if the statistical data are not sufficiently dense for example, then G is all the more interesting because the utilities of the answers are more strongly differentiated.

In particular if $\mathcal{P}$ is sensibly uniform, then the role of $\mathfrak{U}$ can become predominant: the order of the preferences stays close to the order of the utilities and an error in the evaluation of $\mathcal{P}$ causes only a small modification in the structure of the preferable questionnaire.

Moreover there exists a great variety of problems where the utility of the different answers can become predominant; the case of medical diagnosis furnishes a good example of this because here it is important to deal more often with the "most dangerous" rather than with the "most probable". Since the values of u(e) are not bounded, the choice of the valuation $\mathfrak{U}$ can be very subjective and it is necessary to fix it with the greatest possible objectivity.

Remark Let $\overline{E}$ be the set of answers with zero utility and let E* be its complementary set in E. Then:
$$p(\alpha) = \sum_{e \in \overline{E}} p(e) \ , \ v(\alpha) = \sum_{e \in E^*} p(e)\ u(e)$$

and $0 \leqslant w(i) \leqslant 1$ for any $i \in F$. E* forms an incomplete system of eventualities but L_u and G can be defined without any difficulty and Huffman's construction reduces to this case: it suffices to start with the w(e) for $e \in E^*$.

9.3 Cost of the Questions

9.3.1 Costs and expenses

It is proposed to construct arborescent questionnaires defined on a triple ($\mathcal{A}$, $\mathcal{P}$, C), where $\mathcal{A}$ is the set of bases of questions, $\mathcal{P}$ is the probability distribution of the answers and C is a set of real numbers.

Definition 2 The *cost of question i* of an arborescent questionnaire is some strictly positive real number c(i); we write

$C = \{c(i) \mid i \in F\}$.

Definition 3 The *expense of a vertex* x of a questionnaire is the sum of the costs of its proper ancestors: $D(x) = \sum_{h \in A(x)} c(h)$,

Since the root α has no proper ancestors we set $D(\alpha) = 0$.

A *questionnaire with costs* is a quadruple (X, Γ, P_Γ, C). We will define several different classes of questionnaires with costs corresponding to various supplementary conditions.

a. For fixed $(\mathcal{A}, \mathcal{P}, C)$ the most general class of questionnaires with costs is generated by the set of $M!$ possible allocations of the elements of C to the M questions of any arborescent questionnaire built on $(\mathcal{A}, \mathcal{P})$. In this case we say that *the costs are free*.

b. A more restricted class is formed by the allocation of a unique cost $c(i)$ to the question with basis $a(i)$ $(i \in F, a(i) \in \mathcal{A})$. In this case we say that the cost is linked or bound to the basis. Let K be this class.

c. A third class is formed by imposing the condition that the cost is a function of the basis. For example we could take $c(i) = \log_2 a(i)$.

These classes, introduced and studied respectively by P. P. Parkhomenko and S. Petolla, G. Petolla, G. T. Duncan, will form the subject of the following paragraphs; Taboy has studied a special case of the third class in the light of certain sorting problems.

The construction of a questionnaire with costs can be achieved in two steps:

1. Construct a questionnaire without cost

2. (a) For free costs we choose any permutation

$$\begin{pmatrix} 1 & 2 & \cdots & M \\ \tau(1) & \tau(2) & \cdots & \tau(M) \end{pmatrix}$$

and we take the M ordered pairs $(a(i), c_{\tau(i)})$.

(b) for binding bases-costs only one permutation will be imposed and for an order on $\mathcal{A}$ such that $a(1) \leq a(2) \ldots \leq a(M)$ the inequality $a(i) \leq a(j)$ will not imply *a priori* any comparability between $c(i)$ and $c(j)$.

(c) for costs as functions of the bases, only one permutation will be admissible and moreover

$$a(i) = a(j) \Rightarrow c(i) = c(j) .$$

Definition 4 The *cost of the questionnaire* $Q = (X, \Gamma, P_\Gamma, C)$ is the

CONDITIONING OF THE QUESTIONS AND ANSWERS

mathematical expectation of the expense of the answers:

$$L_C = \sum_{e \in E} p(e) D(e) . \tag{13}$$

When $c(i)=c_0 (\forall i \in F)$, then $D(i)=c_0 r(i)$ so that $L_C(Q)=c_0 L(Q)$.

If C is not uniform, then a proof analogous to that of §4.5 shows that

$$L_C = \sum_{i \in F} p(i) c(i) . \tag{14}$$

9.3.2 Free costs

The determination of an arborescent questionnaire with minimal cost also called C-optimal, among the questionnaires admitting the same triple ($\mathcal{A}$, $\mathcal{P}$, C) proceeds by reduction rules generalising those of Chapter 6.

We will perform substitutions of arcs as above but it will also be possible here to effect a supplementary transformation preserving the support (X,Γ) and the valuation P_Γ: the transfer T_1 effects a permutation of the costs $c(i)$ and $c(i')$ of two questions:

The cost L_C^T of the questionnaire deduced by T_1 from the questionnaire with cost L_C is:

$$L_C^T = L_C + (p(i) - p(i'))(c(i') - c(i)) \tag{15}$$

which sometimes causes a reduction in the cost.

The transfer T_0 performs substitutions of arcs in such a way as to exchange the bases of two vertices.

The transfer T_2 effects an exchange of arcs with terminal extremities i and i' without modifying the probabilities nor the costs of i and i', it leads to

$$L_C^T = L_C + (p(i) - p(i'))(D(i') - D(i)) .$$

The use of these three transfers enables us to give reduction rules for the cost generalizing the rules given in Chapter 6.

With each question, we associate the sum

$$\overline{D}(i) = D(i) + c(i)$$

which we will call *the total expense*.

Property 4 The vertices of a C-optimal questionnaire satisfy the rules C0, C1, C2 and C3.

C0: questions ordered according to non-decreasing total expenses have non-increasing bases,

C1: questions ordered according to non-increasing probabilities have non-decreasing costs.

C2: vertices ordered according to non-increasing probabilities have non-decreasing expenses.

C3: the difference between the expenses of vertices with the same probability is never greater than c_1 the maximum cost in C.

The transfer operations T1 and T2 enable us to reduce L_C; if it is not possible, we can proceed to a reordering of the vertices in such a way as to group together, for any sub-set formed with vertices with the same expense, the vertices of smallest probabilities. We operate in such a way that the vertices with the smallest probabilities are the outcomes of the question with maximal rank and maximal total expense.

Thus the necessary conditions for C-optimality generalize those for L-optimality. In a questionnaire without cost we had considered the minimal question with maximal rank σ: $a_\sigma = \min_{a \in \mathcal{A}} a$ and p_σ is the sum of the a_σ smallest elements of $\mathcal{P}$. In the case of costs we will use the minimal question with maximal rank and with maximal cost σ:

$$c_\sigma = \max_{c \in C} c, \quad a_\sigma = \min_{a \in \mathcal{A}} a .$$

We speak of condensation (cf. Def. 6.3) when we contract the set of vertices Γ_σ, where σ is a minimal question with maximal rank and maximal cost. The Sufficient Condition (Property 6.7) for L-optimality can be generalized as follows:

Property 5 A sufficient condition for a questionnaire Ω to be C-optimal is that any sub-questionnaire deduced from Ω by a series of condensations possesses a minimal question with maximal cost.

These properties of C-optimality most of which were discovered independently by S. Petolla and Parkhomenko have enabled S. Petolla to generalize Huffman's algorithm to questionnaires with costs.

Property 6 The C-optimal questionnaire with free costs is determined by Huffman's algorithm as follows:

1. Construct the L-optimal questionnaire on ($\mathcal{A}$, $\mathcal{P}$) by Huffman's algorithm for heterogeneous questionnaires.

2. Allocate to the questions ordered by non-increasing probabilities the costs of C in non-decreasing order.

We remark that, by doing this, we obtain:

Property 7 $\min \sum_{i \in F} p(i)c(i)$ is obtained for the same arborescence and the same probabilities of questions p(i), (i ∈ F) as $\min_{i \in F} \Sigma\, p(i)$ when c(i) placed in non-increasing order are allocated to the probabilities p(i) placed in non-decreasing order, the minima being taken on the sets of questionnaires built on ($\mathcal{A}$, $\mathcal{P}$, C) and ($\mathcal{A}$, $\mathcal{P}$).

This property generalizes the inequality pointed out by Hardy, Littlewood and Polya:

Given two series

and
$$p_1 \geq p_2 \geq \ldots \geq p_M$$
$$c_1 \leq c_2 \leq \ldots \leq c_M,$$

then
$$\sum_{j=1}^{M} p_j c_j = \min \sum_{j=1}^{M} p_j c_{\tau(j)},$$

where $\tau = \begin{pmatrix} 1 & 2 & \ldots & M \\ \tau(1) & \tau(2) & \ldots & \tau(M) \end{pmatrix}$ is a permutation of the M indices $\tau(j)$.

Note that the c_j are determined by C whereas the p_j are obtained by Huffman's construction relative to the pair ($\mathcal{A}$, $\mathcal{P}$).

Remarks Whereas the utility function enables us to generate a family of questionnaires starting from the triple ($\mathcal{A}$, $\mathcal{P}$, $\mathcal{U}$), where $\mathcal{P}$ and $\mathcal{U}$ were associated with the answers and $\mathcal{A}$ with the questions, the questionnaires with free cost associated $\mathcal{P}$ with the answers and $\mathcal{A}$ and C with the questions. The fundamental difference between these two extensions of questionnaires is due in reality to the binding imposed on $\mathcal{P}$ and $\mathcal{U}$ (that is, that p_e and u_e are allocated to the same answer $e \in E$) whereas no binding is imposed between $\mathcal{A}$ and C; the extensions of Huffman's algorithm are then made by working on w in one case, on p in the other but we must afterwards, only in the second case, allocate the cost to the questions already constructed.

We can relate the expenses of the vertices to their utilities by noting however that the generalised Huffman's algorithm tends on the one hand to increase the rank of the answers with great expense and on the other to reduce the rank of answers with great utility.

Most of the problems originating from physical data processing or human sciences making use of questionnaires with costs impose a binding between the a(i) and the c(i) (i ∈ F) in such a way that they cannot in general be constructed by Huffman's algorithm. This case does however give a lower bound for the cost of questionnaires built on ($\mathcal{A}$, $\mathcal{P}$, C).

9.3.3 Binding of the Costs to the Bases

The costs like the bases, need not all be distinct and it is necessary to determine a questionnaire with minimal cost in a family admitting a given binding bases-costs. We will still speak of a C-optimal questionnaire and the binding will be characterized by a bipartite graph representing the permutation

$$\begin{pmatrix} 1 & \ldots & j & \ldots & M \\ \tau(1) & \ldots & \tau(j) & \ldots & \tau(M) \end{pmatrix}$$

and having M arcs with no extremities in common.

Given a questionnaire $Q=(X,\Gamma,P_\Gamma)$ belonging to K, it is possible to associate the following three preorderings with the set of questions $\{q_1, q_2, \ldots, q_n\} \subset X$:

0_1 the complete preordering defined according to the bases:

$$q_i \underset{1}{\prec} q_j \Leftrightarrow a_i \leq a_j$$

0_2 the complete preordering defined according to the costs:

$$q_i \underset{2}{\prec} q_j \Leftrightarrow c_i \geq c_j$$

0_3 the complete preordering defined according to the total expense

$$q_i \underset{3}{\prec} q_j \Leftrightarrow D_i + c_i \geq D_j + c_j .$$

In addition it is also possible to obtain a maximal partial ordering compatible with 0_1 and 0_2. This is the partial ordering $0_4'$ included in $0_1 \cap 0_2$ which is obtained by suppressing $q_i \underset{1}{\prec} q_j$ and $q_j \underset{2}{\prec} q_i$ when we have simultaneously $a_i < a_j$ and $c_i < c_j$. Because these two inequalities are incompatible in a C-optimal questionnaire with free costs.

G. Petolla has proved:

Property 8 There always exists a C-optimal questionnaire in K which contains a question q satisfying the following three criteria:

 - q is minimal with maximal cost for the partial ordering 0_4;

 - q is minimal for the preordering 0_3 of the total expenses.

 - the successors of q are the answers with smallest probability.

This leads to the optimality theorem:

Theorem 1 For any probability distribution $\mathcal{P}$, there exists a complete order Ω of the questions compatible with the partial order 0_4 included in $0_1 \cap 0_2$ such that any questionnaire built by an algorithm derived from that due to Huffman by a choice of minimal questions according to the three criteria of Property 8, is C-minimal.

Thus it is possible to reduce the exhaustiveness of the search for the C-optimal questionnaire by using the ordering 0_4. However we must carry out the following four steps:

1. determine the preorderings 0_1 and 0_2 and an ordering 0_4 included in $0_1 \cap 0_2$,

2. determine all the complete orderings Ω on F compatible with 0_4,

3. construct a questionnaire derived from that due to Huffman by using the concept of minimal question in the sense of Property 8 for each of the complete orderings,

4. select the questionnaire (or questionnaires) with minimal cost L_C.

Remarks The ordering 0_4 can be represented by an order graph (F, φ) whose set of vertices is the set of questions and in which the number of arcs can be very small: we can conceive intuitively that a question with basis a will often have a greater cost than a question with basis b<a, which corresponds to a contradiction between 0_1 and 0_2.

If the number of arcs of (F, φ) is large, then the number of distinct complete orders Ω is small. This can be the case if the number of distinct bases is small, for example when the order of $\mathcal{A}$ is 2, 3, 4 or 5, because for $a_i = a_j$, there is no incompatibility between 0_1 and 0_2 in such a way that $0_1 \cap 0_2$ generates a sub-graph admitting, for constant a, a large number of arcs.

A heuristic - based in reality on the consideration of questionnaires with cost function of the basis (cf. further §9.3.4) - has been suggested by Y. Cesari and G. Petolla with a view to determining a complete order ω :

$$q_i \underset{\omega}{\leq} q_j \Leftrightarrow \frac{c_i}{\log a_i} \geq \frac{c_j}{\log a_j} . \qquad (16)$$

This order ω which proceeds also from the same ideas as heterogeneous information (Definition 8.5) does not ensure a direct determination of the C-optimal questionnaire but it does enable us to obtain directly and without the least exhaustive research an upper bound for the minimal cost L_C by taking the binding into account.

G. Petolla's example

$$\mathcal{A} = (2, 2, 3, 3, 3, 4)$$

$$\mathcal{P} = (10^{-2}, 3.10^{-2}, 5.10^{-2}, 8.10^{-2}, 8.10^{-2}, 10^{-1},$$
$$10^{-1}, 10^{-1}, 11.10^{-2}, 11.10^{-2}, 11.10^{-2}, 12.10^{-2})$$

$$C = (8, 5, 4, 4, 2, 1)$$

Binding: (a_1, c_2), (a_2, c_5), (a_3, c_1), (a_4, c_3), (a_5, c_4), (a_6, c_6).

The questions will take the same indices as the bases (for example q_1 will have basis $a_1=2$ and cost $c_2=5$).

The heuristic gives the order ω :

$$q_3 \underset{\omega}{<} q_1 \underset{\omega}{<} q_4 \underset{\omega}{<} q_5 \underset{\omega}{<} q_2 \underset{\omega}{<} q_6 \ ;$$

but we could also have taken $q_5 < q_4$ without any other modification.

This leads to the questionnaire built by the algorithm derived from that due to Huffman, the successive minimal questions being q_3, q_1, q_4, q_5, q_2 and q_6 (root) and their probabilities 9.10^{-2}, 16.10^{-2}, 29.10^{-2}, 32.10^{-2}, 23.10^{-2} and 1 (Fig. 9.3).

The cost of this questionnaire is $L_C=5.42$.

Five different orders can be explored by starting from $0_1 \cap 0_2$. They give the costs:

5.90 ; 5.52 ; 5.42 ; 5.50 ; 5.42

and the heuristic has enabled us to obtain directly one of the C-optimal questionnaires (which is far from being the general case). A study based on the same heuristic considerations is developed in Problem 7.

9.3.4 Logarithmic costs

The costs introduced by Duncan are essentially increasing functions of the basis. We are then in the case of binding where $O_1 \cap O_2$ has no arcs except those joining two vertices (a_i, c_i) and (a_j, c_j) such that $a_i = a_j$. The sub-graph generated by the sub-set of vertices with the same basis is moreover complete. G. Petolla's general method is thus applicable in good conditions if the number of distinct bases is not very great. It is clear that Duncan's costs do not permit any extension in the case of homogeneous questionnaires. On the other hand choosing the logarithmic function enables us to make an informational study of questionnaires with the costs as a function of the basis.

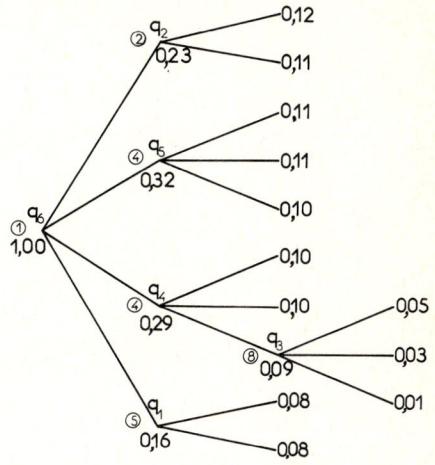

Figure 9.3

C-optimal questionnaire with binding. The questions bear the indices of the bases, the costs are encircled.

Definition 5 A *questionnaire with logarithmic costs* is a heterogeneous arborescent questionnaire in which the costs of the questions $(i \in F)$ are $c(i) = \log_2 a(i)$.

Let Q be a questionnaire with logarithmic costs, routing length $L(Q)$, and cost $L_c(Q)$ processing information $I(Q)$, expressed to base 2 and with heterogeneous information $H(Q)$.

By Property 8.5

$$H(Q) \leq L(Q) \text{, that is}$$

$$\sum_{i \in F} p(i) \sum_{j \in \Gamma i} \frac{p(j)}{p(i)} \log_{a_i} \frac{p(i)}{p(j)} \leq \sum_{i \in F} p(i)$$

and in particular for any i

$$p(i) J_{a_i}(i) = p(i) \sum_{j \in \Gamma i} \frac{p(j)}{p(i)} \log_{a_i} \frac{p(i)}{p(j)} \leq p(i)$$

that is

$$p(i) \sum_{j \in \Gamma i} \frac{p(j)}{p(i)} \log_2 \frac{p(i)}{p(j)} \leq (\log_2 a_i) p(i)$$

Summing over F gives:

$$I(Q) \leq L_c(Q) \qquad (17)$$

that is, we have

Property 9 The information with base two of a questionnaire with logarithmic costs is less than or equal to the cost of the questionnaire: it is equal if the condition of Property 8.5 is satisfied.

In the case when the minimal basis is $a_1 > 2$, we could take as cost function $c(i) = \log_{a_1}(a_i)$ and the logarithm with base a_1 in the calculation of $I(Q)$.

Consideration of heterogeneous quasi-questionnaires with logarithmic costs has enabled Duncan to generalize Kraft's inequality (Theorem 8.4) to heterogeneous questionnaires, which can be shown by taking Property 8.5 into account, the notation of that property being used again here.

Property 10 A heterogeneous quasi-questionnaire is a questionnaire if and only if

$$\sum_{i \in E} \prod_{j:=0}^{r-1} \frac{1}{a_{x(j)}} = 1 ;$$

otherwise the left hand side is less than 1.

Writing $r_a(i)$ for the number of questions with basis a which are proper ancestors of vertex i, such that

$$r(i) = \sum_{a \in \mathcal{A}'} r_a(i)$$

we can express the above equation in the form given by Duncan

$$\sum_{i \in E} \sum_{a \in \mathcal{A}'} a^{-r_a(e)} = 1, \qquad (18)$$

$\mathcal{A}'$ being the set of different bases.

9.4 Questionnaires in the Sense of Campbell

9.4.1 Questionnaires and Renyi's Information

In the same way as a comparison was made between the routing length and Shannon's information, Cesari has studied a type of general questionnaire relative to the concept of coding length introduced by Campbell and with which it is possible to associate Renyi's information.

Consider a polychotomic questionnaire with basis a and probability distribution $\mathcal{P}$ on E.

Campbell has defined a coding length L_t by:

$$L_t = \frac{1}{t} \log_a \sum_{e \in E} p(e) a^{tr(e)} \quad \text{for} \quad t > 0 \qquad (19)$$

and

$$L_0 = \lim_{t \to 0} L_t = \sum_{e \in E} p(e) r(e) \quad \text{for} \quad t = 0$$

which can be compared with Renyi's information of order α (defined in Theorem 7.6) for a complete distribution.

Cesari has extended the definition of L_t for $-1 < t < 0$ which corresponds to $\alpha > 1$ for $\alpha = \frac{1}{1+t}$.

We recover Shannon's information for $\alpha = 1$ and the routing length for $t = 0$.

Definition 6 An *optimal questionnaire of order* t or a t-*optimal questionnaire* is a polychotomic questionnaire minimizing L_t for given $(a, \mathcal{P}, t)$.

The study of t-optimal questionnaires requires the introduction of the value Π_t defined for any $z \in E \cup F$ by

$$\Pi_t(z) = a^{-tr(z)} \sum_{y \in E(z)} p(y) a^{tr(y)} . \qquad (20)$$

We verify that for any answer $e \in E$

$$\Pi_t(e) = p(e) \qquad (21)$$

and for any question $i \in F$:

$$\Pi_t(i) = a^t \sum_{j \in \Gamma i} \Pi_t(j) . \qquad (22)$$

For:

$$\Pi_t(i) = a^{-tr(i)} \sum_{j \in \Gamma i} \sum_{y \in E(j)} p(y) a^{tr(y)}$$

that is

$$\Pi_t(i) = a^t \sum_{j \in \Gamma i} a^{-tr(j)} \sum_{y \in E(j)} p(y) a^{tr(y)}$$

because $r(j) = r(i) + 1$ and if $j_1 \neq j_2$ and $j_1, j_2 \in \Gamma_i$, then $E(j_1) \cap E(j_2) = \emptyset$.

Campbell's length, L_t, can then be expressed directly by means of the valuation of the root α in the form

$$L_t = \frac{1}{t} \log_a \Pi_t(\alpha) \quad \text{for} \quad t \neq 0 . \qquad (23)$$

A t-optimal questionnaire is obtained for given $(a, \mathcal{P}, t)$

when $\Pi_t(\alpha)$ is minimal, if $t>0$ and when $\Pi_t(\alpha)$ is maximal, if $t<0$.

A generalization of the conditions of L-optimality for polychotomic questionnaires has enabled Cesari to extend Huffman's algorithm to questionnaires of order t ($t>0$).

To realize this algorithm a minimal question q must be found at each phase by grouping together the a answers (or the $\beta+1$ answers for the first phase of a polychotomic questionnaire in the broad sense) having the smallest values of Π_t; then Π_t is determined according to (22) by taking the product of a^t with $\sum_{j \in \Gamma i} \Pi_t(j)$. This product which must be repeated at each phase is the only alteration of Huffman's algorithm for L-optimal questionnaires. The computation of L_t is then immediate because the algorithm directly produces $\Pi_t(\alpha)$.

When $-1<t<0$ the decrease of Π_t as a function of the rank along a path - the necessary condition for t-optimality - can no longer be ensured because $a^t<1$; however Huffman's algorithm extended by Cesari still gives the optimal questionnaire for $t<0$. But note that $\Pi_t(\alpha)$ may be increased by contraction of a sub-questionnaire: if in a t-optimal questionnaire Q there exist two adjacent vertices x and $y \in \Gamma x$ such that $\Pi_t(\alpha)<\Pi_t(y)$ then suppression of the vertices of $\mu x - \mu y$ and afterwards contraction at $\{x,y\}$ enable us to form a heterogeneous questionnaire Q' of order t having $|E(x)-E(y)|$ answers less than Q and with coding length $L_t'<L_t$.

This phenomenon can be interpreted in the following way:

Suppose that a questionnaire on $(a, \mathcal{P})$ has been constructed knowing that the probability of obtaining an exact answer at each question is a^t. Then the questionnaire maximizing the probability of obtaining the result can be deduced from the t-optimal questionnaire by suppression of vertices followed by contractions as above and repeated as long as there exists an arc (x,y) with extremities related by $\Pi_t(x)<\Pi_t(y)$.

This questionnaire (heterogeneous) possesses less than E answers but gives the result with the greatest probability of the correct answer; for t=0, the *reliability* of the answers is excellent as each outcome obtained has a probability 1 of being exact.

Consideration of these "t-optimal incomplete questionnaires" enables us, for a constant error probability equal to a^t, deliberately to suppress those answers whose probability of occurrence is small and whose rank in Huffman's questionnaire) is large (these vertices would have too large an error probability).

Property 11   A polychotomic questionnaire $(E \cup F, \Gamma, P_\Gamma)$ has a Campbell length L_t which is a continuous function of t for $t \in]-1, +\infty[$.

Property 12 Every t-optimal questionnaire is such that

$$(t > 0) \quad L_t \geqslant I_N^\alpha (p_1, p_2, \ldots, p_N) , \qquad (24)$$

with equality if and only if the questionnaire is polychotomic in the strict sense and if

$$r(x) = r(y) \Rightarrow \Pi_t(x) = \Pi_t(y) .$$

Whatever $i \in F$, we then have

$$\Pi_t(i) = a^{-(t+1)r(i)} \Pi_t(\alpha) .$$

9.4.2 Charges and Expenses

Cesari has defined a generalization of the coding length by substituting a real-valued function of i, $(\gamma_i)^t$, in place of the constant coefficient a^t appearing in (19).

We still suppose the questionnaire to be polychotomic with basis a and moreover with $t \leqslant 0$ and $\gamma_i > 1$.

The new "coding" length will then be denoted by

$$L_\gamma = \frac{1}{t} \log_a \sum_{e \in E} p(e) \prod_{i \in A(e)} (\gamma_i)^t \quad \text{for} \quad t < 0 , \qquad (25)$$

where $A(e)$ is the set of proper ancestors of e.

For $t=0$ we define

$$L_c = \lim_{t \to 0} L_\gamma = \sum_{e \in E} p(e) \sum_{q \in A(e)} \log_a \gamma_q . \qquad (26)$$

If for $t=0$ we then put $D(e) = \sum_{q \in A(e)} \log_a \gamma_q$ and $c(i) = \log_a \gamma_i$, where $e \in E$ and $i \in F$, we will rediscover the costs of questions and expenses introduced by S. Petolla.

In the case $t<0$, we define for each $x \in E \cup F$ a *charge* given by

$$\kappa_t(x) = \prod_{q \in A(x)} (\gamma_q)^t .$$

and for each $e \in E$ and $i \in F$ a *weight* given by

$$W_t(e) = p(e)$$

and

$$W_t(i) = (\gamma_i)^t \sum_{j \in \gamma_i} W_t(j)$$

generalizing the expenses and probabilities from the case $t=0$.

If the preorderings of the increasing $a(i)$ and decreasing

$\gamma(i)$ are not compatible, then a γ-optimal questionnaire, that is, a polychotomic questionnaire minimizing L_γ for given $(a, \mathcal{P}, t, \{\gamma\})$ may be obtained by using an algorithm derived from Huffman's (Theorem 1).

If the preorderings are compatible, then a γ-optimal questionnaire may be found by a direct application of Huffman's algorithm extended to the case L_t.

9.5 Questionnaires in the Broad Sense

In §4.4.4 we discussed a restriction of the theory of latticoid questionnaires obtained by modifying the axioms $(q_1),\ldots,(q_5)$. In the other direction, it is also possible to extend the theory by a modification of the axioms. This has already been done in the case of quasi-questionnaires (§8.3.1) where (q_3) was deleted and (q_4) and (q_5) were slightly modified as a consequence. Here we will discuss some other possible extensions.

9.5.1 Infinite questionnaires

Definition 7 An *infinite questionnaire* is an l.q.s.c. valued graph $Q_\infty = (X, \Gamma, P_\Gamma)$ satisfying the axioms (q_1), (q_3), (q_4) and (q_5) of Definition 4.1.

We will assume that X is *denumerable* and write $X = E \cup F$.

Further we will distinguish infinite latticoid or arborescent questionnaires according as the supports are latticoid or arborescent (cf. §2.3.5).

In the arborescent case we can show that P_Γ induces a probability space $(E, \mathcal{P}(E), P)$ and in the latticoid case we can form the infinite arborescence of paths in order to establish a probability space.

Let $(p_i, (i \in N))$ be the probabilities of the answers.

If $\sum_{i=1}^{\infty} p_i \log \frac{1}{p_i}$ converges, then we will set

$$I(Q_\infty) = \sum_{i:=1}^{\infty} p(i) \log \frac{1}{p_i} \tag{27}$$

and we will say that I_∞ is the *information transmitted* by the infinite questionnaire.

We define the routing length $L(Q_\infty)$ of an infinite questionnaire by the formula bearing on the probabilities of the answers and their rank on the various paths, which may possibly be infinite (cf. 4.3.3).

Using a method similar to that of the Noiseless Coding Theorem (Theorem 8.5) of quasi-questionnaires from which we deduced an upper-bound on L-optimal questionnaires, we can show that if $I(Q_\infty)$ is finite, then there exists a questionnaire Q_∞^0 with finite routing length such that

$$L(Q_\infty^0) < I(Q_\infty) + 1 \ . \tag{28}$$

Moreover if $I(Q_\infty)$ is infinite, then $L(Q_\infty)$ is also infinite. It is possible by using an extension of quasi-questionnaires to form an infinite questionnaire Q^0 satisfying (28) without actually solving the problem of constructing an infinite questionnaire with minimal routing length.

In the case when the series $\sum_{i=1}^{\infty} p_c(i_c) r_c(i_c)$ converges, where the subscript c is used to remind us that we are working with a compatible questionnaire, it may sometimes be convenient to calculate the first B terms for instance when the evaluation of $\lim_{N \to \infty} \sum_{1}^{\infty} p_c(i_c) r_c(i_c)$ is not easy. In that case we perform a *truncation* that is, a contraction such that, for fixed B,

$\sum_{i=B+1}^{\infty} p(i)$ is less than some fixed number η. We then proceed by using restricted products as at the end of §8.1.3 by accepting the risk η of not attaining the answer.

Example (Duncan)

Consider the infinite probability distribution:

$$\mathcal{P} = \{p_i | i \in N\}$$

with $p_i = p^{i-1} q$ and $0<p<1$, $q=1-p$.

We verify that $\mathcal{P}$ is complete, i.e. $\sum_{i=1}^{\infty} p_i = 1$ and, we may then calculate the transmitted information because the infinite series is evidently convergent:

$$I(\mathcal{P}) = \sum_{i:=1}^{\infty} p_i \log \frac{1}{p_i}$$

and we find

$$I(\mathcal{P}) = \frac{1}{q} (p \log \frac{1}{p} + q \log \frac{1}{q}) \leq \frac{1}{q} \ . \tag{29}$$

If $p=\frac{1}{2}$, it is possible to construct an infinite questionnaire which will be L-optimal. Let Q_D be an infinite dichotomic questionnaire having at every rank r a question and an answer with probability $p_r = 2^{-r}$ and such that $\lim_{r \to \infty} p_r = 0$.

The routing length $L(Q_D) = \sum_{i=1}^{N} p_i r_i$ is the sum of the convergent series $\sum_{r=1}^{\infty} r.2^{-r} = 2$.

But (29) gives the same value: we will say that Q_p is an *infinite absolutely optimal questionnaire*. It is the limit as $N\to\infty$ of the absolutely optimal questionnaires with maximal height studied at the end of §8.1.3 and which we may now consider as being deduced from Q_p by *truncation*.

9.5.2 Questionnaires with Circuits

Deletion of axiom (q_1) allows us to form questionnaires admitting circuits. Many attempts have been made with a view to generalising in this direction; they are due particularly to Patris and Schneider.

Consider a questionnaire $Q=(E \cup F, \Gamma, P_\Gamma)$ in which $x_1, x_2 \in F$ where x_2 is a proper descendant of x_1, and the path $[x_1 x_2]$ is unique.

For a graph G_c by adjoining a new arc (x_2, x_1) and let $\Gamma_c = \Gamma \cup \{(x_2, x_1)\}$. We form a questionnaire with circuits Q_c by imposing a valuation $p(x_2, x_1) > 0$ and modifying the valuation of the arcs of $[x_1 x_2]$ in such a way that Q_c satisfies the axiom of conservation of flow (q_4) and (q_5).

The adjunction of (x_2, x_1) has created a unique elementary circuit (Property 2.5) but also an infinity of distinct paths corresponding to going round the circuit $[x_1 \ldots x_2 x_1]$ a varying number of times. The development of the paths of Q_c is an infinite arborescence.

The same procedure may be carried out in the case of many circuits by "cutting a return arc" of the type (x_2, x_1) to effect the development (in a non exhaustive manner because of the infiniteness).

Let B be the arborescent infinite questionnaire, which is developed from Q_c keeping the hypothesis of proportionality of flow always valid in order to calculate the routing length $L(Q_c)$.

We show that the length $L(Q_c)$ is finite in the case of a unique circuit γ of length ℓ.

Suppose we are required to evaluate

$$L(Q_c) = \sum_{e_B \in E_B} p_B(e_B) r_B(e_B)$$

where E_B is the set of answers of B and $p_B(e_B)$ is the probability of an answer of B with rank $r_B(e_B)$.

Corresponding to an elementary path μ - which must be finite because $E \cup F$ is finite - linking α to a vertex i descending from x_1 there is an infinite number of paths in B, μ, $\mu*\gamma$, $\mu*\gamma^2, \ldots, \mu*\gamma^k, \ldots$, where the symbol $*$ and the power indicate the concatenation of μ with $1, 2, \ldots, k, \ldots$ times the elementary circuit γ.

Let $r_B(i)$ be the length of the elementary path μ; the other paths considered such as $\mu*\gamma^k$ will then have length $r_B(i)+k\ell$.

Let $p_B(i)$ be the probability of i along the path μ; the probability $p_B(i^k)$ of the image of i corresponding to the path $\mu*\gamma^k$ is $p_B(i^k)=v^k p_B(i)$, where v is the conditional probability for the circuit being closed. v is obtained as the product of the conditional probabilities of the arcs of γ, that is $p((a,b)|a)$ for every vertex a along circuit γ:

$$v = \prod_{(a,b) \in \gamma} \frac{p(a,b)}{p(a)} < 1 . \tag{30}$$

Let ν be the answer of E_B which corresponds to an elementary path.

Then we can write:

$$L(Q_c) = \sum_{\nu \in E_B} \sum_{k:=0}^{\infty} (r_B(\nu) + k\ell)v^k p_B(\nu) .$$

The simple series $S = \sum_{k=0}^{\infty} (r_B(\nu) + k\ell)v^k p_B(\nu)$ converges and its sum is

$$S = \frac{1}{1-v}(r_B(\nu) + \frac{\ell v}{1-v} p_B(\nu))$$

and consequently

$$L(Q_c) = \sum_{\nu \in E_B} \frac{1}{1-v}(r_B(\nu) + \frac{\ell v}{1-v} p_B(\nu)) \tag{31}$$

which is strictly finite, because v is strictly less than 1 and E_B is finite.

Another method of attack for questionnaires with circuits consists in introducing states related to each vertex in such a way as to form a Markov chain. We must then introduce transition probabilities. From a state i to a state j, it will be $p((i,j)|i)$ which we assimilate to the probability of arc (i,j) conditioned by i. The hypothesis of proportionality of flow will then be made implicitly as in the case of the questionnaires K (§4.4.4). By means of a trick to prevent the terminals from blocking the system (viz, by adjoining a loop (e,e) at any answer with $p((e,e)|e)=1$ without changing the set e of paths (without loops) entering into E) we may produce again a formulation of axiom (b3) for K-questionnaires:

$$p_2^{(t)}(i) = \sum_{h \in \Gamma^{-1}i} p_1(h,i) p_2^{(t-1)}(h)$$

where the indices t-1 and t refer to the time introduced with the states of the Markov chain.

This method enables us to find (31) again when there is a circuit, to show the convergence of $L(\Omega_c)$ when there are many circuits and to form an algorithm for evaluating $L(\Omega_c)$ within ε (Patris).

9.5.3   Flow Charts and Circuits

Various different formalisations of computation processes involve graph concepts. As examples we may cite Bovett who puts the operations on the arcs (costs or time in the arcs) and Cooper who has defined a family of transformations of graphs some of which have been used here as well. A method of introducing circuits into questionnaires gives a link with the theory of Markov chains and the theory of automata whereas problems of coding have been linked more with homogeneous or arborescent questionnaires. The german school of logic has introduced the so-called algorithmic graph which we will call a flow chart.

Definition 8   A *flow chart* is a graph representing the elementary operations of an algorithm having only one *entrance* and at least one *exit*.

Thus it is an l.q.s.c. graph which may have circuits.

Thiele then considers the so-called *interpretative flow charts* by forming the triple (Ω, ϕ, ψ), where Ω is a flow chart ϕ is a "logic" mapping from Γ to the set $\{0, 1, \ldots, a-1\}$, a being the maximum out-degree of Ω and ψ is a "functional" mapping from the vertices of transmitter type into a set Q.

Whereas ϕ is the usual coding of the arcs coming out of a vertex of a graph, ψ can be replaced by a mapping from the arcs onto Q. This mapping gives a cost ψ_{ij} or an "operating time" which may be allocated to the untested arithmetical operations which are represented in this model by the arc (i,j).

For the presence of a transmitter must be eliminated from any flow chart (if we want to link it to a questionnaire) according for example to the procedure used in §5.3.1.

In data processing it is customary to use tests with two, and sometimes three, outcomes which enables us to generate dichotomic questionnaires (a doubling of tests with three outcomes into two tests with two outcomes is frequent).

But the proper aim of an algorithm is to use the smallest possible number of different instructions in order not to take up too much storage space. An algorithm therefore certainly involves "loops" which translates into the language of questionnaires as: the flow chart always has at least one circuit.

Suppose therefore that it is required to evaluate the

occupation time of an algorithm in a computer. According to custom, "random" variables corresponding to the outputs of the tests of the algorithm (outcomes of the flow chart) must be given a probability - the limit of the frequency for a sufficiently large number of experiments - which can allow an evaluation of a routing length. In fact the routing length of a questionnaire with circuits deduced from the flow chart by the knowledge of the probabilities of the answers of each test will give a data processing expert sufficient information and (31) will not generally be used directly.

We generate on the flow chart a questionnaire having the following three properties: the presence of circuits, of costs and utilities. Such a questionnaire must be represented by a quintuple

$$(X, \Gamma, P_\Gamma, C, \mathcal{U}).$$

The utilities will serve two ends:

1. The answers may have very different significances and the programmer may want to favour certain of them with a strong utility and to penalize others.

2. Since the probability distribution is often not well known, it is interesting to mitigate this deficiency by reducing its effects through the introduction of the preference $\nu_i = u_i p_i$. We know that from the point of view of useful length - which the practical man will choose to consider rather than information - u_i and p_i play symmetrical roles.

The costs will in fact be operation times which are often obtained as follows:

Let t_i be the time for the operations of comparison or decision (dichotomic questions)

and Ψ_{ij} the time for computing or processing operations taking place between the vertices i and j.

Then $\tau(i,j) = t_i + \Psi_{ij}$ will be called the *time for the arc* (i,j) of the flow chart.

Further the quantity

$$L_\tau = \frac{\sum_{e \in E} \sum_{\mu \in \mathcal{C}_e} \sum_{(i,j) \in \mu} \tau(i,j) p(e) u(e)}{\sum_{e \in E} p(e) u(e)}. \qquad (32)$$

will be called the *mean operating time*.

If the questionnaire was arborescent and τ_{ij}=constant, then L_τ would reduce to a useful length. The presence of circuits and simultaneously of $\tau(i,j)$ indicates that this application is an extension of questionnaires with costs, utilities and circuits. In particular even if the questionnaire was arborescent, we could have some difficulties in determining a τ-optimal questionnaire because the $\tau(i,j)$ play a role such that

$$\sum_{(i,j)\in \mu} \tau(i,j) = \tau(e)$$

can be assimilated to an expense; but two vertices e_1 and e_2 which come out from the same question i do not have the same τ.

Without considering τ-optimization, C. Prost has studied a class of flow charts where it is possible to reduce L_τ by operations of the transfer of sub-questionnaires type (defined in the case where there are circuits). Furthermore a questionnaire with circuits and having a smaller routing length can be obtained in a more immediate manner. Finally we should point out even though there are only rare cases in which the routing length of a questionnaire with circuits is infinite, they can certainly be encountered in procedures corresponding to undecidable problems.

9.6 Realizable Questionnaires

9.6.1 Constraints in questionnaires

The questionnaires which have been our main preoccupation up to now have assumed a great flexibility for the realization of the questions which in general had to provide a finer partition of the answers: in the arborescent case the questions appeared as bound to a partition operator to be specified by some semantic concept.

In the latticoid case the set $E(i)$ attached to the question i must be decomposed into a union of non empty sets $E(j)$, not necessarily distinct, and ensuring the covering of $E(i)$. If $E(j)$ and $E(k)$ were included in $E(i)$, then $E(j) \cap E(k) = \emptyset$ if j and k have no common descendants.

No other restriction has arisen to reduce the possibilities of L- or C-optimization. We have seen that the strategy of formulating an equiprobable dichotomic questionnaire (§8.2.2) leads to the observation that the outcomes of a question are associated with events whose number had to be confined between two consecutive powers of 2. However we have not yet had to face a constraint phenomenon prohibiting certain partitions or certain coverings. The detection problem studied below has the object of exposing the scope and the limitations of the applications of questionnaires in the present context. It will also be possible to use a simple *relation operator* to make the partition (the case of the question formulated by y<-1?) and a composite operator involving arithmetical operations and a relation operator (the question formulated by (x+1)(y+1)<0?). These questions are formulated and constructed in this problem according to a process operating upstream to downstream.

9.6.2   A Detection Problem

Consider a moving point whose position in space is given by its co-ordinates X, Y, Z relative to rectangular axes. The 6 planes making up a (2 x 2 x 2) cube centered at the origin and with edges parallel to the axes determine 27 regions in space. It is asked to determine the region of space where the moving point has been detected by asking on average a minimal number of questions with 2 possible answers. The 27 regions determined by the 6 planes containing the sides of the cube will be denoted as shown in the following table

↑ Y	Z<-1			-1<Z +1			Z>1		
+1 →	6	7	8	15	16	17	24	25	26
-1 →	3	4	5	12	13	14	21	22	23
	0	1	2	9	10	11	18	19	20
	↑	↑		↑	↑		↑	↑	
	-1	+1		-1	+1		-1	+1	X →

1. The probability of detecting the moving point in one or the other region is 1/27.

If the questions were trichotomic it is evident that 3 independent questions localising in X, in Y or in Z would enable us to obtain the answer in every case. It would involve a product of 3 trichotomic equiprobable questionnaires of one question each.

But the problem states precisely that the questions should be dichotomic, the minimal average number of comparisons related to $N=2^4+11$ that is

$$L_H = 4 + 2 \cdot \frac{11}{27}$$

will be obtained if no answer has a rank greater than 5. The product $A_{1 \times 2 \times 3}$ of three questionnaires of $N_1=N_2=N_3=3$ answers is therefore excluded (in contrast with the case of trichotomic questions); it would give a routing with length

$$3(1 + \frac{2}{3}) = 5 .$$

To study the questions to be asked one can begin by trying to optimize a questionnaire with 9 answers corresponding to the regions Z<-1; it is necessary to separate the answers into two classes having respectively 4 and 5 elements. This can be realized by a relation operator of the type X<-1? or X>1?. We are led to put the regions 1, 2, 3 and 6 into the same class (Fig. 9.4) corresponding to the criterion:

$$(X + 1)(Y + 1) < 0.$$

```
6  | .  .
3  | .  .
   |_____
   .| 1  2
```

The questions of rank 1 and 2 and the question of rank 3 are then immediate. We can now study the three dimensional problem by recalling that the first question will have to separate the answers into two classes which can have:

    either   11 and 16 elements

    or   12 and 15 elements

    or   13 and 14 elements

otherwise the arborescence would not be balanced.

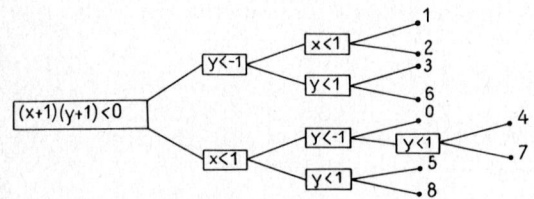

Figure 9.4
L-optimal question-
naire for the plane
(z<-1)

The first question analogous to that of the two-dimensional case $(X+1)(Y+1)(Z+1) > 0$?, would enable us to separate the answers into two classes having 13 and 14 elements respectively as shown in the following table:

```
   Z<-1              -1<Z<1            Z>1

  6 |              16   17           25   26
  3 |              13   14           22   23
    |_____      _____         _____
    | 1  2         9|                18|
```

then the criterion $(X+1)(Y+1)(X-1) > 0$ is satisfied by the seven answers 1, 3, 6, 14, 17, 23, 26 for the class with 14 elements, whereas the criterion $Z<-1$ is satisfied by the five answers 0, 4, 5, 7, 8 for the class with 13 elements.

From this we may obtain the balanced questionnaire answering the detection problem (Fig. 9.5) when the regions where the moving point is detected are equiprobable and

$$L_H = 4 + \frac{22}{27}.$$

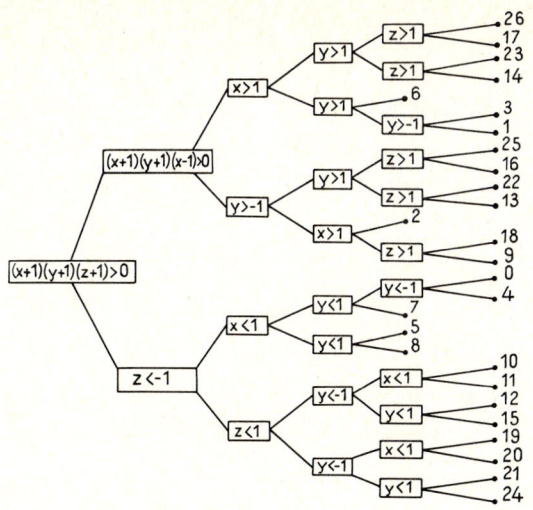

Figure 9.5
Detection problem.
Optimal equiprobable questionnaire.

2. For each independent variable X, Y, Z we suppose that the probabilities are distributed in such a way that the two planes perpendicular to the corresponding axis separate the first and the last quartile.

For X only, the probabilities are then

$$P_X^0 = P\{X < -1\} = \frac{1}{4},$$

$$P_X^1 = P\{-1 < X < 1\} = \frac{1}{2},$$

$$P_X^2 = P\{X > 1\} = \frac{1}{4}.$$

The probability for the moving point to be in the region X_i, Y_j, Z_k ($i,j,k := 0,1,2$) is

$$P_{ijk} = P_X^i P_Y^j P_Z^k.$$

We may now write out the table of probabilities opposite the the table of regions

298   GRAPHS AND QUESTIONNAIRES

6	7	8		15	16	17		24	25	26
3	4	5		12	13	14		21	22	23
0	1	2		9	10	11		18	19	20

$\frac{1}{64}$	$\frac{1}{32}$	$\frac{1}{64}$		$\frac{1}{32}$	$\frac{1}{16}$	$\frac{1}{32}$		$\frac{1}{64}$	$\frac{1}{32}$	$\frac{1}{64}$
$\frac{1}{32}$	$\frac{1}{16}$	$\frac{1}{32}$		$\frac{1}{16}$	$\frac{1}{8}$	$\frac{1}{16}$		$\frac{1}{32}$	$\frac{1}{16}$	$\frac{1}{32}$
$\frac{1}{64}$	$\frac{1}{32}$	$\frac{1}{64}$		$\frac{1}{32}$	$\frac{1}{16}$	$\frac{1}{32}$		$\frac{1}{64}$	$\frac{1}{32}$	$\frac{1}{64}$

We see that the answers are grouped according to the probabilities:

1 answer with probability 2^{-3},

6 answers with probability 2^{-4},

12 answers with probability 2^{-5},

8 answers with probability 2^{-6}.

The optimal questionnaire must be built in such a way that the answers characterized by the probability 2^{-r} have rank r if possible. In particular region 13, the interior of the cube, must have rank 3. Since the interior of the cube is the only region corresponding simultaneously to $X^2<1$, $Y^2<1$, $Z^2<1$ the questionnaire of rank 0 can be $X^2>1?$. A question of rank 1 will be $Y^2>1?$; we see then that the 3 answers 4, 13 and 22 are the only answers which can belong to a sub-questionnaire determined by $X^2<1$, $Y^2<1$.

4 and 22 being separated from 13 by the question $Z^2>1?$, the question $Z>0$ will allow questions 4 and 22 with probability 2^{-4} to have rank 4.

The arborescence is very easily built from these first elements (Fig. 9.6). It possesses the following property:

The vertices with rank r have probability 2^{-r}.

It is therefore an absolutely optimal questionnaire.

The routing length is

$$1.3.2^{-3} + 6.4.2^{-4} + 12.5.2^{-5} + 8.6.2^{-6},$$

that is

$$L_H = 4.5 .$$

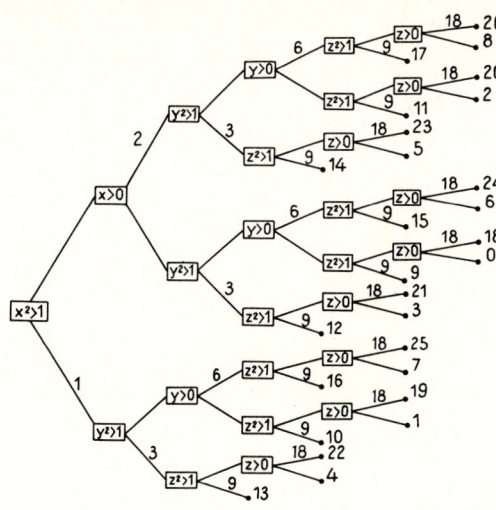

Figure 9.6
Detection problem.
Non equiprobable
questionnaire without
inefficiency

$$p(r,s) = 2^{-r}.$$

This questionnaire is the double product of three optimal questionnaires without inefficiency A_X, A_Y and A_Z where each of the three arborescenees has three answers with probabilities ($\frac{1}{2}$, $\frac{1}{4}$, $\frac{1}{4}$,) with ranks (1,2,2) as shown in the graph of Fig. 9.7,

Quotations We can valuate the outcomes of each question. In the elementary questionnaire A_X in which the answers correspond to regions whose numbers differ by 1, the outcomes of the first question can take the value 0, if $X^2>1$ and 1, if $X^2<1$ and those of the second question 0, if $X<0$ and 2, if $X>0$. In this way the numbers of the regions will be made up by summation of the answers to the questions of rank 0 and 1. We operate in the same way for the questionnaires A_Y and A_Z but for a progression of numbers of regions going 3 by 3 (for X and Z constant) or 9 by 9 (for X and Y constant). The notation for the outcomes of the questions will be related uniquely to their formulation and not to the rank at which these questions are asked along the various paths linking x_0 to (E).

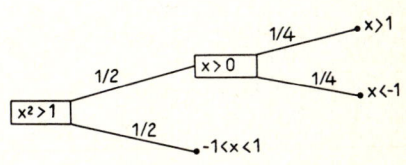

Figure 9.7

As well as the probabilities linked to each vertex of the questionnaire, we can also allocate coefficients to each arc of the questionnaire, two arcs corresponding to an identical answer coming out of a question posed in an identical form being allocated the same coefficients. The N paths coming out of the root will be characterized by a final quotation which will be the sum of the quotations of the arcs used.

We verify that the quotations shown in the table allow an effective enumeration of the 27 regions of space according to the diagram given when posing the problem.

Question	Outcome	Quotation	Rank
$X^2 > 1$	Yes / No	0 / 1	0
$X > 0$	Yes / No	2 / 0	1
$Y^2 > 1$	Yes / No	0 / 3	1 or 2
$Y > 0$	Yes / No	6 / 0	2 or 3
$Z^2 > 1$	Yes / No	0 / 9	2, 3 or 4
$Z > 0$	Yes / No	18 / 0	3, 4 or 5

In particular this quotation can be applied systematically step by step at each phase of the experiment. The summation of the quotations along a path then gives an automatic determination of the quotation of the terminal vertex (answer). We can then say, following Yaglom and Yaglom, that the questions are "directly oriented" or that the information treated by the consecutive questions has been memorized directly with a view to its further utilisation.

This method of quotation is not dependent on the special properties of the particular example met with, but nevertheless it can not always be employed in a systematic way.

9.6.3 Partitions and formable questions

Let Q be an arborescent questionnaire, $\mathcal{A}$ the set of bases of its questions and E the set of its answers.

Then we will introduce a set R(E) related to the set E of answers of this questionnaire.[1] Let $E(i) \subset E$ be a subset such that $|E(i)| \geq a_i > 1$ and let $\mathcal{R}_i$ be the set of all the partitions of E(i) into a_i non-empty sets, $E(j_u)$, for $u=1,2,\ldots,a_i$.

Let $R_i \subset \mathcal{R}_i$ be a set to be called the *set of admissible partitions for E(i)*.

---

[1] we recall that E(i) is an event of $(E, \mathcal{P}(E), P)$ if and only if Q is arborescent, cf. Properties 4.7 and 4.13

Definition 9 The ordered pair $(E(i), R_i)$ will be called the *situation* of i and denoted by $S(i)$.

We will also write $S(i)=S(\lambda_1,\ldots,\lambda_t)$ whenever $E(i)$ and R_i are expressed in terms of the parameters λ_t which may be real or Boolean for example.

Definition 10 The *set* of *admissible partitions* for a set of independent elements E is the set $R(E) = \cup R_i$ where the union is taken over all non-empty sets $E(i)$ of order at least 2.

Definition 11 In a questionnaire Q we say that a question i with basis a_i is R_i-*formable* if the partition of $E(i)$ by the events $E(j)$ associated with the outcomes of i is an admissible partition for $E(i)$. If all the questions of the questionnaire are R_i-formable we say that Q is $R(E)$-*realizable*.

We will sometimes shorten the terminology by speaking simply of *formable questions and realizable questionnaires*.

If the event $E(j_u)$ associated with an outcome of i admits only one element, then j_u will be an answer to the questionnaire.

Definition 12 We say that a question i is *formulated* if there exists an operator allowing the construction of an admissible partition for $E(i)$.

This operator, external to the questionnaire, will be linked to the semantic, that is to the use of the questionnaire itself; it is this which will enable us to interpret the questionnaire. This operator ψ must belong to a domain which needs to be specified.

Note that $R(E)$ is defined by the set of bases, $\mathcal{A}$, and results from the compatibility conditions (4.1 and 4.2) of arborescent questionnaires.

Example Let $E(i)=\{a,b,c\}$ be an event corresponding to three answers a, b, c.

The set of partitions of $E(i)$ into two non-empty subsets is

$$\mathcal{R}_i = \{\Pi_a, \Pi_b, \Pi_c\}$$

where

$$\Pi_a = \{b,c\} \cup \{a\}, \quad \Pi_b = \{a,c\} \cup \{b\},$$

$$\Pi_c = \{a,b\} \cup \{c\},$$

This may be illustrated diagrammatically as shown in Figure 9.8

The homogeneous questionnaires with basis 2 which we can construct with the aid of the partitions in $\mathcal{R}_i$ are necessarily arborescent and will have two questions, the semantic of which will be distinct (example for Π_a, Fig. 9.9).

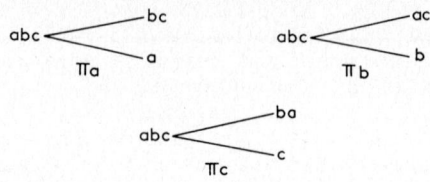

Figure 9.8

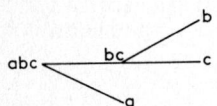

Figure 9.9

The questionnaire shown in Fig. 9.9 is realizable if and only if R(E) contains the partitions Π_a and $\{b\} \cup \{c\}$.

Definition 13 The set R(E) of the admissible partitions is said to be *compatible* if there exists at least one realizable questionnaire.

Let Q be a realizable questionnaire and let $(E(i), R_i)$ be a situation to which will be associated:

- the complete system of eventualities

$$\left\{ \frac{p(e)}{\sum_{e \in E(i)} p(e)} \middle| e \in E(i) \right\}$$

- the set of admissible partitions

$$R(E(i)) = \bigcup_{E(j) \subset E(i)} R_j \subset R(E)$$

R(E(i)) includes the sets of admissible partitions relative to all the possible events which may be taken sequentially or in parallel[1] that is:

sequentially they are the increasingly fine partitions:

$$\{e\} = E(e) \subset E(x_1) \subset E(x_2) \ldots \subset E(j) \subset E(i)$$

in parallel they are disjoint partitions

---

[1] This reasoning of a purely set theoretical nature appeals to the two orderings α and γ established on arborescences (2.3.4) and justifies the systematic use of graphs, in particular, arborescences, to describe the diagrams of decisional processes using questionnaires.

$$\{e\} = E(e) \subset E(x_1) \quad \text{and} \quad \{e'\} = E(e') \subset E(y_1)$$
$$\ldots \ldots \ldots \ldots \ldots \ldots \ldots \ldots \ldots \ldots \ldots$$
$$E(x_j) \subset E(x_{j+1}) \quad \text{and} \quad E(y_k) \subset E(y_{k+1})$$

An arborescent questionnaire separating this complete system of eventualities by formable questions is called a *realizable sub-questionnaire associated* with the situation S(i) of order $|E(i)|$.

If $|E(i)|=1$, it is said to be *degenerate*; if $E(i)=E$, then it is a realizable questionnaire of order $N=|E|$.

Conversely we will say that a situation S(i) is possible if there exists a realizable sub-questionnaire associated with S(i).

It is possible that the set of R(E)-realizable questionnaires, Q_R, say, does not contain certain questions which are however R_i-realizable: for example in the detection problem some questionnaires contained the question $y<-1$? others the question $y^2>1$? and none included the question which was not forbidden and was therefore formable, $(x-1)(y+1)<0$? We will make the following

Convention In order to reduce the exhaustive nature of the discussion in some way we will suppose that R(E) is restricted to the set of formable questions belonging to at least one questionnaire of Q_R but we will still use the notation R(E).

Let $(E(i), R_i)$ be a possible situation; then there exists a realizable questionnaire separating the complete system E such that for $e \in E$, $p(e)>0$ and $\sum_{e \in E} p(e)=1$ by means of formable questions.

9.6.4 Arborescent Realizable L-Optimal Questionnaires

Definition 14 An *L-optimal R(E)-realizable questionnaire* (respectively, *preferable, C-optimal*) is an arborescent questionnaire with minimal routing length (respectively useful length, cost) among the R(E)-realizable questionnaires with given ($\mathcal{A}$, $\mathcal{P}$) (resp. with given ($\mathcal{A}$, $\mathcal{P}$, $\mathcal{U}$), ($\mathcal{A}$, $\mathcal{P}$, C)).

Property 6.4 can be generalized:

Theorem 2 Any L-optimal realizable questionnaire consists of L-optimal realizable sub-questionnaires.

This theorem due to F. Dubail, can be extended to other cases of optimality (C-optimality, minimal heterogeneous information, minimal noise) which leads to just as many construction algorithms for extremal realizable questionnaires.

Let $(E(i), R_i)$ be a possible situation of order n_i.

. If $n_i=1$, the associated sub-questionnaire is degenerate, and is therefore L-optimal realizable with routing length written
$$L(\{e_i\}, \emptyset) = 0 .$$

. If $n_i>1$, then R_i is not empty and there exists a set $\{i\}$ of R_i-formable questions. Suppose that for any possible situation of order
$$1 \leqslant m < n_i \leqslant N$$
we are able to construct the associated optimal realizable questionnaire and let
$$E(j_1),\ldots,E(j_u),\ldots,E(j_{a_i}) .$$
be the partition of $E(i)$ into a_i disjoint non-empty subsets.

To any u there corresponds the situation $(E(j_u),R_u)$ for which we know how to construct an L-optimal questionnaire with routing length $L_0(E(j_u),R_u)$.

The questionnaire Q_R of which the first question is a question i and of which the sub-questionnaire obtained by suppression of i are realizable L-optimal is a sub-questionnaire associated with $(E(i),R_i)$ such that

$$L(Q_R) = 1 + \sum_{u:=1}^{a_i} \frac{p(j_u)}{p(i)} L_0(E(j_u), R_u) .$$

An L-optimal realizable questionnaire will be one of those for which:

$$L(Q_{R_o}) = 1 + \min_{i \in I} \sum_{u:=1}^{a_i} \frac{p(j_u)}{p(i)} L_0(E(j_u), R_u)$$

It is then possible to construct recursively downstream to upstream all the realizable L-optimal sub-questionnaires. We must take care however always to preserve *all* the L-optimal realizable sub-questionnaires as long as the situation $(E(i),R_i)$ remains strictly included in $(E,R(E))$.

Among various realizable L-optimal questionnaires of order $n<N$, perhaps only one will be part of a realizable L-optimal questionnaire of order N.

This algorithm due to F. Dubail which facilitates the construction of an L-optimal questionnaire with the same routing length as the one proceeding from Huffman's algorithm (in the sense of realizability) involves essentially dynamic programming without allowing in general a direct sequence as in the case where all the arborescent questionnaires are *a priori* realizable (cf. §6.4).

9.6.5 The Equivalence of Constraints and Costs

Let R(E) be the set of admissible partitions.

If R(E) is a proper subset of $\mathcal{P}$(E), then R(E) can be given in Dubail's form by a list of authorized partitions or in the implicit form of formulation of the type of alphanumerical comparison with two outcomes, before (or less than) and after (or greater than).

We will say that the constraints bear on the realization of certain questions. These constraints can also be given in a special form as in a problem posed by Katona and solved by Cesari "construct a questionnaire of given height having minimal length of routing". Lastly, the cost C of the questions is also a kind of constraint which it is interesting to present by means of situations and formable questions.

Let P_C be a problem of C-optimality with free costs for the questionnaires defined by ($\mathcal{A}$, $\mathcal{P}$, C), supposing that the costs are given on the set of questionnaires corresponding to the situation (E, $\mathcal{P}$(E)) of order N.

Let P_R be a problem of realizable L-optimal questionnaires on ($\mathcal{A}$, $\mathcal{P}$). Then G. Petolla has established certain implications.

Property 13 Given a problem of realization P_R, we can construct a problem of cost P_C such that any L-optimal realizable questionnaire is also C-optimal.

Let K* be an L-optimal realizable questionnaire and L(K*) its routing length.

Let $c(q) = c > 0$ be a cost attributed to any formable question.

Let $c(q) > c\dfrac{L(K^*)}{p_o}$ be a cost attributed to any question which is not formable where p_o is the minimal probability of such a question ($p_o > 0$).

If K is a realizable non-optimal questionnaire, then

$$L(K) > L(K^*) \Rightarrow L_C(K) = cL(K) > cL(K^*) = L_C(K^*) .$$

If K is not a realizable questionnaire, then there exists an event E(y) associated with a question y of K for which no partition is admissible and such that

$$L_C(K) = \sum_{x \in F} c(x)p(x) \geq c(y)p(y) > cL(K^*) = L_C(K^*) .$$

Property 14 Given a problem of cost P_C, we can construct a realization problem P_R a solution of which is a C-optimal questionnaire solution of P_C.

Let $\overline{K}$ be a C-optimal questionnaire on $(\mathcal{A}, \mathcal{P}, C)$.

If $\overline{K}$ is also L-optimal on $(\mathcal{A}, \mathcal{P})$, then the problem without constraint of realization is "weakly equivalent" to the problem with cost.

If $\overline{K}$ is not L-optimal, let $\mathcal{B}$ be the set of questionnaires B on $(\mathcal{A}, \mathcal{P})$ such that

$$B \in \mathcal{B} \Rightarrow L(B) < L(\overline{K}).$$

Put $R'(E) = R(E) - R_b$,

where R_b is the set of formable questions of B which are not in $\overline{K}$ whatever $B \in \mathcal{B}$.

Then $\overline{K}$ is the L-optimal questionnaire solution of the problem with constraint of realization.

We remark that as well as this theoretical analogy between L-optimal realizable questionnaires and C-optimal questionnaires, there is also an analogy in the type of optimizing procedure because for logarithmic costs there exists an algorithm due to Duncan which proceeds as before by means of dynamic programming.

9.7 Questionnaires in Practice

9.7.1 The Dynamic Aspect of Interrogation

So far the general study of questionnaires has been undertaken in the form of constructing models which are to be used in reality in a dynamic manner.

The practical use of a questionnaire can be systematized as follows:

We are given, in the form of a program for example, the support (X, Γ) of a questionnaire Q with a view to studying a random phenomenon of which we are looking for an elementary realization e belonging to a set of elementary events with distribution $P_E = \{p_1, \ldots, p_N\}$; only one event of this distribution will participate at the terminal $e \in X$.

Suppose (X, Γ) is an arborescence.

The random phenomenon characterized by $\mathcal{P}$, and participating in the definition of the questionnaire by the distribution P_E deduced from $\mathcal{P}$ by the choice of a permutation, is presented in the form of data introduced at the entrance (root). Partition operators facilitating the formulation of the R_i-formable questions act according to the data and when one (and only one) path of e has been described by the operating of the questionnaire solicited from the introduction, then the information has been transmitted by the questionnaire (in the quantitative sense of Theorem 7.4); from the practical point of view, we therefore dispose of one and only one of the

N answers which we can consider as validated by a succession
of finer and finer partitions or of "recognitions" of events.
On average L(Q) of these partitions will have been effected.

Note that this dynamic process equally describes a flow
chart as a questionnaire.

If the operator having to realize the partition E(i) imposed
by (X,Γ) enters into action this signifies that the data
introduced was associated with an eventuality $e \in E(i)$. If
the outcome j_u is validated then $e \in E(j_u)$ and $E(j_u) \subset E(i)$,
$E(j_u) \neq E(i)$. The random phenomenon T studied may be different
from the set of validated answers but it is possible to define
a one-to-one mapping between the set of answers E and T.

If (X,Γ) is a latticoid, then, according to the operator
used, some unique one of the paths leading to an answer of Q
will have been validated. This can allow in particular
(see Ludde and Thiele as well as the example of the Soumax
program §10.2.2) the prevention of the process of building a
priori a compatible questionnaire.

In reality the random phenomenon T studied has been analyzed
too finely by the compatible questionnaire and we effectively
dispose only of the answers of a latticoid questionnaire.
There will then be a surjection of $E_c \rightarrow T$ which we can decompose
into the product of a surjection $E_c \rightarrow E$ with a bijection $E \rightarrow T$,
where E and E_c are the sets of answers of Q and a compatible
questionnaire C.

Aggarwal has studied the problem of adapting an arborescent
questionnaire (X,Γ,P_Γ) to a phenomenon whose events
$E(i) \in P(E)$ are well described by the support $(E \cup F, \Gamma)$ but
whose probability distribution P' is distinct from the distribution P_E induced by P on E: he has opened the way to the
solution of the *adaptability* of questionnaires.

9.7.2   A Random Experiment

Let N be an integer greater than or equal to 2 and let x be
an integer, $1 \leq x \leq N$ determined by a random experiment such
that $P\{x_i = i\} = p_i$, the system of eventualities e_i with probability p_i being complete. Determine x with the aid of a dichotomic realizable L-optimal questionnaire in which all possible
partitions are imposed by the comparison operator "$x \geq k$?"
without any limitation on the integer k.

Consider the two situations studied by F. Dubail,

a.   S(a,b) when $1 \leq a < b \leq N$ and $a < x \leq b$.

Let $E(i) = \{a, a+1, \ldots, b\} = E(a,b)$ and let R_i be the set of
partitions denoted by R(a,b):

$$R(a,b) = \{\rho_k / \rho_k = \{a, a+1, \ldots, k-1\} \cup \{k, k+1, \ldots, b\}.$$
$$\text{and} \quad k := a+1, \ldots, b\} \quad .$$

The outcome "yes" to the question $x \geqslant k$ indicates:

$$x \in \{k, k+1, \ldots, b\} \quad ;$$

$R(a,b)$ admits $b-a$ formable questions.

b. $S(a,a)$ where $E(i)=\{a\}$ and $R_i = \emptyset$ that is $S(a,a)$ is a terminal situation of order 1.

The only possible situations are of the form:
$S(a,b)$ where $1 \leqslant a < b \leqslant N$, in such a way that

$$R(E) = \bigcup_{1 < b \leqslant N} \bigcup_{1 \leqslant a < b} R(a,b)$$

Then we find that $L(a,a)=0$ and

$$L(a,b) = 1 + \min_{a+1 \leqslant k \leqslant b} \left\{ \frac{\sum_{i:=a}^{k-1} p_i}{\sum_{i:=a}^{b} p_i} L(a, k-1) + \frac{\sum_{i:=k}^{b} p_i}{\sum_{i:=a}^{b} p_i} L(k,b) \right\}$$

where $L(a,b)$ is the routing length of the realizable questionnaire having $b-a+1$ answers. Let $K(a,b)$ be the smallest integer such that the minimum is realized. Then the sequence of integers $K(a,b)$ enables us to construct an L-optimal questionnaire the first question of which will be formulated by $x \geqslant K(1,N)$?

Example

$N = 6$, $\mathcal{P} = (0.30 \; ; \; 0.15 \; ; \; 0.15 \; ; \; 0.05 \; ; \; 0.30 \; ; \; 0.05)$.

The routing lengths of the L-optimal questionnaire and the realizable L-optimal questionnaire are:

$L_H = 2.35$ and $L_R = 2.55$. (Fig. 9.10)

However the distribution $\mathcal{P}' = \{0.30; 0.30; 0.15; 0.15; 0.05; 0.05\}$ leading to two answers of rank 2 with probability 0.30 would enable us to realize Huffman's questionnaire. We would obtain $L_R' = L_H$.

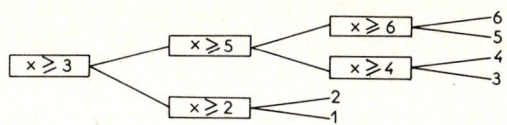

Figure 9.10

9.7.3 Absorption Tests

Kumar and more especially F. Dubail have studied the problem of absorption tests:

N individuals belonging to a community have a blood which can react in three different ways to a certain reagent. It is possible to take individual members to realize one test per person but we can also make tests by grouped samplings. In the first case we must realize N-1 "questions". In the second we may be satisfied with only one test if the reaction is "good"; if it is bad, we will have made an effort for nothing.

The three types of reaction are:

R1 No reaction

R2 Minor reaction

R3 Major reaction

If there is a mixture, then the strength of reaction of some single individual is stronger than all the others: character R3 absorbs the other characters; character R2 absorbs R1 which in the presence of another R1 would not react. There are 3^N answers with probabilities deduced from the trinomial law.

$$(q_1 + q_2 + q_3)^n .$$

For example the N answers will be

(R1,...,R1) with the probability q_1^n;

. . . .

(R2,R3,...,R3) with probability $q_2 q_3^{n-1}$

. . . .

and there are C_N^{N-1} ways to obtain it;

. . . .

(R3,...,R3) with probability q_3^n

and there is only one way to obtain it;

The tests will have basis 3 or 2 according to the situation determined by the preceding question.

There then exists an arborescent realizable optimal questionnaire.

In the case when $N \geqslant 2$ and $q_1^N \geqslant \frac{1}{3}$, Huffman's trichotomic questionnaire is not realizable. Dubail has completely solved this problem - generalizing the problems of tests where R3 and R2 coincided, studied by Sobel and his pupils: the situation is not formal but is given in the form of an ALGOL procedure; however Kumar has obtained by different methods formulas usable in a great number of cases.

9.7.4 Weighings

A well-known problem is that of identifying the one bad coin among 13 using a simple pair of scales. We can ask moreover to discover whether the bad coin is too light or too heavy. The usual formulation is to make it into a minimax problem (minimum height of an arborescent questionnaire). L-optimality turns this into a maximin problem (reduction of the difference between ranks). We will substitute N for 13 (with $N \geqslant 3$).

We dispose here also of dichotomic or trichotomic questions and we can show that the trichotomic L-optimal questionnaire is not realizable because the necessary questions are at most of basis 2 when an outcome has not been "equilibrium". Since there can be two kinds of imbalance - whether the plate goes up or goes down - there will therefore be two questions with basis 2. But there are 2N possible answers (what coin among N and what is the direction to the error in weight of this coin), Furthermore a heterogeneous questionnaire with all questions of basis 3 except two questions of basis 2 must have 2p+1 answers.

Finally the L-optimal realizable questionnaire cannot be a trichotomic Huffman questionnaire (in the broad sense).

Masson has shown that if $N \neq \frac{1}{2}(3^h-5)$ - respectively $N=\frac{1}{2}(3^h-5)$ - there exists a heterogeneous realizable questionnaire having three - respectively five - questions with basis 2 and all the other questions with basis 3 which is L-optimal. Since there are 2N possible answers, let us write $2N=3^k+2\alpha+1$; whereas the L-optimal trichotomic questionnaire is such that $L_H = k + \frac{3\alpha+2}{2N}$
the heterogeneous questionnaire of Masson, Ω_M is such that:

$$L_M = L_H + \frac{1}{N} \text{ if } N \neq \frac{1}{2}(3^h - 5)$$

or

$$L_M = L_H + \frac{3}{2N} \text{ if } N = \frac{1}{2}(3^h - 5) .$$

However if we are not trying to determine in every case whether the bad coin is lighter or heavier, then we can construct a questionnaire (Ω') with routing length $L_M - \frac{1}{N}$; it enables us to determine whether the bad coin is too light or too heavy except in the case when it produces with a probability 1/N where all the weighings have given rise to equilibrium of the scales. The height of Masson's questionnaire is h=k+1:

it is what Yaglom and Yaglom have called the minimal number of weighings necessary to determine "in every case" a bad coin.

We find for example:

For
$$N = 3 \qquad h = 2$$
$$4 \leqslant N \leqslant 12 \qquad h = 3$$
$$13 \leqslant N \leqslant 39 \qquad h = 4$$
$$40 \leqslant N \leqslant 120 \qquad h = 5$$
$$121 \leqslant N \leqslant 363 \qquad h = 6 \ldots$$

The height of questionnaire (Q') is less by 1 than the height indicated here when $N=\frac{1}{2}(3^h-1)$, that is for N=4, 13, 40 etc.

Remarks

1. The apparent analogy between the properties of optimality of realizable (arborescent) sub-questionnaires and the sub-questionnaires of Chapter 6 must not be taken too literally. Without constraint the L-optimality of an arborescent questionnaire constructed on ($\mathcal{A}$, $\mathcal{P}$) implies L-optimality relative to all the questionnaires arborescent or not. It could well not be the same for the generalization to realizable questionnaires.

2. An example was given in §4.4.3 which enabled us to interpret latticoid questionnaires by semantics. A theory (semantic) has been developed in which interpretative questionnaires are quite directly related to interpretative flow charts; Ludde and Thiele have obtained results in this formal setting which seem to be quite general.

The realization model studied in §9.6 enables us in general to solve the real-life problems posed by the practical user of questionnaires; certain types of questions (as for example 9.7.2 and 9.7.3) enable us to use the given pairs $(E(i),R_i)$ very directly. In other cases these pairs will not be made explicit and we often look for an algorithm proceeding upstream to downstream. The significance of the questions apparently allows us to do without the tools of admissible partitions and possible situations.

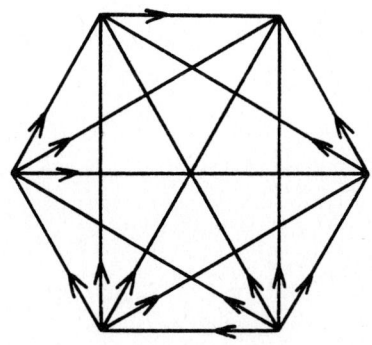

Chapter Ten

INTERROGATIONS, COMPARISONS, SORTINGS

10.1 Indirect Interrogations and Pseudoquestionnaires

10.1.1 Direct and Indirect Interrogations

Realizable arborescent questionnaires involve a restriction of the set $\mathcal{R}_i$ of partitions of the event $E(i)$ and imply that the outcome j of a question i depends only on the probability of realization of the elementary event $e \in E(i)$ introduced as input at the root. At the introduction of the given data, there corresponds only one state e of the phenomenon T under study, whatever its nature, (human, biological, social, economic, physical, data processing, etc.). It is within this framework that it has been possible to introduce a bijection $E \rightarrow E'$ from the set of answers of the arborescent realizable questionnaire onto the set of elementary events of T.

If this bijection maps $E(i)$ to $E'(j')$ we will say that the probability of the event $E'(j')$ in question conditioned by the outcome i is $p(j'|i)=1$. Any such procedure for studying T will be called a *direct interrogation*.

In some cases, such as for example that pointed out in §9.4.1 for $t<0$ or that of the inhabitants of town C in Problem 5, it can happen that the phenomenon being studied does not furnish any realiable answer - it is no longer possible to define a bijection but at most a surjection from E onto E'. That is, the probability of $E(j')$ knowing outcome i is $p(j'|i) \leqslant 1$. It may also happen that the answers of the questionnaire are infinite in number.

We say that the phenomenon T is *indirectly interrogated* by means of the experimental system Q and it will be necessary to introduce a family of auxiliary bipartite graphs D in order to be able to identify the answers of T by means of those of Q.

Such methods of indirect interrogation may be studied by means of the generalization that D. Chenais and Terrenoire have called pseudoquestionnaires. We will formulate a question i of a questionnaire Q whose outcome will enable us to modify the information obtained by the experimenter from T and according to the value of the mutual information of Q and T we will be led to look for one or the other partition. The certainty of the state of T will be obtained at the end of a process which might converge very slowly.

10.1.2 Pseudoquestionnaires

Let (X,Γ) be a finite or infinite arborescence with root α and suppose, to simplify the exposition, that it is homogeneous with basis a. Let $E \cup F$ be a partition of X into answers

$y \in E$ and questions $x \in F$: the paths $[\alpha y]$ are all finite.

We are given;

- a *denumberable set* Ω;

- a *measure* P enabling us to form the probability space $(\Omega, \mathfrak{P}(\Omega), P)$;

- a particular complete set of events on $(\Omega, \mathfrak{P}(\Omega), P)$ comprising N non-empty events:

$$T = T_1 \cup T_2 \cup \ldots, \cup T_j, \ldots \cup T_N, \; T_j \in \mathfrak{P}(\Omega) \; ;$$

T is called the *set of events under study;*

- a *probability distribution* on T:

$$\mathcal{P} = (p(T_1), p(T_2), \ldots, p(T_N)), \; \sum_{j:=1}^{N} p(T_j) = 1 \; .$$

- a set D of *complete systems of events* q^i on $(\Omega, \mathfrak{P}(\Omega), P)$ each comprising a non-empty events

$$D = \{q^i | q^i = q^i_1 \cup q^i_2 \cup \ldots \cup q^i_a, \; q^i_k \in \mathfrak{P}(\Omega)\} \; ;$$

D will be called the *detector* and associated with it will be a set of a x N stochastic matrices enabling us to interpret the events studied by means of D:

$$\nu_{kj}(i) = \frac{P(q^i_k \cap T_j)}{P(T_j)} = p(q^i_k | T_j) \; . \tag{1}$$

These matrices are supposed to be stochastically independent and without any zero column. Given these ideas of sets, complete systems, partitions and probabilities, we may now introduce the concept of a pseudoquestionnaire.

Definition 1 A *pseudoquestionnaire* constructed on D is the quintuple $K = (X, \Gamma, \mathcal{P}, u, B)$ in which

- u is a mapping $F \to D$ such that

$$|u(i)| = a \quad (\forall i)$$

and that its restriction u^τ to any walk $\tau = [\alpha \ldots i]$ of (X, Γ) is an injection;

- $B = \{b_i | i \in F\}$ is a set of bijections b_i between the a arcs coming out of question i and the a events of the complete system $q^i = u(i)$.

The vertices of F and E are the questions and the answers of the pseudoquestionnaire; each bijection attached to a question

INTERROGATIONS, COMPARISONS, SORTINGS   315

ensures that the coding of the arcs of a path conforms with the coding of the sequence of complete systems q^i, considered as the successive partitions of Ω performed at each of the vertices of the path.

Consider an event $\omega \subset \Omega$;

$$p(\omega) = \sum_{j:=1}^{N} p(T_j) p(\omega | T_j)$$

and for a subset q_k^i of the partition q^i such that

$$\forall \ell \quad q_\ell^i \cap q_k^i = \emptyset \quad \text{and} \quad \bigcup_{k:=1}^{a} q_k^i = q^i :$$

$$p(q_k^i) = \sum_{j:=1}^{N} p(T_j) p(q_k^i | T_j) . \qquad (2)$$

Let $x_0 = \alpha, x_1, \ldots x_\ell, \ldots x_h, x_{h+1} = i$ be the vertices of a path μ_i. For $\ell \in \{0, \ldots, h\}$ we will write

$$u(x_\ell) = q^\ell \quad , \quad b_{x_\ell}((x_\ell, x_{\ell+1})) = q_{i(\ell)}^\ell$$

and the event $E(i)$ associated with question i of K will be

$$E(i) = \bigcap_{\ell:=0}^{h} q_{i(\ell)}^\ell .$$

The probability of this event of $(\Omega, \mathcal{P}(\Omega), P)$ is:

$$P(E(i)) = \sum_{j:=1}^{N} p(T_j) p(E(i) | T_j) \qquad (3)$$

and consequently

$$p(i) = \sum_{j:=1}^{N} p(T_j) \prod_{\ell:=0}^{h} \nu_{i(\ell)j}(q^\ell) . \qquad (4)$$

The conditional probabilities

$$\Lambda_j(i) = \prod_{\ell:=0}^{h} \nu_{i(\ell)j}(q^\ell)$$

depend only on the detector D and the path μ_i and are preserved for a family of pseudoquestionnaires differing only by their distributions $\mathcal{P}$.

We will verify the condition of "conservation of flow":

$$p(i) = \sum_{k \in \Gamma i} p(k) \ . \tag{5}$$

Let σ be a section of (X,Γ) and σ' a subset of σ. According to the Definition given in §4.2.2, the probability of the section σ is $p(\sigma) = \sum_{i \in \sigma} p(i) = 1$ and similarly the probability of

σ' is
$$p(\sigma') = \sum_{i \in \sigma'} p(i) \leq 1 \ .$$

The laws of the probabilities $p(q_k^i|T_j)$, defined for any $q^i \in D$ and any system T of N events whatever the N-dimensional distribution $\mathcal{P}$, generalize the concept of a formable question: with any vertex $i \in X$, we associate the sequence of complete systems q^ℓ relative to its ancestors (including itself) as well as the probabilities of the subsets conditioned by the events under study.

For any non-empty event $E(i) = \bigcap_{\ell=0}^{h} q_{i(\ell)}^{\ell}$ we can calculate the probability:

$$p(T_j|i) = \frac{p(T_j) \prod_{\ell:=0}^{h} \nu_{i(\ell)j}(q^\ell)}{p(i)} \ . \tag{6}$$

If $h\Gamma i$ we obtain

$$p(T_j|i) = \frac{p(T_j|h)\nu_{i(h)j}(q^h)}{\sum_{j:=1}^{N} p(T_j|h)\nu_{i(h)j}(q^h)} \tag{7}$$

and

$$\frac{p(i)}{p(h)} = \sum_{j:=1}^{N} p(T_j|h)\nu_{ij}(q^h) \ . \tag{8}$$

We will say that the element $\Pi(T|x)$ of $\mathbb{R}^N$ with components $p(T_j|x)$ is the probability vector associated with x; this vector enables us to characterize the probability of state T_j of the system under study at any vertex i of a pseudoquestionnaire.

Definition 2 We will say that a pseudoquestionnaire K is *trivial* if it admits a section σ such that

$\forall i \in \sigma$, $\Pi(T|i)$ has one 1 and $N - 1$ zeroes.

It will be the same for any questionnaire obtained by a change of $\mathcal{P}$, according to (7).

The detector of K must be such that there exists a partition q associated with each pair (j,j'), $j \neq j'$ such that $\nu_{ij}(q) \neq 0 \Rightarrow \nu_{ij'}(q) = 0$.

INTERROGATIONS, COMPARISONS, SORTINGS   317

We will then say that D is *perfectly reliable*; each question associated with such a detector effects a partition included in the preceding one and we can associate a questionnaire Q, with K whose answers

$$e_1,\ldots,e_{N'} \quad (N' \geqslant N),$$

are such that:
$$\forall k \; \exists j \quad p(T_j|e_k) = 1 \tag{9}$$

The vertices satisfying (9) are called the *interpretable vertices* (whether K is trivial or not).

If $N'=N$ the pseudoquestionnaire will have operated exclusively by direct interrogations.

If $N'>N$ the arborescent questionnaire Q is compatible with a latticoid questionnaire obtained for example by contractions of the sets of vertices for which the same value of j satisfies (9).

We will say that a *trivial pseudoquestionnaire reduces to a questionnaire* (X,Γ,P_Γ) and we will determine $\mathscr{P}$ by means of the probabilities of the answers; if $N'=N$, then the T_j are the elementary events of Ω.

Definition 3 Consider two pseudoquestionnaires

$$K = (X,\Gamma, \mathscr{P}, u, B) \quad \text{and} \quad K' = (X',\Gamma', \mathscr{P}, u', B').$$

We will say that K is an *extension* of K' and that K' is a restriction of K if

- (X',Γ') is a sub-graph of (X,Γ) with the same root;

- u' is a restriction of u;

- for every question i of K', $\Gamma i = \Gamma' i$ and $b'_i = b_i$.

If (X',Γ') is the maximal sub-graph of (X,Γ) of finite height R, we will say that K' is the restriction of K with height R: the vertices of rank $r<R$ are of the same type in K and K' (terminal or non-terminal) and those of rank R are terminal in K'. We will then write $K'=K(R)$.

Definition 4 A pseudoquestionnaire K" is a *sub-pseudoquestionnaire* of K if (X'',Γ'') is a sub-arborescence of (X,Γ), u extends u", and $b''_i=b_i$ for any question of K".

10.1.3   Routing, information, convergence

The routing length of pseudoquestionnaires is defined, as for arborescent questionnaires, by:

$$L(K) = \sum_{i \in F} p(i)$$

and $L(K)$ can be finite or not whereas $L(K(R))$ is always finite.

In the case of the questionnaire Q, arborescent or latticoid, we introduced transmitted information $I(Q)$ and processed information $J(Q)$; it would have been easy to interpret $I(Q)$ as an information to be transmitted by the direct interrogation system Q, but in practice this was not necessary. By contrast, it is necessary in the case of the indirect interrogation system constituted by the pseudoquestionnaire K.

Definition 5 The *information to be transmitted* by a questionnaire Q or by a pseudoquestionnaire K is the information of the random system T with probability distribution $\mathcal{P}$:

$$I(\mathcal{P}) = \sum_{j:=1}^{N} p(T_j) \log \frac{1}{p(T_j)} .$$

We know (cf. Exercise 7.5) that the information processed by a question i of an arborescent questionnaire can be written in the form

$$J(i) = I(E(i), E(\Gamma i)) .$$

We therefore denote by

$$I(\Pi(T|i)) = \sum_{j:=1}^{N} p(T_j|i) \log \frac{1}{p(T_j|i)}$$

the information that must be transmitted by the sub-questionnaire with root $i \in X$ whose terminals are the terminals of the pseudoquestionnaire, $I(\Pi(T|i))=0$ if and only if i is an interpretable vertex.

Definition 6 The *information processed by a question i* of a pseudoquestionnaire is the quantity

$$J(i) = I(\Pi(T|i)) - \sum_{k \in \Gamma i} \frac{p(k)}{p(i)} I(\Pi(T|k)) . \qquad (10)$$

But the sum in the right-hand side of (10) is the mean information that the a sub-pseudoquestionnaires with roots $k (k \in \Gamma i)$ must transmit; it is the information to be transmitted by the subset with vertices Γi conditioned by i. We can therefore write

$$J(i) = I(\Pi(T|i) , \Pi(T|\Gamma i)) . \qquad (11)$$

Definition 7 The *information processed* by a pseudoquestionnaire K of finite height is the quantity

$$J(K) = \sum_{i \in F} p(i) J(i) . \qquad (12)$$

We can develop (12) as follows:

$$J(K) = \sum_{i \in F} p(i) \left[I(\Pi(T|i)) - \sum_{k \in \Gamma i} \frac{p(k)}{p(i)} I(\Pi(T|k)) \right]$$

$$J(K) = \sum_{i \in F} p(i) I(\Pi(T|i)) - \sum_{k \in E \cup \bar{F}} p(k) I(\Pi(T|k))$$

$$J(K) = p(\alpha) I(\Pi(T|\alpha)) - \sum_{e \in E} p(e) I(\Pi(T|e))$$

that is

$$J(K) = I(\mathcal{P}) - \sum_{e \in E} p(e) I(\Pi(T|e)) . \qquad (13)$$

Since K is of finite height, the vertices are interpretable and (13) reduces to $J(K) = I(\mathcal{P})$.

Theorem 1 The information processed by a pseudoquestionnaire of finite height K is equal to the information to be transmitted.

If K is of finite height we evaluate:

$$\lim_{R \to \infty} J(K(R)) = I(\mathcal{P}) - \lim_{R \to \infty} \sum_{e \in E_R} p(e) I(\Pi(T|e))$$

which in general is less than $I(\mathcal{P})$.

The method of using a pseudoquestionnaire K consists in indirectly questioning the random process T (with distribution $\mathcal{P}$) by going along a path as long as it passes through a question x with $p(T_j|x) \neq 1$ for all j.

If there is no path which allows us to make a definite decision, then it will be necessary to reduce the requirements and to impose some stop on the continuance of the questioning.

We say that a vertex x is *interpretable to within* ε, if there exists a j_0 such that $p(T_{j_0}|x) \geq 1 - \varepsilon$.

Let $\mathcal{K}(D)$ be the set of all pseudoquestionnaires constructed on D in which the terminal vertices are interpretable - their supports are (finite) arborescences. The restriction of K in which the vertices interpretable to within ε become terminals (so none of their ancestors are interpretable to within ε) will be denoted by $K(\varepsilon)$.

The restriction to height R of $K(\varepsilon)$ will be denoted by $K(R, \varepsilon)$ and the set of vertices of rank R in $K(R, \varepsilon)$ which are not interpretable to within ε by $E_{R\varepsilon}$. These latter are terminals in $K(R, \varepsilon)$ but only questions in $K(\varepsilon)$. We will denote the probability of the portion E_{R_ε} of the section by

$$p(E_{R_\varepsilon}) = \sum_{x \in E_{R_\varepsilon}} p(x).$$

The study of stopping criteria for indirect interrogation processes involves various concepts of convergence analyzed in detail by Terrenoire. They are connected with the height, with the routing length of pseudoquestionnaires or with $p(E_{R_\varepsilon})$.

Definition 8 A pseudoquestionnaire constructed on D will be said to be *strongly convergent*,

   if $\forall\ \varepsilon > 0$, $K(\varepsilon)$ is of finite height

L-*convergent*

   if $\forall\ \varepsilon > 0$, $\lim_{R\to\infty} L(K(R;\varepsilon)) < \infty$

H-*convergent*

   if $\forall\ \varepsilon > 0$, $\lim_{R\to\infty} p(E_{R_\varepsilon}) = 0$

Property 1 The different types of convergence of pseudoquestionnaires are connected by the following two implications:

1. strong convergence implies L-convergence

2. L-convergence implies H-convergence.

The first implication follows directly from the definitions. For the second we note that

$$L(K(R;\varepsilon)) = \sum_{r:=0}^{R-1} p(E_{r_\varepsilon}),$$

Thus, if $L(K(R;\varepsilon))$ is finite, the series with general term $u_r = p(E_{r_\varepsilon})$ converges for any $\varepsilon > 0$ and $\lim_{r\to\infty} p(E_{r_\varepsilon}) = 0$.

If D is finite, then the three concepts are equivalent.

H-convergence implies

$$(\forall\ \varepsilon)\ \lim_{R\to\infty} \sum_{e \in E_R} p(E_{R_\varepsilon})\ I(\Pi(T|e)) = 0 \qquad (14)$$

and this relation remains true for strongly convergent or L-convergent pseudo-questionnaires; we will also consider the trivial pseudoquestionnaires as particular cases of strongly convergent pseudoquestionnaires.

Theorem 2 The information processed by a pseudoquestionnaire K is less than or equal to the information to be transmitted and there is equality if and only if K is H-convergent.

Thus a kind of duality between questionnaires, with

$J(Q) \geq I(\mathcal{P})$ and pseudoquestionnaires with $J(K) \leq I(\mathcal{P})$ appears with equalities holding in the arborescent case or for H-convergence.

The concept of L-convergence has enabled Terrenoire to generalize the problem of towns (Problem 5) in the following manner:

The inhabitants of A always tell the truth, those of B tell lies and those of C randomly tell the truth or its contrary with probability $\frac{1}{2}$. It is asked to identify the type of inhabitant by means of dichotomic questions.

Whatever the partition q, that is the formable question used, the matrix $\nu(q)$ will be:

$$\nu(q) = \begin{pmatrix} 1 & 0 & \frac{1}{2} \\ 0 & 1 & \frac{1}{2} \end{pmatrix}.$$

As soon as there is no limitation of the questions, we dispose of an infinite detector for which a sufficient condition for L-convergence is satisfied: there exists a matrix s in which the columns are all distinct.

Consequently there is always a possibility of deciding on the town; at least on average, because L-convergence allows infinite paths for which beyond a certain rank the probabilities of the vertices tend to zero.

10.1.4 Applications to Pattern Recognition

Pseudoquestionnaires are well adapted to certain processes of structure recognition for which they facilitate the generation of efficient algorithms.

10.1.4.1 Diagnostic Aid

A project for giving a diagnostic aid in toxicology has been realized by means of pseudoquestionnaires (D. Chenais, Faure, Terrenoire). The choice of toxicology as the first field of medicine to receive this attention was indicated by the amount of statistical evidence which has been accumulated through international and interdisciplinary collaboration which is sufficiently precise to enable us to construct a detector which satisfactorily represents the real situation. The data may be represented by the actual answers given by the patient to a set of questions formulated in accordance with the partitions effected by the detector. In medical terms these answers are the symptoms whose combination permits the elaboration of the syndrome and then the diagnosis; the poisoning has been caused by toxin T_j. As in the theoretical case, by (1), the real detector consists of the set of matrices $\nu(i)$ giving the probabilities of the symptoms (outcomes q_k^i) when toxin T_j has been absorbed.

The practical application presents two main but on the face

of it contradictory requirements: to obtain a quick (paths of minimal length) and confident (minimal probability of error) answer. Corresponding to the first requirement is the idea of trying to construct a pseudoquestionnaire K(R) of limited height, and to the second the idea of trying to perform a truncation to within ε . We stop the questioning algorithm whenever for each vertex i of a section of the pseudoquestionnaire K there exists a toxin $T_{j(i)}$ such that

$$p(T_{j(i)}|i) > 1 - \varepsilon$$

However, as in the case of realizable questionnaires (cf. §9.7.1), the pseudoquestionnaire in toxicology is used step by step, that is for a patient symbolized by a given set of data. In this case we will be satisfied merely to construct step by step the itinerary formed by the consecutive outcomes.

Starting with an initial situation for which the vector Π is not undetermined - but related to general conditions or local conditions which may be fixed "once for all" - we have the choice between all the elements of the detector D. This detector is indeed finite but it is not possible to use more than R questions and there is a decision to be taken: which element of D must we choose? We will choose that which maximizes the processed information $I(\Pi(T|\alpha), \Pi(T|\Gamma\alpha))$ among all the q^1 of D. By means of (7) it is easy to determine the probabilities of the various toxins conditioned by the outcomes of the question, the formulation of which has thus been chosen. We are then in a situation analogous to the initial situation. We will pursue the indirect interrogation and the construction of the path of the arborescence while keeping the components of Π up to date. In practice we have realized a conversational system permanently giving the six greatest components of Π; we can stop the process as soon as these components are sufficiently differentiated without waiting for R to be attained.

The convergence of this type of pseudoquestionnaire seems to be very good in the case where the poisoning is due to a unique toxin already having a reference - multiple causes have not yet given rise to statistics or complete clinical tables -. We may compare the algorithm sketched here with that of Fano which also bore on the maximization of the information processed in the direct interrogation processes.

In practice, we do not use the fact that D is finite, limiting ourselves to an R relatively small as soon as we know how to determine the first toxins selected: the rest is the doctor's job.

The hypothesis that the mapping u is an injection is expressed by the fact that we do not have recourse twice to the same test. It is important to note that the clinician, G. Cau, has compared the systematic of this heuristic with the train of thought of the physician forming his diagnosis.

INTERROGATIONS, COMPARISONS, SORTINGS   323

10.1.4.2 Segmentation in a population

Consider a population (Ω) which we can question indirectly by means of formable questions according to a detector D. It is required to segment Ω by forming a "small" number of families characterized by the profile of their answers without imposing beforehand a series of type profiles. The value retained for R is fixed in advance to be very small and we try to build a pseudoquestionnaire $K(R, \varepsilon)$ such that the processed information is greater than a value fixed beforehand without separating the answers by more than ε. The same strategy of maximization of processed information $I(\Pi(T|i), \Pi(T, \Gamma i))$ is proposed once more here; but the inputs no longer correspond to only one individual: they correspond to the whole population under study. The maximization of the processed information will require, for R fixed, to choose the "most significant questions" and the analysis of the survey leading to segmenting Ω will be based only on the questions of the detector thus selected: we will say that these questions explain the behaviour of the individuals to within ε.

The inconvenience of these methods lies essentially in the number of logarithmic calculations to be effected, all the greater when the detector possesses more elements, but the rapidity of the increase of the processed information enables us to justify this heuristic which does not present any programming difficulty.

10.1.4.3 Word Recognition

Simon and Roche have used the same heuristic in the case of the recognition of five letter words (at most). A preliminary treatment is effected on the crude data introduced into a computer in the form of zeroes and ones arranged in a rectangular array. We then form a figure consisting of various segments of line or elementary primitives indexed from 1 to p (p being smaller than 8): the zeroes represented by dots (Fig. 10.1) indicate the absence of line; the numbers 1 to p refer to different "line segments" (of the letters) for which small angular differences are accepted; lastly each segment possesses an origin and an extremity and we will attach to each letter coded values indicating direction (by sector of $\pi/4$), the relations between segments, etc.).

The recognition of the words requires two (non-homogeneous) pseudoquestionnaires intervening one at the level of the word and the other at the level of the letter.

The *recognition of the word* uses a detector comprising 5 systems of events formulated by "what is the jth letter $j=1,\ldots,5$.

The program searches sequentially for the j for which the processed information is the greatest. For a dictionary with 50 French words chosen "at random", it suffices on average to find two letters to determine a word with a probability 0.8.

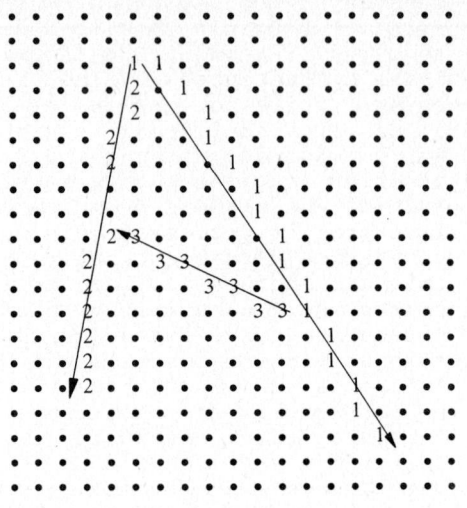

Figure 10.1

The recognition of the letter uses a detector having 9 systems of events such as those corresponding to:

- number of elementary segments, (for example 2 for V, 3 for A, 8 for 0).

- number of contact points of two segments,

- direction of the third segment,

- direction of the 1st (of the 2nd) segment; number of independent cycles (2 for B for example).

The double pseudoquestionnaire is processed in alternate fashion: that relative to the word operates once, then that relative to the letter operates one or more times. It has been established that it suffices to ask the first three questions of the detector to determine the letter. We then pass to the pseudoquestionnaire relative to the word.

A variation of this method consists in asking alternatively a question relative to one and to the other pseudoquestionnaire because it is not always necessary to completely identify a letter. For example if MARDI and MARIE are the only words of the dictionary in which the third letter has a cycle, then the answer to the question "number of segments of the 5th letter" (1 or 2 for mardi according to the linearity of the line, 3 to 5 for Marie in the same way) immediately completes the recognition if the cycle of the R has already been recognized.

10.2 Comparisons and Questions

Here we will study some cases in which the admissible partitions of the realizable questionnaires are obtained by means of the comparison operator with two outcomes > and < neglecting the case of equality, the occurrence of which will be considered as having zero probability.

Consider a complete system of events and suppose that there is a first stage of computation leading, according to the values of the indices or of known functions of the variables, to N eventualities. The second stage of computation or of processing is supposed to be independent of the eventuality, while the third stage is dependent on the eventuality which imposes a given type of sub-program.

Using a well-known classical method, we can put a code with value 0, 1, 2 or N-1 at the end of stage 1, then at the beginning of stage 3 compare the code to 0, then to 1, then to 2 ... and branch towards the sub-program to be executed as soon as the number of the eventuality has been identified to be that of the code. The theory of questionnaires shows that the progression must be made not from the beginning but from the middle, then by way of dichotomic separation. By middle we must understand a number comprised between the minimal and maximal values N_1 and $N-N_1$ (formula 8.30). So that, if the eventualities are equiprobable then the number of tests will be L_H whereas the classical method of progression gives L_{max} which can be much greater.

This method can be adapted to some machines by putting a code 1 in one of the 32 or 64 binary positions of a memory word, with the exclusion of the others. The different eventualities are then indexed by their abscissa in this memory word.

The consultation of tables for the search of a function $y(x)$ or the formation of a program in the technique called "table programs" can benefit from the same method and the questionnaire asked to resolve the problem of search in a minimal time will be dichotomic (equiprobable or not) because all the questions are comparisons with two possible outcomes.

Another classical problem in data processing is that of carrying out the minimum number of computations or more precisely to avoid carrying out redundant computations. It is sometimes useful to codify certain results in an automatic manner according to the outcomes of a sequence of questions as it was done in §9.6.2 for the problems of detection in regions of different probabilities.

10.2.1 Compatible realizable latticoids and arborescences

Suppose we want to find the maximum and the sub-maximum or second greatest. The search for the maximum of N numbers corresponds to the formulation of a questionnaire with N answers and therefore leads a priori to asking N-1 dichotomic questions. If these questions can be asked freely, then we can realize a sequence of questions involving *mutual exclusion*, as for example "does the maximum belong to the first N_1 numbers?" (where N_1 takes the value defined above). Whatever the outcome, yes or no, it will be possible to realize a questionnaire enabling us to ask either k or k+1 questions to determine the maximum of $N=2^k+\alpha$ numbers.

But in fact a question making use of such a mutual exclusion does not correspond to an elementary operation in a computer. The *restriction of the* field of a question must be restricted to only one comparison: is A greater than B? having two possible outcomes. In these conditions, J. Ville has shown that it is necessary to ask in any case at least N-1 questions because we cannot be sure of the result if the number had only been compared with N-2 others. He has constructed a questionnaire for determining the maximum of N numbers in the form of a latticoid in which the questions can have two predecessors.

Condensation of this type of latticoid facilitates the immediate compilation of a program for finding the maximum of N numbers (Fig. 10.2) but without using all the information processed by the N-1 questions of a path.

In fact in the case of 5 numbers, the two paths (i) using alternatively an arc yes (towards the top) and an arc no (towards the bottom) enables us to classify the five numbers, whereas the two extreme paths (e) corresponding to 4 outcomes yes or to 4 outcomes no do not enable us to obtain any information on the order of the 4 remaining numbers.

Let us form the arborescence of the paths. Since there are 5 numbers, 5!=120 permutations of them correspond to 120 eventualities for the classification of the numbers. A path (i) which will order the 5 numbers in 4 comparisons has a probability 1/120 of being taken while an extreme path (e) has a probability 1/5 of being taken.

The compatible questionnaire (Fig. 10.3) shows the 16 distinct paths of the latticoid and indicates the preordering relations obtained at the outcome of the 4 comparisons as well as the probabilities of the 16 eventualities.

Note that the comparison operator used here uniquely determines the compatible questionnaire C without having to calculate the $\lambda:(\cdot)$; in fact it is C which enables us to evaluate the arcs of the latticoid shown in Fig. 10.2.

The questionnaire C enables us to find the maximum of 5 numbers in 4 questions. To obtain the sub-maximum, that is the greatest of the 4 remaining numbers, the latticoid questionnaire requires 3 comparisons. By taking account of the preordering obtained by means of the arborescent questionnaire, what on average is the number of comparisons necessary to obtain the sub-maximum?

Denoting the paths linking the root to the answers according to the outcomes of the questions of rank 0 to 3 by a number

INTERROGATIONS, COMPARISONS, SORTINGS   327

with 4 binary digits, we can construct the table relative to the 120 cases.

In 48 cases

 (Paths 0000 and 1111)   3 comparisons are required

in 24 cases

 (Paths 0001, 0010, 0101, 0110
 and 1110, 1101, 1010, 1001)   no comparison is necessary

in 24 cases

 (Paths 0011, 0100 and 1100,
 1011)   1 comparison is necessary

in 24 cases

 (Paths 0111, and 1000)   2 comparisons are necessary

For, the first two groups require the maximal number of comparisons or no comparison whereas in the third group it is required to make one comparison: for example d>e for 0011 and b>c for 0100; the preorder obtained by path 0111, b>a, b>c, b>d, b>e and e>a imposes the comparisons c>d and either c>e, or d>e.

On average:

$$\frac{24.1 + 24.2 + 48.3}{120} = 1.8$$

comparisons are necessary.

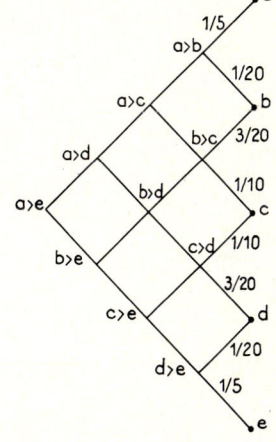

Figure 10.2

An arborescent questionnaire enables us to determine the maximum and the sub-maximum of 5 numbers with 5.0 comparisons on average instead of 7 with two latticoid questionnaires. But such a questionnaire is not optimal.

Huffman's algorithm applied to the questionnaire whose eventualities have probabilities equal to the values indicated in Fig. 10.3 shows indeed that the routing length could be reduced to 3.508 (as against 4) in what concerns only the finding of the maximum of 5 numbers.

But such a questionnaire is not realizable: it would require the use of questions from outside the range of comparisons available in a computer.

It should be noted that we have deliberately left aside the possibility of equality cases which in any case will have negligible probability when the range of the possible values of the numbers is important (9 significative figures for example) and if the numbers a, b, c, d, e all have the same probability 10^{-10} of taking any one of the values in the interval of definition.

The information processed by the latticoid is:

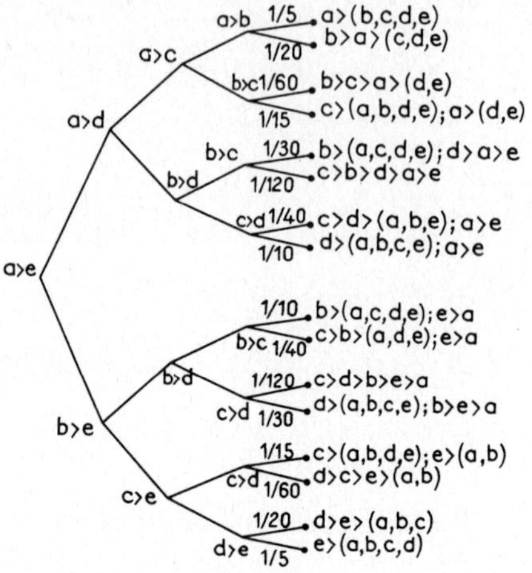

$$J(\Lambda) \leqslant I(B) = 3.639$$

Figure 10.3

the maximum value obtained for the compatible questionnaire B, with proportional flows and determined from the latticoid questionnaire in which the flows are themselves deduced from C. We note that $I(C)=3.452$ for questionnaire C (Fig. 10.3), that $I(\Lambda)=\log_2 5=2.322$ and finally that $I(D)=2.847$ for the arborescent questionnaire having 8 answers with the same probabilities as the 8 arcs entering into the set of answers of Λ.

10.2.2 Products of Arborescent Questionnaires

The finding of the sub-maximum of n numbers by means of the auxiliary results obtained in the course of the finding of the maximum is relatively economical but application of products of questionnaires will be more powerful because it will reduce the routing length even more.

The maximum of n numbers can be obtained by the product of n-1 questionnaires with only one question

$$a_i > M \quad (\text{for } 1 = 2,3,\ldots,n).$$

If yes, then we attach the value a_i to M.

Otherwise we will preserve M. At the start, $M:=a_1$.

The maximum and the sub-maximum will be obtained simultaneously by a product of n-1 questionnaires.

• First questionnaire equiprobable, with only one question: $a_1 > a_2$.

Let M be the greatest and S the other number.

- n-2 successive questionnaires for

$$i := 3, 4, \ldots, n.$$

1st question:

$a_i > S \rightarrow$ No (M and S are preserved)
↓
Yes. (then ask the second question).

2nd question:

$a_i > M \rightarrow$ No (so $S < a_i < M$ and thus a_i is the new sub-maximum)
↓

Yes (a_i is the new maximum and the value given earlier to M becomes the sub-maximum).

The probabilities of the answers in each of these elementary questionnaires are:

$$P\{a_i < S\} = \frac{i-2}{i} \qquad (15)$$

and

$$P\{S < a_i < M\} = P\{a_i > M\} = \frac{1}{i}, \qquad (16)$$

from which the elementary routing length is

$$L_i = 1 + \frac{2}{i}$$

for a processed information:

$$I_i = \log_2 i + \frac{i-2}{i} \log_2 \frac{1}{i-2} \qquad (17)$$

The noise of the product questionnaire is therefore:

$$B = \sum_{i:=3}^{n} \left[1 + \frac{2}{i} - \log_2 i + \frac{i-2}{i} \log_2 (i-2) \right] \qquad (18)$$

In particular for n=5:

$$I = 5.456 \quad \text{and} \quad L = 5.566.$$

We remark that transposing the order of the two questions $a_i > S$ and $a_i > M$ (for any i) would give, for the same structure of questionnaires, a routing length

$$L' = I + \sum_{i:=3}^{n} (2 - \frac{1}{i}), \qquad (19)$$

that is L'=6.217 for n=5, which is still greater than the value 5.8 found in paragraph 10.2.1.

The difference L-I is small, but the *transmitted information* I is the sum of two informations:

1. *information to be transmitted* corresponding to the isolated determination of the maximum ($\log_2 n$) and of the sub-maximum [$\log_2(n-1)$] that is I($\mathcal{P}$)=4.322 for n=5;

2. a *subsidiary information*, to a certain extent extraneous because it is not asked for, this information I_s corresponds to what the questionnaire has enabled us to detect about the order of the non-maximal elements. This preordering appears from the outcomes of the questions when an element is taken as maximum or sub-maximum before being rejected. The procedure does not use this information. However a similar procedure, TID, which will be studied later is in a certain sense the generalisation of the procedure SOUMAX in the case of complete ordering: I_s is then fully used and the noise is indeed negligible.

The ALGOL procedure can be written as:

```
procedure SOUMAX (m, s, j, k, a, n); value n;

real m, s; real array a; integer j, k, n;
```

comment: the integers j and k indicate the position of the maximum m and the sub-maximum s in the table a which consists of n elements, the finding of the minimum and of the next to minimum can be obtained in the same way by simple substitution of comparisons < for comparisons >;

```
begin : integer i : boolean e ;
   if n<2 then
      begin m : = s = a [1] ; j : = k : = 1; go to ALONE
      end;

e : = a [2] > a [1] ;
m : = if e then a [2] else a [1] ;
s : = if e then a [1] else a [2] ;
j : = if e then    2 else    1 ;
k : = if e then    1 else    2 ;

   for i : = 3 step 1 until n  do
      begin if a [i] > s  then
         begin if a [i] > m  then
            begin s : = m ; m : = a [i] ; k : = j ; j : = i
            end ;                     else
            begin s : = a [i] ; k : = i
            end ;
         end ;
      end ;
ALONE : end
```

Sobel has systematically studied the determination procedures for the maximum and sub-maximum and has suggested algorithms quicker than SOUMAX but they are very delicate to program.

10.2.3 Sequential Questionnaires

The programming of the Maximum procedure gives rise to a certain problem. Whether we consider Λ or C (Figures 10.2 and 10.3) we are led to formulate comparisons with ranks 1, 2 or 3 differently according to the result of the preceding comparisons.

For example we will compare b and d if d>a>e or b>e>a but a and c or c and e in the other cases. Stated otherwise, the probability of using each of the three comparisons b>d, a>c and c>e is 1/3 (which we can verify according to the probabilities of the answers of C).

This requires the program 1) to preserve in its memory the result of each comparison; 2) to provide as many branchings as the latticoid has questions.

If a memory is not provided, then the transmitted information is less than the processed information. If as many branchings are provided as the compatible arborescent questionnaire has questions, then the memory can be dispensed with without losing information. However it will be better to have recourse only to one flow chart "in line" where all the instructions would have to be effected systematically.

Definition 9 A *sequential questionnaire* is a realizable questionnaire whose associated partitions can be commanded by only one operator depending only on the rank.

If the sequential questionnaire is of height h, then as a rule it has a^h answers. However it can happen that certain operators are inefficient, that is, there is some question j to which there corresponds a pseudo-partition $(E(j), \emptyset, \ldots, \emptyset)$. In reality this will be a quasi-questionnaire in which certain internal vertices are quasi-questions or vertices with zero probability.

In a sequential questionnaire of height h, there are a^h distinct paths linking the root to the answers. But certain answers as well as certain questions could be the same and then the information processed would be greater than the information transmitted; the questionnaire will have support equal to a Dedekindian in which the only vertices with more than one entering arc are the terminals; such a questionnaire will still be called sequential (Fig. 10.4). The a^h paths can be preserved by memorizing certain quantities (for example we will preserve the name or the value of the maximum in a question formulated as a comparison) which as a rule are compulsory in this type of realizable questionnaire. The principal interest of sequential questionnaires lies in the fact that the only questions to be formulated, comparisons for example, are those of the questionnaire itself and thus it is not necessary to do service operations to test if there is need to ask a question and what question; these service operations can often double the number of questions determined by the finest of theories, when the questionnaire is not sequential.

For example the finding of the maximum of 5 numbers gives rise to a sequential questionnaire Q represented in latticoid form in Fig. 10.4. The compatible questionnaire defined by the semantic associated with the comparisons is of type B; it is shown in Fig. 10.5. We remark that the answers of this compatible questionnaire have the same probabilities as those of C (Fig. 10.3) for a distinct semantic.

The probabilities of the arcs conditioned by the initial vertices are:

$$\frac{p(i,j^+)}{p(i)} := \frac{1}{2}, \frac{2}{3}, \frac{3}{4}, \frac{4}{5},$$

where the arc (i,j^+) corresponds to the affirmative outcome (that is to the arc going up Fig. 10.4 and 10.5) to the question $i := \alpha, \beta, \gamma, \delta$.

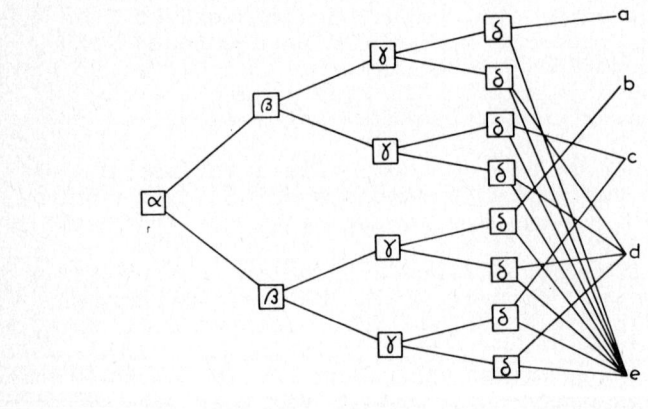

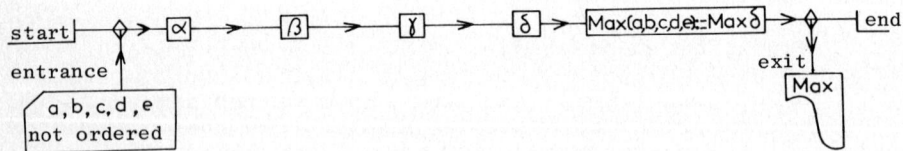

Figure 10.4 The comparisons effected are:

(α) $a > b$; (β) $M_\alpha > c$; (γ) $M_\beta > d$; (δ) $M_\gamma > e$,

where M_i is the greatest among the elements compared at question i.

INTERROGATIONS, COMPARISONS, SORTINGS   333

The purely sequential
flow chart (Fig. 10.4)
shows the interest of sequential questionnaires when
they exist.

We calculate:

L(Q) = 4 > J(Q) = I(C) = 3.452

J(Q) > I(Q) = $\log_2 5$ = 2.322.   (20)

In these realization
problems where the comparison
operator is used, as in many
problems calling for particular operators (weighing,
problems for example), we are
not able to realize a
questionnaire processing only   Figure 10.5
the information to be trans-
mitted. Whether it is required to select the greatest from a
list of N numbers or the bad coin among N coins, it is most
often impossible to answer exclusively the problem posed
without giving some subsidiary information, preordering on the
non-maximal items, or weight of the bad coin. We can therefore
suggest a:

Conjecture. There exists a class of problems for which the
obtaining of two informations relative to two phenomena A and
B are related in such a way that all the realizable questionnaires enable us to obtain I(AB) and not I(A) only.

Without being able to define the content of this class (and
it is precisely there that the preceding proposition is a conjecture) we therefore reverse the classical proposition
requiring the solution of a problem to transmit an information
I(A) without transmitting the subsidiary information I(B|A)
and we will say that to impose I(A) only, means to set the
problem badly or to truncate the element of information I(A|B)
obtained necessarily by any realizable questionnaire.

In the case of the sequential questionnaire Ω, the subsidiary information, I(C)-I(Q) is far from being negligible.
This does not signify that Q is the "best" sequential questionnaire enabling us to obtain the maximum of 5 numbers. We note
that this questionnaire enables us to generate a family of
sequential questionnaires for 2 ≤ N, and we may realize
various corresponding programs as exercises.

10.3 <u>Questionnaires for Sorting</u>

10.3.1   <u>L-optimal sorting</u>

We will try to resolve the problem of sorting of numbers
placed in the same set which could be for example the fast

access memory unit of a computer. We will operate with purely numerical words because ordering of alphabetic words presents itself in the same way except for the richness of the code, whereas the case of groups of words related to a key of storage can be reduced to the preceding ones.

We will study the optimal sorting only in the sense of the routing length: only the questions enable us to act on the information and in the first place we will neglect the transfer of numbers which the method will lead us to make.

The sorting of n numbers in the fast access memory leads us to build a questionnaire with n!=N answers which we can consider as equiprobable: they are the n! possible orders given by the ordering relation <. In practice the order corresponding to ≤ would give a number of answers greater than N but would lead us to recommend questions in which one outcome (=) would have a probability considerably less than the other two (< and >).

The operator enabling us to formulate the questions will be the comparison "A<B?" whose outcomes will be *yes* (A<B) and *no* (A>B).

We will deal with this problem in the context of classifying by increasing order.

Dichotomic tests relative to a system of $n!=N=2^k+\alpha$ answers correspond to a mean minimal number of comparisons, $L_o = k + \frac{2\alpha}{N}$.

The sorting problem consists in the realization of a questionnaire with routing length as near as possible to L_o enabling us to classify n numbers in the natural order.

For n=3, L_o=2.66 and the questionnaire is immediate: two questions enable us to determine the maximum and, if one of the numbers has been recognized to be included between the two others, they enable us to find the order. In 4 cases out of 6, it is necessary to use 3 comparisons. This questionnaire can be interpreted as the product of two arborescent questionnaires with N_1=2 and N_2=3 answers (Fig. 10.6).

For n=4 the number of answers is $4!=2^3 \times 3$. We will still consider the questionnaire as a product with $N_1=2^3$ and N_2=3. The L-optimal questionnaire, for which L_o=4.66 is still immediate if the third question compares the smallest elements determined by the preceding questions (Fig. 10.7).

The questions are only shown explicitly in the case when the

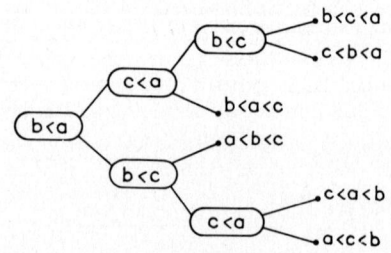

Figure 10.6

first two outcomes are yes, the paths leading to the 18 other answers are obtained by symmetry.

For n=5 the number of answers $5! = 2^3 \times 15$ enables us to consider the product with $N_1 = 2^3$ and $N_2 = 15$. The first three questions will be identical with those of the case n=4 and the fifth number to sort will be compared only starting from rank 3. An L-optimal arborescence A_2 is constructed by considering among the possible comparisons those giving a partition of the 15 answers into 8 and 7. If b<a, d<c and b<d, the 7 formable comparisons give rise to the following cases (Fig. 10.8)

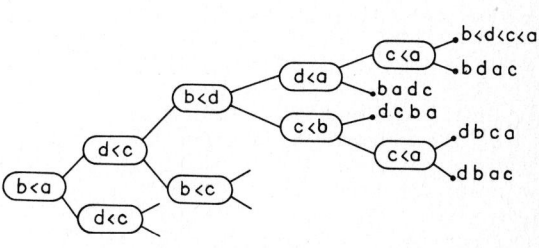

Figure 10.7

Then the comparison with rank 3 is d<e?. The other comparisons can then be written without any difficulty (Fig. 10.9).

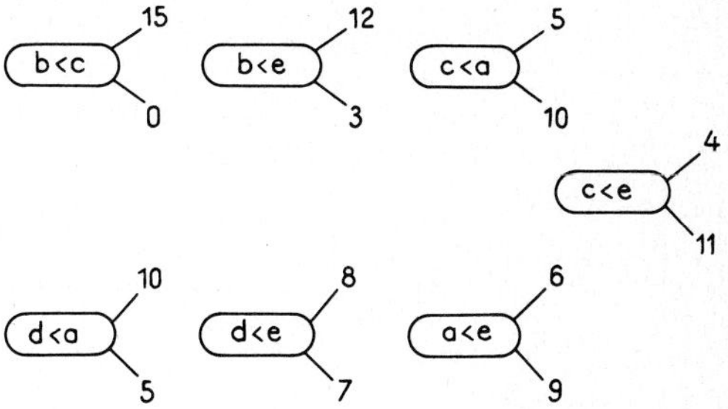

For n=6 the number of answers $6! = 2^4 \times 45$ enables us to consider the product with $N_1 = 2^4$ and $N_2 = 45$. The first four comparisons, the outcomes of which are equiprobable can be written:

⓪   b < a   ?
①   d < c   ?
②   f < e   ?
③ min ⓪ < min ②   ?

Figure 10.8

In the case when the 4 successive outcomes have been for example, four yes (Fig. 10.10), a table of all the 11 questions formable a priori shows that the partition of the set of 45 eventualities into two events comprising N_1 and $45-N_1$ eventualities such that

$$16 \leqslant N_1 \leqslant 32 \quad \text{and} \quad 16 \leqslant 45 - N_1 \leqslant 32$$

can only arise for

$$N_1 = 19 \quad \text{and} \quad 45 - N_1 = 26.$$

Many questions enable us to obtain 26 yeses and 19 noes. One of these questions leads in the final stage of the experiment to resolving a system of eventualities presenting themselves in the form of a product of dichotomic questionnaires with 3 answers of the type (ABC) (αβγ) of questionnaire 10 of Problem 5. Since the only type of question admissible is comparison and since the product of two equiprobable questionnaires with three answers is a generator of noise it is impossible to construct an L-optimal questionnaire if the question of rank 4 is not a>c? The choice of this question enables us to continue the construction of this L-optimal sorting questionnaire for 6 numbers (Fig. 10.10) with the decomposition of the cardinalities of events (Fig. 10.11).

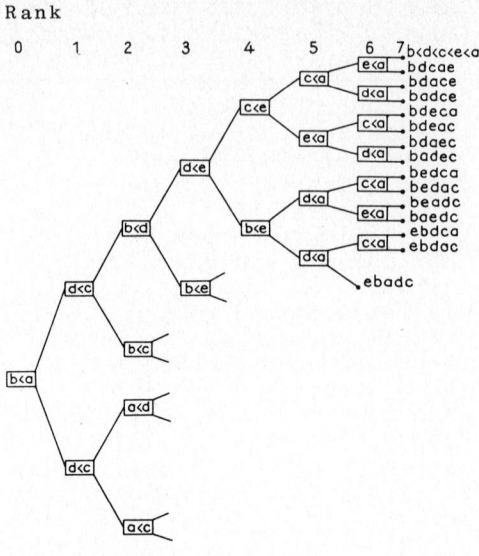

Fig. 10.9 L-optimal sorting of five numbers.

Figure 10.10 L-optimal sorting of six numbers.

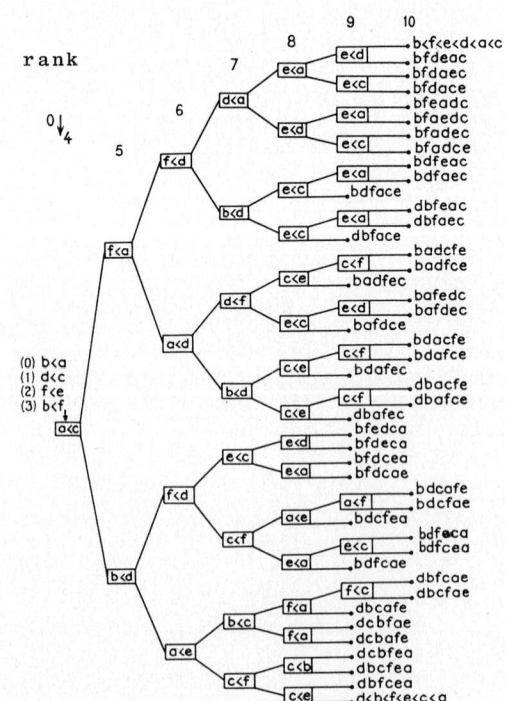

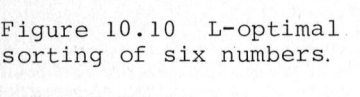

Figure 10.11

For n>6 the number of answers N=n! can be put in the form of a product of powers of prime numbers. The power p of 2 is obtained easily by summation of the integer quotients of n by the successive powers of 2.

We may then effect a decomposition of the eventualities and try to construct a product of questionnaires in which $N_1=2^p$, N_2 is odd and the first questionnaire is sequential.

The numbers taken two by two enable us to make [n/2] comparisons, the minimum of these pairs, taken two by two enabling us to make [n/4] comparisons: by continuing step by step with the smallest numbers belonging to the pairs formed earlier it is easy to realize p comparisons. But to continue further it is necessary to proceed by means of the evaluation of the partitions which can be obtained by the different questions to be asked. What is evident for n=3, immediate for n=4, and reasonable for n = 5 or n = 6 is yet feasible for 7 and 8 but does not seem to be systematizable for a value of n of the order of 100. A first algorithm should enable us to decide, among the different admissible partitions for each rank, which one is to be preserved: and since we must continue until the maximum rank is reached in order to be sure that a set of questions is optimal, the application of this method would be heavy going. Moreover the orders of the comparison being related to the paths followed, a sorting program could not possibly be made efficient because of the number of transfer operations and branching operations neglected up to now.

In fact it is useless to continue the search for sorting questionnaires which are both L-optimal and realizable beyond 6: Cesari has tried to find if such a questionnaire existed for n=7. A necessary condition for the arborescence to be balanced is that any vertex with rank k is associated with a partial order compatible with at least 2^{k-r} total orders and at most with 2^{k+1-r} total orders (with $7!=2^k+\alpha$ that is k=12). To show that such a questionnaire is not realizable, Cesari has used a computer to consider the finer and finer realizable partitions obtained by following the path giving the least probable partial order. The heuristic thus avoided exhaustivity. In all the cases there exists a rank ρ and a vertex ξ with rank ρ such that the number of total orders compatible with the partial order defined by ξ is strictly less than $2^{12-\rho}$. He has operated in the same way for n=8, with the same result, which is *a priori* not surprising because 8! has three more powers of 2 and the same odd factor 315. However Cesari has tried all the partitions, including those where the questions with two equiprobable outcomes were not all among the questions of rank 0 to 6.

Nevertheless Cesari has realized L-optimal questionnaires for sorting 9 and 10 items. For 11 and 12 his first researches on a computer have led to an impasse as in the cases 7 and 8.

Property 2 There exist L-optimal realizable questionnaires with the comparison operator to sort 2, 3, 4, 5, 6, 9 and 10 items but not 7 and 8 items.

10.3.2 Realizable Sortings

We can next try to construct questionnaires which without being L-optimal, will at least have the same height as L-optimal questionnaires: only the second part of the preceding necessary condition must then still be satisfied (not more than 2^{k+1-r} total orders); we will say that they are questionnaire solutions of a minimax problem. Cesari and Hadjan[3] have realized such questionnaires presenting a proper noise $B_p = \frac{1}{7}$ and Sobel conjectures that there do not exist any, for n=7 or n=8 possessing a proper noise of 1/7 or 2/7.

Ford and Johnson have suggested a fairly efficient sorting algorithm. This algorithm facilitates the realization of questionnaire solutions of the minimax problem for $2 \leqslant n \leqslant 11$ and for n:=20,21 but we do not know of any solution beyond that. Ford and Johnson's algorithm gives at the same time the solution of the maximin problem (therefore L-optimal) for n=5 but not for n=6. By an exhaustive method recalling that of Cesari, Wells has shown that the Ford-Johnson method leads for n=12 to a height greater by 1 than the solution of the minimax problem.

Finally these researches lead to the following conclusions:

1. It is legitimate to match the numbers to be sorted to find the minimum (or the maximum) of each pair, then for each group of
$$2^2, 2^3, \ldots, 2^h$$
numbers. Some sorting algorithms studied for example by Sobel are governed according to this rule, but because of the very low limitation for h (h=2 or 3) lead to routing lengths which are quickly prohibitive.

2. By substituting a homogeneous questionnaire for the semi-homogeneous questionnaire we are led naturally to recommend the use of questions with basis 2, 3, 5 for the factor decomposition of n!. If this procedure was possible it would be more powerful than that of dichotomic comparisons and would lead to defining a questionnaire in which L and the acquisition would be considerably less than the information expressed actually in base 2 (for n=100, the reduction would exceed 50%). Of course it is not possible to have at hand questions having as basis the greatest prime number less than n as soon as n exceeds 10 or 20. But this procedure would lead very naturally to simulate a semi-homogeneous questionnaire by proceeding by successive products of arborescences.

The product of n L-optimal questionnaires is a questionnaire in which the noise is equal to the sum of the inefficiencies of the n arborescences. If we realize a questionnaire with n! answers by a simulation method it would be possible to evaluate with precision the upper limit of the noise generated.

More simply still the sorting of n numbers could be decomposed into two stages:

a. sort n-1 numbers, that is order them in increasing (or decreasing) order;

b. insert the nth number ($n=2^k+\alpha$) by placing it between the two nearest numbers. Since there are n possible places, on average, by operating through dichotomy, $L_0 = k + \frac{2\alpha}{n}$ comparisons are necessary.

The number of comparisons required by a sorting through dichotomic insertion is

$$L = \sum_{p:=2}^{n} L_p ,$$

where L_p is the routing length of the optimal questionnaire with p eventualities.

The noise of this questionnaire is $B = L - I$, that is

$$B = \sum_{p:=2}^{n} (L_p - \log_2 p) ,$$

limited by

$$n B_{max} = 0.0861 \, n .$$

Hence

n	I	nB_{max}
5	6.9	0.4
20	61.1	1.7
100	524.8	8.6
10 000	118 460	861

This method was proposed simultaneously by Iverson, Kislitsyne and Picard in 1963 and leads to a practically negligible noise. This classification by insertion according to a dichotomic scale must not be confused with the simple insertion by comparisons in increasing order in which the routing length is

$$L' = \sum_{p:=2}^{n} \frac{p-1}{2} - \frac{1}{p} ,$$

which increases very rapidly.

This method of sorting by insertion is easily applicable in the case of manual sorting. But the actual technology of electronic computers does not readily lend itself to insertion in the fast access memory. That is why Hibbard, who had seen the possibility of this method, has preferred to renounce its use. It seems that the power of this method is based on;

- the reduction of the noise due to the extraneous comparisons in considerable proportions (the method proposed finally by Hibbard requires 1.4 times more comparisons on average);

- the reduction to n of the number of extraneous comparisons for the longest paths,

- the possibility of using the preliminary sorting of a part of the data and that of inserting the numbers one after the other step by step as they arrive eventually in the course of data processing;

- the non-use of auxiliary memories because the pth number will be inserted in a group of p numbers which do not overlap beyond the pth storage unit.

The algorithm of sorting by dichotomic insertion (TID) will be obtained by a double product of sequential questionnaires: on the one hand for the insertion of the ith number and on the other for the description of all the values of i from 3 to n.

We use only one auxiliary memory which we will consider as the $(n+1)^{st}$ memory of the table where the numbers to be ordered are placed.

Procedure: TID (x,n) ; real array x ; integer n ;

Comment: Sorting by increasing order of the numbers through dichotomic insertion, the only memory necessary for intermediate storing requires that the array x be declared with n+1 elements of which the last is fictitious.

```
begin : integer i, j, k, m, s ;

   if n = 1 then go to EXIT ;
   if x [n]  < x [n - 1] then
      begin  x [n + 1] : = x [n] ;
             x  [ n ]   : = x [n - 1] ;
             x [n - 1] : = x [n + 1] ;
      end ;
   if n = 2 then go to EXIT ;
   for i : = n - 2 step - 1 until 1 do
   begin x [n + 1] : = x [i] ; m : = i ; s : = n + 1 ;
      A : k : = (m + s) ÷ 2 ;
             if x [i]  < x [k] then s : = k else m : = k ;
             if s - m ≠ 1  then go to A ;
             for j : = i step 1 until m - 1 do
             x [j] : = x [j + 1] ; x [m ] : = x [n + 1] ;
   end ;
   EXIT : end
```

In the case where we have at hand the information concerning the *a priori* probability of the elements of the file of being already classified, a substantial economy of time is obtained by a generalization of the TID program.

Although these methods are nearly L-optimal, they reveal themselves to be scarcely fit for use in practice because of the number of shiftings in memory core which they impose at each insertion. Among the best sorting techniques of the fast access memory we must point out those due to Flores and Gamzon which use direct addressing. Having determined the smallest and the greatest elements of the file of items to be ordered and having allocated to them the extreme positions of the memory, the direct addressing consists in determining by only one function of the first degree the theoretical place of any item by supposing equipartition: the definitive arrangement takes into account the possible double allocations. We then operate with questions of basis n.

Whereas the number of comparisons, required by the TID method of sorting or Ford and Johnson's method, is of the order of $\log_2 n!$, that is: $n \log_2 n$ except for a multiplicative constant, the number of "operations" - comprising both arithmetical operations and service instructions and integrating the search for the maximum and the minimum - increases linearly with n in Gamzon's method which reveals itself to be very efficient: the routing length tends to $L(n) = \frac{e}{2e-3}(n-2)$ as n tends to infinity.

It is to be remarked finally that the methods of sorting on magnetic tapes by fusion can be studied by means of equiprobable realizable questionnaires. According to the fusion basis (basis of the questionnaire for which $x \Gamma y$ if the file figuring at the "step" x proceeds from the fusion of the files figuring at the different "steps" $y \in \Gamma x$), it is possible to determine if a method of the polyphase type is quicker than another of the "balanced" type for example. Cesari has given fairly complete tables enabling us to decide on the method according to the characteristics of the data processing material used.

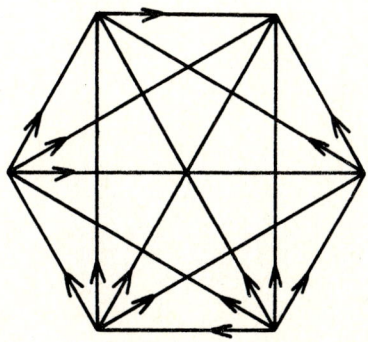

PROBLEMS

Problem 1 Incidence-Matrix and Connectivity

Let $A=(X,\Gamma)$ be a graph of order 8 defined by the adjacency matrix:

1. Determine the number of arcs m and the number of weakly connected components p.

   Discuss the properties of symmetry, reflexivity and transitivity in A.

2. Find the incidence matrix of the graph obtained from A by suppression of the loops and the incidence matrices of the two largest connected components C and D (which define them uniquely).

	1	2	3	4	5	6	7	8
1	.	1	.	.	.	.	.	.
2	1	.	1	.	.	.	.	.
3	1	.	.	.	.	.	.	.
4	.	.	.	1	1	.	1	.
5	.	.	.	.	.	1	.	.
6	.	.	.	.	.	.	1	.
7	.	.	.	1	.	.	.	.
8	.	.	.	.	.	.	.	1

Figure 1

3. Which elementary cycles are also circuits? Which arcs do not belong to elementary circuits? Find a simple cycle which is not elementary.

4. Show that C and D, weakly connected by hypothesis, are also lower quasi-strongly connected.

   Do they have any stronger form of connectivity? If so, what is it? If not, why not?

5. Algorithm for the transformation between the incidence and adjacency matrices.

   Find an algorithm for constructing the adjacency matrix of an s-graph $\{a_{ij}\}$ from the incidence matrix $\{s_{ij}\}$.

   We suppose that:

   $s_{ij} = 1 \Rightarrow$ arc j comes out of vertex x_i

   or arc j is a loop at x_i.

   $s_{ij} = -1 \Rightarrow$ arc j enters into vertex x_i

   $s_{ij} = 0 \Rightarrow x_i$ is not an extremity of arc j,

   and that:

   q_1 arcs come out of x_i and enter into $x_k \Rightarrow a_{ik} = q_1$

q_2 arcs come out of x_k and enter into $x_i \Rightarrow a_{ki} = q_2$

q_3 loops come out of $x_i \Rightarrow a_{ii} = q_3$.

The given matrix S has n rows and m columns, and the matrix A to be found has n rows and n columns the sum of its non-zero entries being m.

Problem 2   Eulerian and Hamiltonian Paths

Let A be the graph discussed in Problem 1, where C and D were also defined.

1.   Determine the cycle rank k and the cocycle rank ℓ of A. Enumerate the elementary cycles of C and D. Which elementary cycles constitute a basis of independent cycles?

2.   Let A_j be the partial graph of A obtained by the suppression of j arcs. For what value of j is A_j circuitless? How can we be sure that j is minimal? Let i be the minimal value of j. What kinds of connectivity are possessed by C_i and D_i (obtained from C and D by suppression of i arcs)? Are these kinds of connectivity dependent on the choice of arcs belonging to A and not to A_j?

3.   Discuss the elementary cocycles of A. Find a basis of elementary independent cocycles in A.

In view of Problem 1, question 3, we know that A has at least one cocircuit (why?). Enumerate the cocircuits of A.

4.   Same as question 2 but A_j is to be cocircuitless.

5.   Let B_h be the graph obtained from C ∪ D by adjunction of h arcs. Find a graph B_2 which is strongly connected. Let W and Z be the arcs of B_2 not in Γ. Since B_2 is strongly connected, it necessarily has a circuit passing through all its vertices. Is there a simple circuit passing once and only once over each arc of B_2 (Eulerian cocircuit)? If not, is there an Eulerian cycle or an Eulerian path (passing once and only once over each arc)? Are these answers dependent on the choice of the arcs W and Z?

6.   Show that it is possible to choose W and Z in such a way that B_2 possesses an elementary circuit passing through the n vertices of B_2 (Hamiltonian circuit). With this choice does B_2 have a cycle or an Eulerian path?

Problem 3   Paths in a Circuitless Graph

Let G = (X,Γ) be a graph of order n(>6) with adjacency matrix given by:

$$a(i,j) = \begin{cases} 1, & \text{if } i < j \leqslant n \text{ and } j \leqslant i + 3 \\ 0, & \text{otherwise} \end{cases}$$

where a(i,j)=1 signifies the existence of an arc going from vertex i to vertex j and the first index is the index of the rows.

1. Evaluate the indegrees and outdegrees of all the vertices and the numbers of arcs and circuits in G.

2. Evaluate the number of independent elementary cycles.

3. Show that this graph has a Hamiltonian path and also a Hamiltonian cycle.

4. Let G^p be the graph on the same set X which expresses the existence of paths of length p in G. Indicate the form of the adjacency matrix of G^p for p=1,2. Give the limiting form as p tends to n.

Problem 4   Dissection of Parallelepipeds

Consider a p x q x r rectangular parallelepiped consisting of pqr small 1cm cubes.

We make a "vertical" cut wherever we wish, then arrange the resulting pieces however we wish and make another "vertical" cut and so on. The problem is to minimize the number C of cuts necessary to separate the pqr elementary cubes.

We will assume that a "vertical" cut will allow us to cut several of the pieces at once by placing them suitably.

Consider the following particular cases:

1. A 5 x 8 x 1 parallelepiped - or a block of chocolate

2. A p x p x p parallelepiped - or a cube

3. A 3 x 4 x 5 parallelepiped.

Finally examine the case when a cut is only allowed to split one piece into two (r=1), with application to a 5 x 8 x 1 parallelepiped.

Remark: This "very pleasant and delectable" problem was suggested by C. Berge.

Problem 5   Towns and Liars

Consider three towns A, B and C which are such that the inhabitants of A always tell the truth, those of B always tell lies and those of C alternately tell lies and the truth. Observer X wants to find out which town he is in and also which town his informant comes from. How can we reduce to the minimum the number of questions he must ask knowing that his informant will answer only "yes" or "no"?

The questionnaire which the observer must submit to know where he is (3 possible outcomes: A, B, or C) and to whom he is speaking (α from A, β from B or γ from C, i.e. three possible outcomes) has 9 answers. However the solution has on the face of it 12 answers. For each inhabitant of C may give true answers to odd questions and false answers to even questions or vice versa. Thus there are two types of inhabitant of C:

γ those beginning with the truth, (TLTL)

$\bar{\gamma}$ those beginning with a lie, (LTLT)

We see that there are three possible cases:

When the informant comes from C he may be either one of the types γ or $\bar{\gamma}$ or he may surely be of type γ or surely of type $\bar{\gamma}$. We could say in the last two cases that the inhabitants of C give as their first answer to a stranger (knowing the date) a true answer on odd days and a false answer on even days. We propose to study the questionnaires corresponding to the following cases:

Equiprobable dichotomic questionnaires:

1.   12 answers (A, B, C) (α, β, γ, $\bar{\gamma}$). (This problem was suggested by Yaglom and Yaglom.)

2.   9 answers (A, B, C) (α, β, γ).

3.   9 answers (A, B, C) (α, β, $\bar{\gamma}$).

deduced directly from 1 by the suppression of questions which have become useless.

4.   Questionnaire 2 optimized (γ).

5.   Questionnaire 3 optimized ($\bar{\gamma}$).

Non-equiprobable dichotomic questionnaires ($p(\alpha)=p(\beta)$).

6.   Questionnaire 1 optimized to take into account the probabilities $p(\gamma)=p(\bar{\gamma})=\tfrac{1}{2}p(\alpha)$.

7.   Questionnaire 4 optimized to take into account the probabilities $p(\gamma)=p(\alpha)$; $p(\bar{\gamma})=0$.

8.   Questionnaire 5 optimized to take into account the probabilities $p(\bar{\gamma})=p(\alpha)$; $p(\gamma)=0$.

Supposing that the inhabitants of A, B and C always tell the truth, discuss

9.   the product of equiprobable questionnaires.

10.   the equiprobable questionnaire 9 optimized.

Determine the routing length and the information transmitted in each of the 10 cases.

Problem 6   Routing Length, Products and Ordering

1. Realize an arborescent questionnaire which will facilitate the ordering (sorting) of 4 distinct numbers X, Y, Z, T.

1.1 Evaluate the number N of answers (all equiprobable). Evaluate the information I_N and the routing length L_a of the polychotomic optimal questionnaire with basis:

$$a = 2 \, , \, a = 3 \, , \, a = 5 \, .$$

(Use a as the base of the logarithms in the evaluation of I_N.)

Is it necessary to use one of the formulae established in §6.6.2 for the value of L_5?

1.2 For the effective realization of the questionnaire we have at hand one of the following systems of questions $\{Q_a\}$.

   1st system

$\{Q_2\}$ consists of questions with basis $a=2$ constituted by algebraic comparison.

   V<U having two outcomes   V<U   (upper arc)
                             V>U   (lower arc).

   2nd system

$\{Q_a\}$ consists of questions with basis $a=3$ and $a=2$.

The first question α separates the N answers into 3 equiprobable classes:

   {X is the smallest element or (X and T) are the two smallest.}

   {Y is the smallest element or (Y and T) are the two smallest.}

   {Z is the smallest element or (Z and T) are the two smallest.}

Then at rank 1 the question is formulated relative to the first of the preceding classes:

$$(X < Y < T) \quad \text{or} \quad (X < T < Y) \quad \text{or} \quad (T < X) \quad ? \; ;$$

and similarly for the other classes (by cyclic permutation of X Y Z). Further a final series of questions, whose bases are to be specified, facilitates the separation of the answers which are still grouped.

   3rd system

$\{Q_5\}$ consists of questions with basis 5 and we suppose that this system enables us to realize a polychotomic optimal questionnaire with routing length $M_5=L_5$.

Whatever $\{Q_a\}$, it is assumed that the permitted transfers are without penalty and we will determine the routing length by considering questionnaires realized with the aid only of the questions of the given system $\{Q_a\}$.

Schematize the realizable questionnaires and evaluate the routing lengths M_2 and M_3 of the questionnaires corresponding to Q_2 and Q_3; justify M_3.

2. Suppose it is required to determine the two greatest numbers (Ω,π) and the two smallest (ω,ρ) in a list of n+4 distinct numbers in disorder, where $\Omega>\pi$ and $\omega>\rho$.

2.1 Evaluate the information J_a obtained to base a by the determination of (Ω,π,ω,ρ).

Calculate J_2 and J_5 when n=12.

2.2 We have at hand a set of questions with basis a=5 which are capable of detecting the position of a given number among four other numbers already ordered.

Construct a product of questionnaires enabling us to determine (Ω,π,ω,ρ) by using in the first place the ordering of the four numbers with the aid of $\{Q_5\}$.

Determine the routing length C_5. Computation for n=12.

2.3 We have at hand a set of questions with basis a=2 which are capable of detecting whether a given number belongs to an interval (A,B) of the type A<X<B?, with at least one of the two extremities being not equal to infinity.

Construct a product of questionnaires enabling us to determine (Ω,π,ω,ρ) by using in the first place the ordering of the 4 numbers with the aid of $\{Q_2\}$. Will the product be optimal?

This product can be realized in at least three different ways by starting from questionnaires facilitating the comparison of the ith number with the two smallest and the two greatest of the i-1 first numbers. Is this procedure appropriate for i=5 and i=6?

Determine the routing length C_2 corresponding to these products of questionnaires. Computation for n=12.

3. Can we apply the methods of paragraph 2 to select the four greatest numbers in a list of n numbers?

Problem 7 Heterogeneous Questionnaires and Insertion

We have at hand questions with bases 6, 3, 2 to realize a heterogeneous questionnaire which will enable us to sort t numbers.

We assume that these questions are permutation instructions with 3 addresses u, v, w, or input addresses. Three numbers to be ordered are placed in u, v and w. At the exit of the instruction, the outcomes of the questions are the 6 compatible orders: the minimum M among the three numbers is placed in u, the intermediate I in v, the supremum S in w. Such an instruction is then a question with 6, 3 or 2 outcomes according as before executing it

. we have no information on the order of (u), (v), (w).

. we know that the content (u) of u is less than the content (v) of v or more generally two of the three numbers are already classified.

. we know the minimum, the intermediate or the supremum but we do not know the relative order of the other two.

1. Study the questionnaires obtained by insertion for t := 3, 4, 5, 6.

Supposing the t numbers are ordered in a random sequence from the start, evaluate the routing length

$$L(t) = \sum_{j \in Q} p_j$$

where the j's are the questions.

It should be noted that several methods enable us to order the tth number when t-1 numbers have already been ordered.

Find an approximate value of the information

$$I = \sum_{i \in E} p_i \log_2 \frac{1}{p_i} ,$$

where E is the set of eventualities (to be specified).

2. A weight πa_j is attached to each question with basis a_j. Evaluate the weighted routing length

$$N(L) = \sum_j p_j \pi a_j ,$$

where a_j is the basis of the question j.

Applications

$\pi_2 = 1 \qquad \pi_3 = 1 \qquad \pi_6 = 1.5$

$\pi_2 = 1 \qquad \pi_3 = 1.5 \qquad \pi_6 = 2.5$

Compare I, L and N(t).

3. We suppose that t=6, but we will start by merging the ordered two sub-lists a<b<c and d<e<f with a view to

establishing the order of the set $\{a,b,c,d,e,f\}$.

Determine L and $N(6)$ under the same hypothesis as before, compare with I.

Problem 8   Equivalent Questionnaires

Two questionnaires are said to be *equivalent* if each is obtained from the other by permutation of the vertices of the same ranks and probabilities. The equivalence class of a questionnaire Q is determined by the set of questionnaires which are equivalent to Q, including Q. Two questionnaires are considered to be *distinct* if they are not equivalent.

Let B be the set of questionnaires having four answers of probabilities p_1, p_2, p_3, p_4 constructed on a question with basis 3 and a question with basis 2;

Let C be the set of dichotomic questionnaires having 4 answers with probabilities

$$p_1, p_2, p_3, p_4 \;;\; p_1 + p_2 + p_3 + p_4 = 1.$$

1.   Can the questionnaires of B be considered as homogeneous, semi-homogeneous, heterogeneous?

Distinguish the two types of arborescences (for B as for C) supporting the questionnaires of B and C; we will say that the questionnaires are of type B_1, B_2, C_1 or C_2. Enumerate the equivalence classes when all the answers have different probabilities.

2.   Let $p_1 > p_2 > p_3 > p_4$. Determine the questionnaires of minimal routing length (L_o for B, L_m for C) and maximal routing length (L_p for B, L_M for C). Say how they depend on the values of p_1, p_2, p_3, p_4.

3.   Suppose $p_1 = p_2 > p_3 = p_4$. Enumerate the equivalence classes. Same questions as in 2.

4.   Application

Suppose
$$p_1 = p_2 = \frac{1}{3} \;;\; p_3 = p_4 = \frac{1}{6}.$$

Calculate the routing lengths of the distinct questionnaires of types B_1, B_2, C_1, C_2.

Calculate the informations I_2 and I_3 of the questionnaires (with logarithms to the bases 2 and 3) and compare these values to the different routing lengths.

Problem 9   Reduction of the circuits in a flow chart

Consider the programme:

```
BEGIN INTEGER M, X; READ (M); X := 0 ;
BEGIN REAL ARRAY A [1: M, 1: M] ; INTEGER I, J, K ;
   FOR I := 1 STEP 1 UNTIL M DO
   FOR J := 1 STEP 1 UNTIL M DO
   READ (A [I,J] ) ;
TRICYCLE: FOR I := 1    STEP 1 UNTIL M-2 DO
          FOR J:= I + 1 STEP 1 UNTIL M-1 DO
          FOR K:= J + 1 STEP 1 UNTIL M   DO
X = IF ((A [I,J] ≠ 0 V A [J,I] ≠ 0) ∧ (A [J,K] ≠ 0 V A [K,J] ≠ 0)
       ∧(A [I,K] ≠ 0 V A [K,I]≠ 0))  THEN X + 1 ELSE X
       WRITE (M,X)
   END
END
```

We look for the mean computation time T of the instruction TRICYCLE as a function of the order M of the matrix A and the elementary time τ_s when $M \geq 3$. For that we take into consideration only the time τ_s necessary for evaluation when the boolean expression takes on the value TRUE or the value FALSE and the probabilities of these answers; these probabilities are considered to be independent of A and M.

1. Give the number XMAX of the 3-element sets (I, J, K) considered by TRICYCLE (XMAX is the maximal value which X can attain at the end of the programme).

τ_1 is the computation time of the complete boolean expression B appearing in the allocation instruction of X in TRICYCLE; evaluate T as a function of M and τ_1.

2. τ_2, the time of computing the boolean expression C:

$$(A [I,J] \neq 0 \text{ V } A [J,I] \neq 0) ,$$

is the time taken by a dichotomic elementary questionnaire with two answers (C TRUE, C FALSE) with probabilities p_T and p_F.

2.1 For fixed I, J, K, what sort of product can one realize to obtain the N_2 answers corresponding to the computation B? Value of N_2?

Calculate the routing length L_1, as a function of p_T:

Applications:

a. $p_T = 1 - p_F = \frac{1}{2}$

b. $p_T = 1 - p_F = \frac{3}{4}$.

2.2 For variable I, J, K, what sort of product is it necessary to realize to carry out the instruction TRICYCLE?

Evaluate T as a function of M, τ_2 and p_T.

3. τ_3, the time of computation of the boolean expression C, is the time taken by a trichotomic questionnaire with 3 answers:

Y_1, if (A [I,J] ≠ 0) = TRUE ;

Y_2, if (A [I,J] = 0 ∧ A [J,I] ≠ 0) = TRUE ;

Y_3, if (A [I,J] = 0 ∧ A [J,I] = 0) = TRUE ;

the probabilities are respectively p_1, p_2 and p_3.

3.1 What questionnaire will enable us to calculate B?

What is the number N_3 of answers and the routing length L_3?

Applications:

a. $p_1 = p_2 = p_3 = \frac{1}{3}$

b. $p_1 = \frac{1}{2}$, $p_2 = p_3 = \frac{1}{4}$.

3.2 Calculate T as a function of M_1, τ_2 and the probabilities.

4. τ_4, the time of computation of the boolean expression D:

$$(A [I,J] \neq 0),$$

is the time taken by a dichotomic questionnaire with two answers with probabilities q_T and q_F.

What questionnaire will enable us to calculate B? What is the number of answers? For $q_F = \frac{1}{2}$, determine the routing length L_4 and evaluate T as a function of M and τ_4.

5. Show that the information I_4 processed by the product of questionnaires easily reduces to L_4.

Now let

$$p_3 = \frac{1}{4}, \quad p_1 = \frac{1}{2} \quad \text{and} \quad q_F = \frac{1}{2}.$$

Show that I_3 (evaluated with logarithms to the base 3) is then proportional to I_4 (evaluated with logarithms to the base 2).

Express I_3 and determine k such that $I_4 = kI_3$.

Indicate the value of the ratio $\tau_3/\tau_4 = h$ necessary to obtain $T(M\tau_3) = T(M\tau_4)$.

Interpret the significance and the value of h/k.

6. After the preceding study, it has been decided to modify the programme starting from the label TRICYCLE by substitution of:

```
BICYCLE: FOR I:= 1  STEP 1 UNTIL M - 2  DO

         FOR J:= I + 1  STEP UNTIL M - 1  DO

         IF (A [I,J] ≠ 0 V A [J,I] ≠ 0)

         THEN FOR K = J + 1  STEP 1 UNTIL M  DO

         BEGIN IF (A [J,K] ≠ 0 V A [K,J] ≠ 0)

         THEN BEGIN

           X = IF (A [I,K] ≠ 0 V A [K,I] ≠ 0)

               THEN X + 2  ELSE X

           END

END;

             WRITE (M,X)

             END

END
```

Evaluate the time T_6 by means of the duration τ_3 and show that there exists a finite value of M starting from which, whatever p (non-zero), the instruction BICYCLE will be executed more rapidly than TRICYCLE.

Take into account the result of §5 to deduce the quickest programme, MONOCYCLE, determining X.

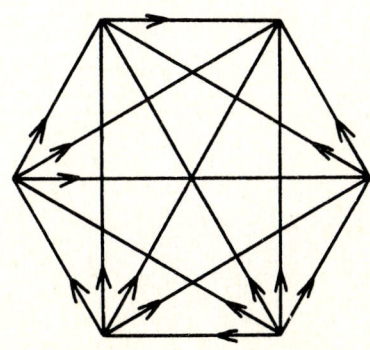

SOLUTIONS TO PROBLEMS

Problem 1 Incidence Matrix and Connectivity

1. The fundamental parameters of A are:

   vertices : n = 8   (given)

   arcs   : m = 11 (given)

   weak connectivity: the adjacency matrix shows the existence of a chain]123[with no vertex of this chain joined to any of the others; there is also a chain]4567[and the vertex 8 is not joined to any other vertex. Hence

   $$p = 3$$

   From the adjacency matrix we see that 1Γ2 and 2Γ1 but 2Γ3 and 3$\not\Gamma$2; 1$\not\Gamma$1 but 8Γ8; 2Γ3, 3Γ1 and 2Γ1 on the other hand and 4Γ5, 5Γ6, 6$\not\Gamma$4 on the other. Consequently A does not have any of the following six properties: antisymmetry, symmetry, reflexivity, irreflexivity, intransitivity, transitivity.

2. The three weakly connected classes are $\{1,2,3\}$, $\{4,5,6,7\}$ and $\{8\}$. Let C be the connected component generated by $\{1,2,3\}$ and D the one generated by $\{4,5,6,7\}$. After suppression of the loops (4,4) and (8,8) from A, we number the arcs from 1 to 9 as follows:

   Let $j=(i,k)$. We look for the smallest ℓ such that, if $\exists \ell > k$ such that $i\Gamma\ell$, then $j+1=(i,\ell)$ and otherwise if $(i+1)\Gamma\ell$, then $j+1=(i+1,\ell)$.

   We will form only one incidence matrix, the first block of 3 rows and 4 columns being the incidence matrix of C and the second block of 4 rows and 5 columns being the incidence matrix of D. To avoid confusion the names of the arcs will be written with a prime. (Fig. 0)

   We note that

$$S_{ik'} = \begin{cases} +1, & \text{if } k' \text{ comes out of } i \\ -1, & \text{if } k' \text{ enters into } i \\ 0, & \text{otherwise (zero is indicated by .).} \end{cases}$$

	1'	2'	3'	4'	5'	6'	7'	8'	9'
1	+1	-1	.	-1	.	.	.	.	.
2	-1	+1	+1	.	.	.	.	.	.
3	.	.	-1	+1	.	.	.	.	.
4	.	.	.	.	+1	+1	.	.	.
5	.	.	.	.	-1	.	+1	.	-1
6	.	.	.	.	.	.	-1	+1	.
7	.	.	.	.	.	-1	.	-1	+1

Figure 0

3. In C there are three elementary cycles given by the arcs; [1',2'] and [1'3'4'] are circuits and]2'3'4'[is not a circuit because 2' and 4' enter into 1.

In D there are similarly]5'7'8'6'[,]5'9'6'[and [7'8'9'] which latter is a circuit.

Among the arcs, only 5' and 6' do not belong to any circuit.

A simple non-elementary circuit is a set of arcs (without repitition) such that one extremity of each arc is an extremity of the next in order (taken cyclically) it being possible for a vertex to be an extremity of more than two arcs in the set.

The chain]1'3'4'2'[is simple, non-elementary and contains a cycle but is not itself a cycle. The same is true of the component D. To find a simple cycle which is not elementary we must start from A. We then find for example]5'7'8'6'10'[by observing the loop 10=(4,4).

4. In C there is a circuit [1'3'4'] passing through all the vertices. C is therefore strongly connected and lower quasi-strongly connected.

In D the vertex 4 is the origin of two arcs and the terminal extremity of none. Thus there is no circuit passing through 4 and D is therefore not strongly connected. However 4 is the origin of paths leading to 5, 6 and 7 and therefore D is l.q.s.c.

In particular the path [5'7'8'] or [4567] passes through every vertex of D which is therefore also u.q.s.c. and therefore quasi-strongly connected.

5. The incidence-adjacency algorithm.

```
'BEGIN'
  'INTEGER' I,J,K,L,C,M,N ;
  'FORMAT' F0=''(20(3X,I2))'' ;
  'FORMAT' F1=''(10(5X,I1))'' ;
  READ (N,M) ;
  'BEGIN'
    'INTEGER' 'ARRAY' G [1:N, 1:N] , S [1:N,1:M] ;
    'FOR' I:=1 'STEP' 1 'UNTIL' N 'DO'
      'FOR' J:=1 'STEP' 1 'UNTIL' N 'DO' G [I,J] :=0 ;
    'FOR' I:=1 'STEP' 1 'UNTIL' N 'DO'
      'FOR' J:=1 'STEP' 1 'UNTIL' M 'DO' READ(S [I,J] ) ;
    WRITE ('' INCIDENCE MATRIX '') ;
    OUTPUT (61,F0,'FOR' I:=1 'STEP' 1 'UNTIL' N 'DO'
        'FOR' J:=1 'STEP' 1 'UNTIL' M 'DO' S [I,J] ) ;
    'FOR' J:=1 'STEP' 1 'UNTIL' M 'DO'
    'BEGIN'
      C:=0 ;
      'FOR' I:=1 'STEP' 1 'UNTIL' N 'DO'
        'IF' S [I,J] ≠ 0 'THEN'
        'BEGIN'
          'IF' S [I,J] = 1 'THEN'
          'BEGIN'
            K:=I ;
    ET1:C:=C+1 ;
```

```
            'IF' C ⩾ 2 'THEN
              'BEGIN'
                G [K,L] :=G [K,L] +1 ;
                'GOTO' ET3 ;
                'END' ;
              'GOTO' ET2 ;
          'END'
       'ELSE'
         L:=1 ;
         'GOTO' ET1 ;
  ET2: 'END' I ;
      'IF' C = 1 'THEN' G [K,K] :=G [K,K] +1 ;
ET3:'END' J ;
    WRITE ('' ADJACENCY MATRIX '') ;
    OUTPUT (61,F1,'FOR' I:=1 'STEP' 1 'UNTIL' N 'DO'
            'FOR' J:=1 'STEP' 1 'UNTIL' N 'DO' G [I,J] );
  'END' ;
'END'
```

INCIDENCE MATRIX

1	1	1	-1	0	0	0	-1	0	0	0	0	0	0	-1	0	0	0	0	0
2	0	0	1	1	1	1	0	0	0	0	-1	0	0	0	0	0	0	0	0
3	0	0	0	-1	-1	-1	1	-1	0	0	0	0	0	0	0	0	0	0	0
4	0	0	0	0	0	0	0	1	1	1	0	-1	0	0	0	0	0	0	0
5	0	0	0	0	0	0	0	0	0	0	1	1	1	0	-1	0	0	0	0
6	0	0	0	0	0	0	0	0	0	-1	0	0	-1	1	1	0	0	0	0
7	0	0	0	0	0	0	0	0	0	0	0	0	0	0	0	1	-1	0	0	0
8	0	0	0	0	0	0	0	0	0	0	0	0	0	0	0	-1	0	-1	-1	0
9	0	0	0	0	0	0	0	0	0	0	0	0	0	0	0	0	1	1	0	-1
10	0	0	0	0	0	0	0	0	0	0	0	0	0	0	0	0	0	0	1	1

ADJACENCY MATRIX

	1	2	3	4	5	6	7	8	9	10
1	2	0	0	0	0	0	0	0	0	0
2	1	0	3	0	0	0	0	0	0	0
3	1	0	0	0	0	0	0	0	0	0
4	0	0	1	1	0	1	0	0	0	0
5	0	1	0	1	0	1	0	0	0	0
6	1	0	0	0	1	0	0	0	0	0
7	0	0	0	0	0	0	0	1	0	0
8	0	0	0	0	0	0	0	0	0	0
9	0	0	0	0	0	0	1	1	0	0
10	0	0	0	0	0	0	0	1	1	0

<u>Problem 2</u>   <u>Eulerian and Hamiltonian Paths</u>

1.   From Problem 1 we already know m=11, n=8 and p=3. Hence the cycle rank k=6 and the cocycle rank ℓ=5.

Among the three elementary cycles [1'2'] , [1'3'4'] ,]2'3'4'[of C and the three elementary cycles]5'7'8'6'[,]5'9'6'[, [7'8'9'] of D only four belong to the same basis of independent cycles because, after the suppression of two loops, the cycle rank reduces to 4. We may take three circuits and a cycle to form a basis. For example [1'2'] ,

[1'3'4'], [7'8'9'] and]5'9'6'[. Each arc of C or D belongs to at least one cycle and each cycle contains at least one arc which is not in any other (see Figure 0).

2. i is at least equal to 4. The loops (4,4) and (8,8) must be suppressed together with at least one arc from C and at least one from D.

The arc 1' belongs to two elementary circuits in C both of which are broken by its suppression. The suppression of 7', 8' or 9' breaks the unique circuit in D. Hence i=4.

The component C_4 also contains the path [3'4'] passing through the three vertices 2, 3 and 1. Hence C_4 is semi-strongly connected.

If we suppress 9', then the path [5'7'8'] still belongs to D_4 which therefore remains semi-strongly connected.

If we suppress 8', then the arc 6'=(4,7) ensures the existence of a path going from 4 to 7. D_4 is therefore l.q.s.c. and only l.q.s.c. because 6 is not the origin of any arcs.

If we suppress 7', then D_4 is u.q.s.c. but it loses lower quasi-strong connectivity because 6 is not the terminal extremity of any arcs and is also not the origin of any path going to 4.

3. A loop cannot belong to a cocycle because its extremity cannot belong both to $S \subseteq X$ and to X-S. The cocycles of A are therefore the same as the cocycles of C and D which have cocycle ranks $\ell_C=2$ and $\ell_D=3$.

Among the elementary cocycles in C,

$$\omega(1) = \{1',2',4'\}, \omega(2) = \{1',2',3'\}, \omega(3) = \{3',4'\},$$

we may take $\omega(1)$ and $\omega(3)$, which has the same arcs as $\omega(\{1,2\})$, to form a basis.

Similarly we may take $\omega(4)$, $\omega(\{4,5\})$ and $\omega(\{4,5,6\})$ as a basis of the elementary independent cocycles of D. These five cocycles constitute a basis for A. For every proper arc belongs to at least one of them and every cocycle contains at least one arc which is not in any of the others. The arcs 5' and 6' do not belong to any circuits (cf. 3, Problem 1) so they each belong to a cocircuit. Since $\omega^+(4)=\{5',6'\}$, they belong to the unique cocircuit of A.

4. Since A has only the one cocircuit $\omega^+(4)$, the suppression of 5' and 6' produces a graph A_2 without cocircuits. C_2 remains strongly connected whereas D_2 is not connected. We find two strongly connected classes in D_2, namely $\{4\}$ and $\{5,6,7\}$.

5. Since the graph $C \cup D$ is not connected, it is necessary to

adjoin an arc coming out of {1,2,3} and entering {4,5,6,7} and an arc coming out of {4,5,6,7} and entering {1,2,3}. Thus $h \geq 2$. Since {1,2,3} and {5,6,7} are equivalence classes in the sense of strong connectivity and since no arc enters into 4 it is sufficient to adjoin two arcs of the type

$$W = (s,4) \quad \text{and} \quad Z = (t,s')$$

where $s, s' \in \{1,2,3\}$ and $t \in \{5,6,7\}$

to form the strongly connected graph B_2.

Counting the non-zero entries of the ith row of the incidence matrix of $C \cup D$ will give the degree of vertex i. Thus we see that the vertices 1,2,5,7 are of odd degree.

Taking s=1 (or 2), s'=2 (or 1) and t=5 (or 7), the vertices 1, 2 and 5 (or 7) will have even degree in B_2, whereas 4 will have odd degree. It follows that, since there are two vertices of odd degree in B_2, there cannot be a circuit or an Eulerian cycle but there is an Eulerian chain in B_2.

Since there is only one arc W entering into 4, we can hope to obtain an Eulerian path by taking 4 as the origin of this chain. Let W=(1,4) and Z=(5,2).

Then the chain beginning with the arcs 5', 7', 8' can be readily extended; [5'7'8'9'Z2'1'4'W6'] constitutes an Eulerian path e_1 comprising the 11 arcs of B_2.

If we now take W=(1,4) and Z_1=(7,2), we find similarly an Eulerian path e_2= [6'9'7'8'$Z_1$2'1'3'4'W5'] .

These paths may also be written in the form:

$$e_1 = [4\ 5\ 6\ 7\ 5\ 2\ 1\ 2\ 3\ 1\ 4\ 7]$$

$$e_2 = [4\ 7\ 5\ 6\ 7\ 2\ 1\ 2\ 3\ 1\ 4\ 5]$$

There is no other way of proceeding for D.

Since, in C, the vertex 1 is of odd outdegree and the vertex 2 is of even indegree, we must choose s=1 and s'=2 and not the contrary.

We see that the Eulerian paths of C are [21231] and [23121] only the first of which is included in e_1 and e_2. Thus we may obtain two more paths e_1' and e_2' by using this variation in C. For example

$$e_2' = [4\ 7\ 5\ 6\ 7\ 2\ 3\ 1\ 2\ 1\ 4\ 5]$$

Hence there are two ways of adjoining two arcs to form B_2. The choice {W,Z} leads to two paths e_1 and e_1' and the choice {W,Z_1} leads to the two paths e_2 and e_2'.

6. To construct a cycle passing exactly once through each vertex of the graph, it is also necessary to adjoin two arcs to A to form B_2.

The four Eulerian paths all contain the elementary circuit [121]. To construct a Hamiltonian cycle, we therefore start from the simple paths

$$f_1 = [4\ 5\ 6\ 7\ 5\ 2\ 3\ 1\ 4\ 7]\quad \text{obtained from } e_1 \text{ and } e'_1$$

and $\quad f_2 = [4\ 7\ 5\ 6\ 7\ 3\ 1\ 4\ 5]\quad \text{obtained from } e_2 \text{ and } e'_2.$

Suppression of the arc (4,7) from f_1 gives a non-elementary circuit, suppression of the arcs (5,6), (6,7) and (7,5) from f_1 gives a non-Hamiltonian circuit whereas suppression of the two first arcs of f_2 gives the Hamiltonian circuit of B_2:

$$h = [5\ 6\ 7\ 2\ 3\ 1\ 4\ 5] = [7'\ 8'\ Z_1\ 3'\ 4'\ W\ 5']\ .$$

Problem 3   Paths in a Circuitless Graph

1. $$d_i = \begin{cases} 2 + i & \text{, for } i \leq 4 \\ 3 + (n-i) & \text{, for } i \geq n - 3 \\ 6 & \text{, for } 4 \leq i \leq n - 3 \end{cases} \text{; degree}$$

indegree	i	outdegree
0	1	3
1	2	3
2	3	3
3	$4 \leq i \leq n - 3$	3
3	$n - 2$	2
3	$n - 1$	1
3	n	0

Circuits: none (upper triangular matrix)

Arcs : from the outdegrees

$$3 + \sum_1^{n-3} 3 = 3n - 6\ .$$

The graph is antisymmetric.

2. $k = m - n + p$. But the graph is clearly connected because of the upper diagonal and therefore $p = 1$ and

$$k = (3n - 6) - n + 1 = 2n - 5\ .$$

k is the number of elementary independent cycles.

3.   The Hamiltonian path is obviously

$$[1,2,3,\ldots,n-2,n-1,n]\ .$$

A cycle would be for example

$$]1,3,5,\ldots,n-1,n,n-2,n-4,\ldots,2,1[$$

or $]1,3,5,\ldots,n-2,n,n-1,n-3,n-5,\ldots,2,1[$

according as n is even or odd.

Neither the cycle nor the chain are unique but there is only the one Hamiltonian path because of the orientation.

4.   The graph G^p has the same vertices as G and its arcs are given by the paths of length p in G.

Let $G=\{a_{ij}\}$ and $G^p=\{p_{ij}\}$ be the adjacency matrices.

Now there exists a path of length p between i and i+p (p>0) and also a path of length p between i and i+3p or i+k for $p \leqslant k \leqslant 3p$.

Hence the adjacency matrix of G^p is a banded matrix of width 2p+1 in which the only non-zero entries are those for which $i \leqslant n-p$ and $i+p \leqslant j \leqslant \min(n,i+3p)$ and these are all equal to 1.

When p<n, it is evident that the non-zero entries are in a triangle placed in the upper right hand corner of G^p. For p=n, $p_{1n}=1$ and $p_{ij}=0$, otherwise. This gives another proof of the existence of only one Hamiltonian path and the absence of circuits.

Further for q>n, $G^q = \emptyset$.

These results can also be established by considering the ordinary matrix products without reference to the interpretation in terms of paths.

Problem 4   Dissection of Parallelepipeds

Let $p=2^{k_1}+\alpha$ be the first dimension of the parallelepiped.
To cut into $2^{k_1}+\alpha_1$ parallelepipeds with the same thickness amounts to realizing a dichotomic equiprobable questionnaire with minimal routing length. The number of cuts will be the height of the L-optimal arborescence. Thus

$$c_1 = k_1, \quad \text{if } \alpha_1 = 0$$
$$c_1 = k_1 + 1, \quad \text{if } 0 < \alpha_1 < 2^{k_1}$$

which we will write as

$$C_1 = k_1 + [\alpha_1].$$

We can operate in the same way "in Y and Z" which gives

$$C_2 = k_2 + [\alpha_2] \quad \text{and} \quad C_3 = k_3 + [\alpha_3]$$

under the same conditions.

But cutting in X, then in Y and then in Z amounts to carrying out the product of the three questionnaires (A_1), (A_2), (A_3).

If the cuttings could be made in a dependent manner, that is, if after having cut $k_2' < k_1$ times perpendicularly to direction X, it was desired to cut $k_2' < k_2$ times perpendicularly to direction Y and (the same for Z), the number of final cuttings would diminish with the routing length which can be minimized in $L_{1.2.3}$ and then

$$k_1 + k_2 + k_3 = L_{1.2.3} \leq L_1 + L_2 + L_3.$$

But we see that in X the number of cuts is not the minimal value of L_1 (That is $k_1 + 2\alpha_1/p_1$) but $k_1 + 1$ if $\alpha > 0$.

At any given phase we can only cut along a plane, whereas for the case 10 of Problem 5, it would be possible to construct the questionnaire of routing $L_{1.2}$ by a process equivalent down to the separation of two "planes" (a cut in the form of an L).

Consequently, the total number of cuts is the height of the product questionnaire of the three L-optimal questionnaires corresponding to p, q and r.

Thus the minimal number of cuts is

$$C = k_1 + k_2 + k_3 + [\alpha_1] + [\alpha_2] + [\alpha_3].$$

Applications:

A cut can cut simultaneously several distinct pieces:

1. Chocolate tablets:

$$p = 2^2 + 1 \quad : \quad k_1 = 2 \; ; \quad [\alpha_1] = 1 \; ;$$

$$p = 2^3 \qquad : \quad k_2 = 3 \; ; \quad [\alpha_2] = 0 \; ;$$

$$r = 2^0 \qquad : \quad k_3 = 0 \; ; \quad [\alpha_3] = 0 \; ;$$

$$C = k_1 + k_2 + [\alpha_1] = 6.$$

2. Cubes:

$p = 2$: $C = 3$;

$p = 3 = 2^1 + 1$: $C = 6$;

$p = 4 = 2^2$: $C = 6$;

$p = 5 = 2^2 + 1$ (and $5 < p \leq 8$) : $C = 9$;

$9 \leq p \leq 16$: $C = 12$;

3. Parallelepiped:

$p = 3, q = 4, r = 5$: $C = 7$;

4. A cut can only split one piece into two: the number of cuts is equivalent to the number of questions of a dichotomic questionnaire with $N = p \times q \times r$ answers, that is $M = p \times q \times r - 1$ (application M=39).

Problem 5 Towns and Liars

1. The solution proposed by Yaglom and Yaglom is determined by the table of questions (columns) and answers (elements of the array) as a function of the nature of the answer. The notation [r,s] of the 12 answers shows the $2^4 - 12 = 4$ cases where only three questions are necessary. The arborescence (Fig. 1) obtained from this is of the balanced type and satisfies the setting of the problem suggested by Yaglom and Yaglom: the questionnaire enables the observer never to ask more than 4 questions. On average, the observer will ask:

$$L_1 = 3 + 2 \cdot \frac{4}{12} = 3.66 \text{ questions}.$$

In the case when the observer has one chance in three of being in each town and, independently, one chance in four of speaking to an inhabitant α, β, γ or $\bar{\gamma}$, this dichotomic arobrescence with 12 equiprobable answers of the balanced type answers the first part of the problem.

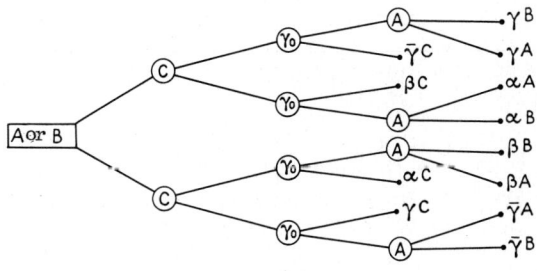

Figure 1 -

12 equiprobable answers

	Questions				
	1	2	3	4	
	Town	Town	Inhabitant	Town	
Answers	A or B?	C?	γ_o?	A?	r,s
αA	1	0	0	1	4,9
αB	1	0	0	0	4,8
αC	0	1	0	(0)	3,2
βA	0	1	1	0	4,6
βB	0	1	1	1	4,7
βC	1	0	1	(1)	3,5
γA	1	1	1	0	4,14
γB	1	1	1	1	4,15
γC	0	0	1	(1)	3,1
$\bar{\gamma}$A	0	0	0	1	4,1
$\bar{\gamma}$B	0	0	0	0	4,0
$\bar{\gamma}$C	1	1	0	(0)	3,6

2. If the inhabitants of C begin by telling the truth (an odd day for example) and the observer knows that the answers of C are in the order TFTF, the last three answers of the preceding array are suppressed and the questionnaire 1 reduces in size to give place to 9 equiprobable answers. Contractions are performed. Indeed the sense of the answer to a question "is it town A?" has been changed as its rank has diminished by 1 between questionnaire 1 and questionnaire 2 (Fig. 2); the routing has length

$$L_2 = \frac{1.2 + 4.3 + 4.4}{9} = 3.33 .$$

3. We obtain an analogous questionnaire when the inhabitants of C can be of type $\bar{\gamma}$ (Fig. 3).

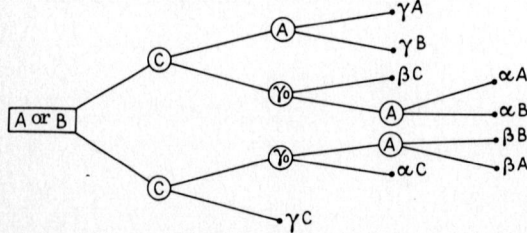

Figure 2 -9 equiprobable answers, γ (TFTF).

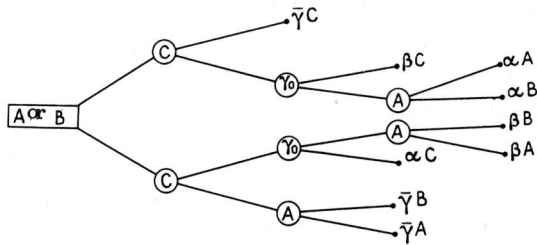

Figure 3. - 9 equiprobable answers, $\bar{\gamma}$ (FTFT)

4. and 5. Since the arborescences 2 and 3 possess answers with difference of ranks equal to 2, they can be modified in such a way as to minimize the routing length. The nature of the question is no longer linked only to the rank but also to the earlier answers or more precisely to the path coming from the first question (root).

The routing of 4 and 5 is L-optimal and has length

$$L_4 = 3 + \frac{2 \times 1}{9} = 3.22 \qquad \text{(Fig. 4 and 5)}$$

6. If we consider the questionnaire 1 again by allocating the same probability to the inhabitants of C to be of type TFTF or FTFT and if we suppose that the chances that the observer will be in A, B or C, in conversation with α, β, or (γ or $\bar{\gamma}$) are equal, then the probabilities of the 12 answers become

$$\left. \begin{array}{l} p(\alpha M) = p(\beta M) = \frac{1}{9} \\ p(\gamma M) = p(\bar{\gamma} M) = \frac{1}{18} \end{array} \right\} \qquad (M := A, B, C)$$

Optimization of the questionnaire requires a modification in the nature of the questions asked in such a way that the 6 answers of smallest probability have a rank at least equal to those of the 6 other answers and that the rules R2 and R3 are also satisfied.

The optimal routing (Fig. 6) of the arborescence 6 takes the value $L_6 = 3.55$

7. and 8. Supposing that half the population of C is of type γ and the other half of type $\bar{\gamma}$, we can take the probability coefficients to be:

$p(\alpha M) = p(\beta M) = 2p(\gamma M)$

and

$p(\bar{\gamma}) = 0 \quad (M := A,B,C)$,

that is

$p(\alpha M) = \frac{2}{15}$

and

$p(\gamma M) = \frac{1}{15}$ (for the arborescence 7).

The questionnaires 4 and 5 are no longer L-optimal because the vertices of rank 4 have stronger probabilities than some vertices of rank 3.

An optimization (rules R1, R2, R3) leads to defining two new questionnaires for which the new questions are replaced by the four varieties of questions due to Yagolm and Yaglom (Fig. 7 and 8). The routing length is then reduced to $L_7=3.133$.

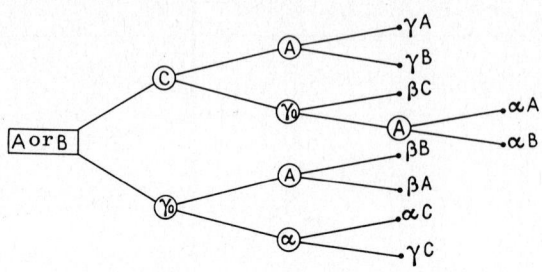

Figure 4 - 9 equiprobable answers, γ(TFTF), balanced type.

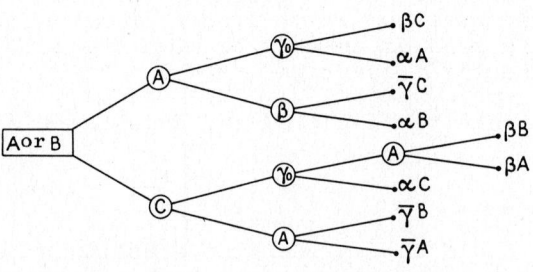

Figure 5 - 9 equiprobable answers, $\bar{\gamma}$(FTFT), balanced type.

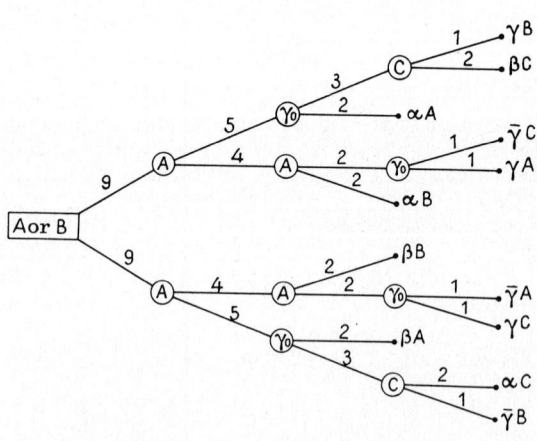

Figure 6 - 12 answers
The arcs (i,j) carry flow proportional to p(i,j).

9. Now we solve the problem of determining the town and the origin of the informant by supposing that the informant, an inhabitant of A, B or C, always tells the truth.

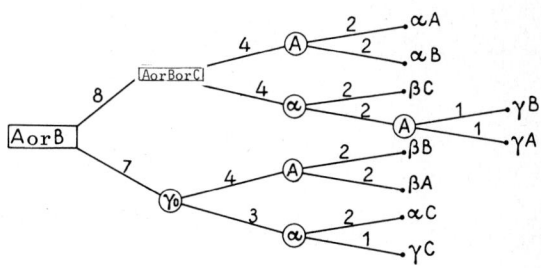

Figure 7 - 9 answers.

Since the possibilities of the place (A, B, C) and the origin (α, β, γ_0) are independent, the questionnaire can be interpreted as a product of two questionnaires with three answers each. The question "Am I in town A or B?" then possibly "Am I in town A?" followed by "Do you live in C?" then possibly by "Do you live in A?" answer the problem of the maximum number of questions which does not exceed 4.

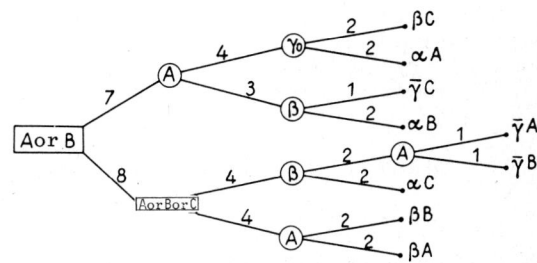

Figure 8 - 9 answers.

The routing of this product of questionnaires (Fig. 9)

$$L_9 = (1 + 2 \cdot \tfrac{1}{3}) + (1 + 2 \cdot \tfrac{1}{3}) = 3.33 ,$$

is greater than that of an equiprobable L-optimal questionnaire with 9 answers

$$L_{10} = 3 + 2 \cdot \tfrac{1}{9} = 3.22 .$$

10. We know that to separate to the best 9 answers it is necessary that the first question leads to two arcs having respectively 4 and 5 answers. Otherwise the theorem on polychotomic equiprobable L-optimal questionnaires (6-4) would put into defect the constructed arborescence.

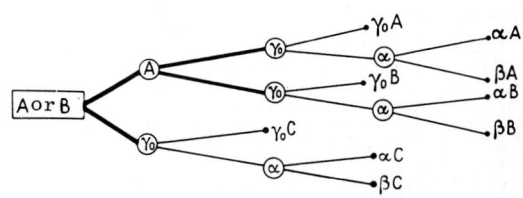

Figure 9 - 9 equiprobable answers. Product of 2 arborescences with 3 eventualities.

To separate the 9 answers in this way

αA	βA	γ_oA
αB	βB	γ_oB
αC	βC	γ_oC

it is necessary to ask in the first place: "do you live in C?" or "Am I in town C?"

The construction of the balanced questionnaire 10 is then immediate (Fig. 10)

Informational study of the town questionnaires

Questionnaire 1 has 12 equiprobable answers, and thus transmits the information

$$I_1 = \log_2 12$$

Questionnaires 2, 3, 4, 5, 9 and 10 have 9 equiprobable answers

$$I_2 = \log_2 9 .$$

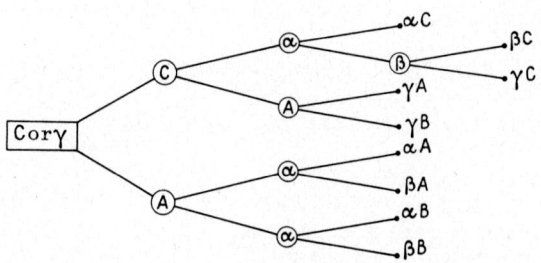

Figure 10 - 9 equiprobable answers (no one tells lies).

Questionnaire 6 transmits the information

$$I_6 = 6 \cdot \frac{1}{9} \cdot \log_2 9 + 6 \cdot \frac{1}{18} \cdot \log_2 18$$

and the questionnaires 7 and 8

$$I_7 = 6 \cdot \frac{2}{15} \cdot \log_2 \frac{15}{2} + 3 \cdot \frac{1}{15} \cdot \log_2 15 ,$$

which leads to the table;

Questionnaire n°	Answers N	Information I	Routing L	Proper noise $L-L_H$	Inefficiency $L_H - I$
1	12	3.585	3.666	-	0.081
2	9	3.170	3.333	0.111	0.052
3	9	3.170	3.333	0.111	0.052
4	9	3.170	3.222	-	0.052
5	9	3.170	3.222	-	0.052
6	12	3.503	3.555	-	0.052
7	9	3.107	3.133	-	0.026
8	9	3.107	3.133	-	0.026
9	9	3.170	3.333	0.111	0.052
10	9	3.170	3.222	-	0.052

The routing lengths and the informations of 6 and 10 shows that the difference

$$L_6 - L_{10} = I_6 - I_{10} = 0.333$$

measures the subsidiary information it is necessary to acquire if the answers of the inhabitants B and C are wrong. We can say that the lie is revealed by a supplement of information of 0.333.

We can then compare the information and the routing lengths of questionnaires 1, 4 and 5 to those of questionnaires 6, 7 and 8 to show the inefficiency of a balanced questionnaire when the probabilities of the answers are or are not all the same.

$$I_1 - I_6 = 0.082 \quad ; \quad I_4 - I_7 = 0.063$$
$$L_i - L_6 = 0.111 \quad ; \quad L_4 - L_7 = 0.089$$

Finally the inefficiency is always less than its maximal value (table §8.1) $B_M = 0.086$ whereas the proper noise is greater than B_M.

Problem 6   Paths, Products and Ordering

1.   Ordering of 4 numbers X, Y, Z, T all distinct.

1.1
$$N = 4! = 24$$
$$I_a = \log_a 24$$

Following the classical formulae, we will write:

$a = 2 \quad N = 2^4 + 8$, $\quad I_2 = 4.58 , L_2 = 4.66$

$a = 3 \quad N = 3^2 + 2 \times 7 + 1$, $\quad I_3 = 2.89 , L_3 = 2.96$

$a = 5 \quad N = 5^1 + 4 \times 4 + 3$, $\quad I_5 = 1.97 , L_5 = 2.00$

In fact $24=5^2-1$ so that we know that the length of L_5 is equal to 2, all the paths coming out of the root and entering into E being of length 2; it is therefore pointless to apply a formula to determine L_5.

1.2 In the case when a=2, the realizable questionnaires are of Huffman's type and $M_2=L_2=4.66$.

For a=5 we suppose also that $M_5=L_5=2.00$ one of the 5 questions of rank 1 having 4 outcomes instead of 5.

Finally, for a=3, the first question is predetermined.

The details are shown in the diagram (Fig. 11) for the case {X is the smallest element or (X and T) are the two smallest ones} indicated by ξ on the arc coming out of the root , all the other analogous cases being relative to the sub-arborescences coming out of the extremities of the arcs η, ζ coming out of α.

At rank 1 the questions of the case ξ lead to only three outcomes:

X < Y < T, X < T < Y, T < X < Y

compatible with the hypotheses.

At rank 2 we must order Z relative to Y and T when X is the minimum (3 possible outcomes) or we must order Z relative to Y when T is less than X (2 possible outcomes).

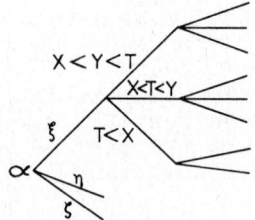

Figure 11

Thus as a total we have:

6 questions with rank 2 and basis 3

and 3 questions with rank 2 and basis 2,

and this questionnaire does not satisfy Huffman's algorithm for heterogeneous questionnaires although it has the same routing length.

We have in fact a product of questionnaires $A_1 A_2$.

A_1 possesses the first question of the product, with root α.

A_2 is isomorphic to the heterogeneous questionnaire with 3 questions of basis 3 and one questions of basis 2 represented downstream from arc ξ on the preceding figure.

Since the questionnaire $A_1 A_2$ has three questions with basis 2 whereas Huffman's homogeneous questionnaire with basis 3 and 24 answers has only one such question, it is evident that $M_3 > L_3$.

In practice we find

for
$$M_3 = 3$$
$$L_3 = 2.96 \ .$$

2. The number of answers is

$$N = (n + 4)(n + 3)(n + 2)(n + 1)$$

supposing $\Omega > \pi$ and $\omega > \rho$ as given. For Ω is one of the n+4 numbers, π one of the n+3 others, etc.

2.1 The information is:

$$J_a = \log_a [13 \cdot 14 \cdot 15 \cdot 16]$$
$$J_2 = \log_2 [2^5 \cdot 3 \cdot 5 \cdot 7 \cdot 13] = 15.4$$
$$J_5 = \log_5 2 \cdot J_2 = 6.6 \ .$$

2.2 a=5

Having used the ordering of the 4 numbers with the aid of $\{Q_5\}: \rho < \omega < \pi < \Omega$, which takes $M_5 = 2$ questions it is sufficient to ask a question for each of the (n+4)-4=n remaining numbers (Fig. 12) and then to make any necessary substitutions.

We find

$$C_5 = n + M_5$$

$$\underline{n = 12 \rightarrow C_5 = 14}$$

(clearly greater than J_5).

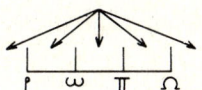

Figure 12

2.3 a=2

Ordering with the aid of $\{Q_2\}$ enables us to consider the five cases denoted by A, B, C, D, E according to the position (of the ith number i) relative to the 4 first numbers studied $\rho < \omega < \pi < \Omega$ (Fig. 13).

For each i we place i relative to the two greatest and the two smallest numbers already pre-ordered and then make the necessary substitutions which will not enter into the enumeration of the routing length.

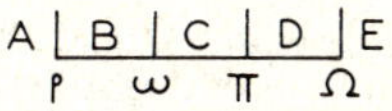

A | B | C | D | E
ρ ω π Ω

Figure 13

An elementary questionnaire will thus be realized for all i from i=5 to i=n+4. The product of $\{Q_2\}$ by the n such questionnaires will answer the problem posed. Since each questionnaire is of the dichotomic type with 5 answers, it necessarily has answers of at least two distinct ranks and the product will therefore include answers situated on n+2 different ranks.

Put

$$p(A) = p(B) = p(D) = p(E) = \frac{1}{i} \; ; \; p(C) = 1 - \frac{4}{i} \geqslant \frac{1}{i}$$

The routing length of such a product of questionnaires will not be optimal - without making any calculation - only if it is very near the information, that is $J_2 = 15.4$ and we will see that it is never the case even for the 4th questionnaire under study.

1st solution

Elementary questionnaire (Fig. 14)

The routing of this questionnaire is

$$L_i = 1 + \frac{9}{i} \, .$$

The product has routing length:

$$C_2 = M_2 + \sum_{i:=5}^{n+4} L_i$$

$$n=12 \quad C_2 = 4.66 + 12 + 9 \sum_{i:=5}^{16} \frac{1}{i} = 28.3$$

with

$$\sum_{i:=5}^{16} \frac{1}{i} = 1.30 \, .$$

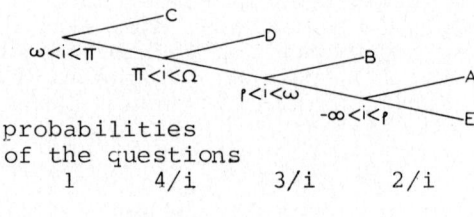

probabilities of the questions
1   4/i   3/i   2/i

Figure 14

2nd solution (Fig. 15)

we obtain

$$L_i = 1 + \frac{8}{i}$$

and

$$n=12 \quad C_2 = 4.66 + 12 + 8 \sum_{i:=5}^{16} \frac{1}{i} = 27.0$$

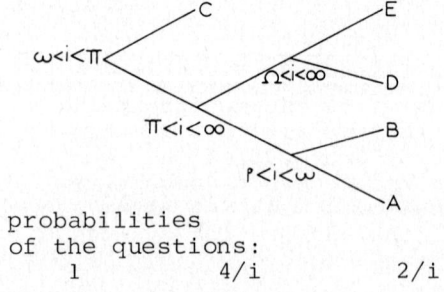

probabilities of the questions:
1   4/i   2/i

Figure 15

3rd solution (Fig. 16)

Elementary questionnaire:

$$L_i = 3 - \frac{1}{i}$$

Hence

$$C_2 = 4.66 + 36 - \sum_{i:=5}^{16} \frac{1}{i} = \underline{39.4}$$

The second solution is more interesting; however if $p(C) < \frac{2}{i}$ it is not recommended, the elementary questionnaire being not optimal.

We must then modify the elementary questionnaire relative to $i=5$.

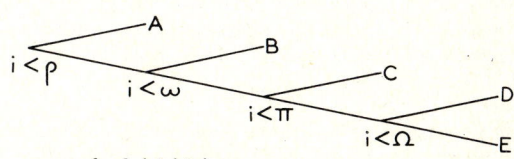

probabilities of the questions
1   1-1/i   1-2/i   2/i

Figure 16

4th solution (Fig. 17)

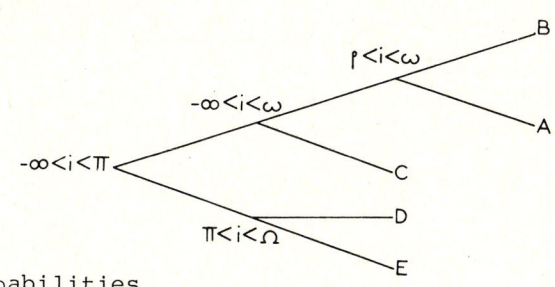

probabilities of the questions:

1   3/i   2/i
           2/i

Figure 17

Consequently

$$L_i = 1 + \frac{3}{i} + \frac{2}{i} + \frac{2}{i} = 1 + \frac{7}{i} < 1 + \frac{8}{i} \ .$$

This questionnaire is optimal and all the other questionnaires, unchanged relative to the second solution and intervening in the product are optimal.

We find

$$C_2 = 27 - \left[\frac{1}{i}\right]_{i=5}$$

$$C_2 = \underline{26.8}$$

and therefore

C_2 is much greater than J_2, which is not unusual because the products of optimal questionnaires are not generally optimal. Moreover we can see directly that the product $\{Q_2\} \times A_4$, where A_4 is the elementary questionnaire of placing the fifth number (with $L_i = 1 + \frac{7}{i1}$), is not optimal because there exist answers with probability $\frac{1}{4i}$ and rank 6 (case C, D and E), 7 (case A, B or C, D, E) 8 (case A, B).

3. The search for the four greatest numbers in a list of numbers can be effected in an easily programmable fashion by using the fourth solution obtained in the preceding paragraph, the problem posed here being equivalent to the preceding one.

The ith number to be obtained can be compared with the preceding ones by setting questionnaires (or sub-programs) enabling us to isolate at first the case $C(\omega < i < \pi)$ which has the strongest probability $1 - \frac{4}{i}$ as soon as $i \geq 6$.

The probabilities of the answers of the questionnaires in the 4th solution ($i \geq 6$) are shown in Fig. 18

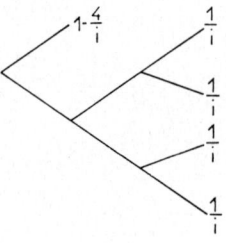

Figure 18

Problem 7 Heterogeneous Questionnaires and Insertion

1. 1.1 Types of questions

The questions enable us to effect the ordering of three numbers (u), (v), (w). In the absence of any other information on these numbers, there are 6 possible orderings and the questions are formed with the basis 6: we will say that they are questions of type 6.

If two of the three numbers (u), (v), (w) are ordered one relative to the other, then the third is determined by a question, said to be of type 3: the usable base being 3.

If we know the minimum, the intermediate or the supremum, there are but two total compatible orderings and the usable basis is 2: the question will be said to be of type 2.

We remark that we are dealing in reality with only one sort of question with basis 6 but it acts in practice according to the type conditioned by the information at the input. According to the hypothesis given when posing the problem (instruction of permutation with three addresses) the question of type 2 cannot be effectively used without knowing the minimum (respectively the intermediate, the superior): if I is the intermediate, we place (u) in u, (v) in v and (w) in w; the two possible outcomes being ((u) in u and (w) in w) or ((u) in w and (w) in u); the system would be blocked if we had left an arbitrary element in v before placing (u) and (w) to command the instruction. This remark will reduce the combinatorics by forbidding the root of a questionnaire to be a question with basis 2, because the first question of an insertion questionnaire cannot judge beforehand the final position of any element.

1.2 Study of the insertion

In an insertion procedure we first order the three numbers by using exclusively the question of type 6; then we insert the 4th number then the tth number and thus there will be exactly t total orders compatible with the order of the first (t-1) numbers.

The questionnaire K_t which is to be realized to order t numbers placed randomly has t! equiprobable answers corresponding each to one and only one total and strict order: we suppose the numbers to be distinct.

But to insert a number in a list of t-1 ordered numbers, amounts to forming a questionnaire $K_{(t-1)t}$ with t answers.

We then write K_t as the product

$$K_t = K_{t-1} \times K_{(t-1)t}$$

which corresponds indeed, to t! eventualities; we will start from K_3 which is a questionnaire with only one question of type 6 such that $L_3 = 1$.

Let a, b, c, d, e, f, be six distinct numbers placed at random in a list.

We will use the following symbolic representation:

(a,b,c)	non-ordered triple or question of type 6
abce...ℓ	ordered n-tuple: $a<b<c<e...<\ell$
d|ab	question of type 3, when $a<b$
c<d	question of type 2, when we know M, I or S.

The arcs coming out of questions indicate the outcomes:

Type 2		type 3	
	at the top $c<d$		at the top $b<d$
c<d		d|ab	in the middle $a<d<b$
	at the bottom $d<c$		at the bottom $d<a$

t=3 The polychotomic elementary questionnaire K_3 having one unique question with basis 6 has routing length $L_3=1$.

t=4 We must determine a questionnaire K_{34}, knowing the partial ordering abc (for example) the 5 other partial orders of the three elements a, b, c, give rise to questionnaires isomorphic to K_{34} (the difference being exclusively related to the formulation of the question); we will speak of the product $K_3 \times K_{34}$ even if we deal in reality with a succession of six restricted products for six of these graphs isomorphic to K_{34}.

The root is necessarily of type 3 and as there are 4 outcomes, a question of type 2 is necessary. Whence (Fig. 19)

with routing length

$$L_{34} = 1.50$$

and consequently

$$L_4 = 2.50 .$$

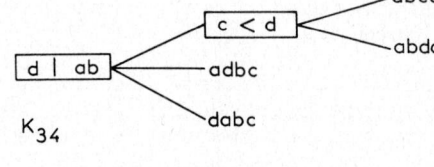

t=5 K_{45} will be trichotomic homogeneous with two questions, two answers of rank 1 and three answers of rank 2, or heterogeneous with one question of type 3 and two questions of type 2. We suppose a<b<c<d (Fig. 20, 21, 22):

with routing length

$$L_{45} = 1.60$$

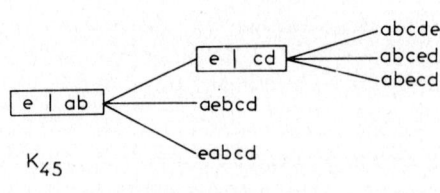

Figure 20

with routing length

$$L'_{45} = 1.80$$

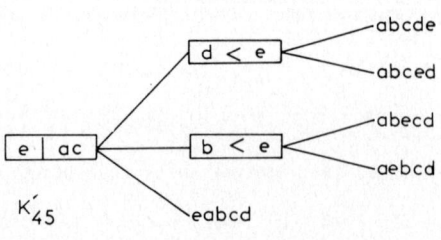

Figure 21

We verify that the
questions of type 2
are indeed consistent
with the hypotheses
for K'_{45} and K''_{45}; this
last questionnaire
has routing length
$L''_{45} = 2.00$.

Finally we obtain

$L_5 = 4.10$

$L'_5 = 4.30$

$L''_5 = 4.50$.

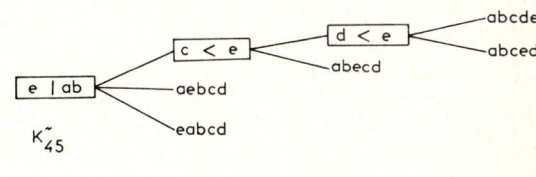

Figure 22

$t=6$ Since there are six answers, we cannot hope to find a trichotomic homogeneous questionnaire in the strict sense but we could have two questions with basis 3 and one question with basis 2 to form an insertion questionnaire of the sixth number.

We will thus form a questionnaire with root of type 3 having three questions with basis 2. We suppose $a<b<c<d<e$ (Fig. 23 and 24). That is

K_{56}

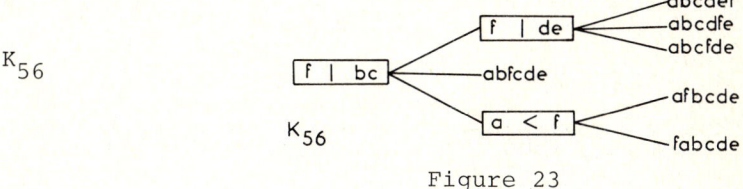

Figure 23

such that

$L_{56} = 1.83$

but also

K'_{56}

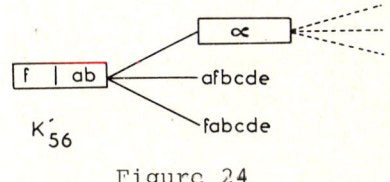

Figure 24

Questionnaire K'_{56} has two answers with rank 1 and one question α with rank 1. This question is the root of a sub-questionnaire where f must be inserted in the ordered cde and which is thus isomorphic to K_{34}.

Effecting the restricted product indicated in α, we find

$$L'_{56} = 1 + \frac{2}{3} L_{34} = 2.00 ,$$

K'_{56} has a dichotomic question of rank 2 whereas

K_{56} has a dichotomic question of rank 1 so that

$$L'_{56} - L_{56} = \left(\frac{1}{t}\right)_{t=6} \text{ , which is verified .}$$

By placing f relative to the ordered pair bd the paths are balanced and the questionnaire obtained is heterogeneous (Fig. 25) and

with routing length

$$L''_{56} = 2.00$$

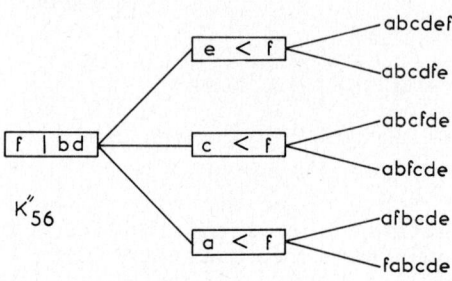

Figure 25

We will then be able to form by combination of the three questionnaires for t=5 and the three questionnaires for t=6, nine questionnaires whose routing lengths (cf. table II further) will lie between

and

$$L_5 + L_{56} = 5.93$$
$$L''_5 + L'_{56} = 6.50$$

1.3 Information

The information can easily be determined because E is the set of permutations of t distinct numbers, thus $|E|=t!$ and all these permutations are equiprobable:

$$(\forall i) \; p_i = 1/t!$$

Hence

$$I(t) = \log_2(t!)$$

that is

$$I(3) = 2.58 \qquad I(5) = 6.91$$
$$I(4) = 4.58 \qquad I(6) = 9.49 \; .$$

2. The routing lengths of the preceding questionnaires are generally less when a maximal number of questions with basis 3 is used. However, when evaluated throughout with logarithms to the base 2, I depends only on t and not on the basis of the questions. To evaluate the difference of routing relative to an optimal questionnaire, the concept of heterogeneous information has been introduced but the problem here concerns only Shannon's information. Here we have recourse to a weighting, a kind of cost function of the questions.

Let N_1 be the weighted routing length with the weights

$$\pi_2 = 1 \quad , \quad \pi_3 = 1 \quad , \quad \pi_6 = 1.5 \quad ;$$

$$N_1(t) = \sum_{j \in Q} p_j \, \pi_{a_j}$$

where Q is the set of questions, π_{a_j} the weight defined by the basis and p_j the probability of the question.

Similarly consider $N_2(t)$ for the weights

$$\pi_2 = 1 \quad , \quad \pi_3 = 1.5 \quad , \quad \pi_6 = 2.5$$

Then we can construct the following tables the first of which is merely for subsidiary use.

We see that for $t \geqslant 5$, the questionnaire with minimal routing length L (and N_1) is distinct from the questionnaire for which N_2 is minimal. The influence of the questions of type 2 which tended to increase L tends in reverse to produce an economy, therefore to reduce N_2; $N_1 - L$ is constant and related to the cost π_6.

$N_2 - I$ is most often positive.

Now let N_3 be the length corresponding to the weights:

$$\pi_2 = 1 \; ; \; \pi_3 = \log_2 3 = 1.58 \; ; \; \pi_6 = \log_2 6 = 2.58 \; .$$

Then we find

$$N_3(K_3) = I(K_3) \quad ; \quad N_3(K_4) = I(K_4) \; ,$$

and

$$i \geqslant 4 \Rightarrow N_3(K_i) > N_2(K_i) + 0.08(i - 2)$$

because the questionnaires K_i ($i \geqslant 4$) have at least $(i-3)$ questions with basis 3 and one question with basis 6.

Table I

Calculation of N_1 and N_2 for the questionnaires facilitating the insertion of t numbers among t-1.

t	Questionnaire	Probabilities of Questions		N_1	N_2
		base 2	base 3		
4	K_{34}	$\frac{1}{2}$	1	1.5	2.0
5	K_{45}	—	$1;\frac{3}{5}$	1.6	2.4
	K'_{45}	$\frac{2}{5};\frac{2}{5}$	1	1.8	2.3
	K''_{45}	$\frac{3}{5};\frac{2}{5}$	1	2.0	2.5
6	K_{56}	$\frac{1}{3}$	$1;\frac{1}{2}$	1.83	2.58
	K'_{56}	$\frac{1}{3}$	$1;\frac{2}{3}$	2.0	2.83
	K''_{56}	$\frac{1}{3};\frac{1}{3};\frac{1}{3}$	1	2.0	2.50

We see then that $N_3(K_i) \geqslant I(K_i)$ ($\forall i$), the inequality being strict as soon as $i \geqslant 4$.

It is a generalisation of the property:

"Information is a lower bound for the routing length of homogeneous questionnaires".

Table II

Computation of N_1 and N_2 for sorting questionnaires (by insertion) of t numbers.

t	Questionnaire	L	N_1	N_2	I
3	K_3	1.00	1.50	2.50	2.58
4	$K_4 = K_3 \times K_{34}$	2.50	3.00	4.50	4.58
5	$K_5 = K_4 \times K_{45}$	4.10*	4.60*	6.90	6.91
	$K_5' = K_4 \times K_{45}'$	4.30	4.80	6.80*	
	$K_5'' = K_4 \times K_{45}''$	4.50	5.00	7.00	
6	$K_5 \times K_{56}$	5.93*	6.43*	9.48	9.49
	$K_5 \times K_{56}'$	6.10	6.60	9.73	
	$K_5 \times K_{56}''$	6.10	6.60	9.40	
	$K_5' \times K_{56}$	6.13	6.63	9.38	
	$K_5' \times K_{56}'$	6.30	6.80	9.63	
	$K_5' \times K_{56}''$	6.30	6.80	9.30*	
	$K_5'' \times K_{56}$	6.33	6.83	9.58	
	$K_5'' \times K_{56}'$	6.50	7.00	9.83	
	$K_5'' \times K_{56}''$	6.50	7.00	9.50	

*Indicates the minimum, for fixed t, in its column.

3. There exist t! orders for arranging a list.

Consequently, to merge two ordered lists of three numbers abc and def into one list of 6 numbers amounts to finding one order among the $\frac{6!}{3!3!}$, that is among the twenty, possible orders.

These orders numbered from 1 to 20 are as follows

1	abcdef	6	adbecf	11	dabcef	16	daefbc
2	abdcef	7	adbefc	12	dabecf	17	deabcf
3	abdecf	8	adebcf	13	dabefc	18	deabfc
4	abdefc	9	adebfc	14	daebcf	19	deafbc
5	adbcef	10	adefbc	15	daebfc	20	defabc

We then try to built a questionnaire as near as possible to a trichotomic optimal questionnaire having 20 answers and therefore 9 questions with basis 3 and 1 question with basis 2. To this end, we establish table III giving the number of answers coming out of the three outcomes of each question of type 3 which it is possible to ask when we know the two partial orders abc and def.

Table III

$\alpha \mid \beta\gamma$	$\alpha < \beta$	$\beta < \alpha < \gamma$	$\alpha > \gamma$
a|de	10	6	4
a|df	10	9	1
a|ef	16	3	1
b|de	4	6	10
b|df	4	12	4
b|ef	10	6	4

It is pointless to construct the other elements of the table, because to compare an element of the first triple with a pair of the second gives the same number of outcomes as to compare an element of the second triple with a pair of the first triple.

Moreover c|ef (for example) will have outcomes with probabilities $(\frac{10}{20}, \frac{6}{20}, \frac{4}{20})$ like a|de, because of the symmetry of the triples.

Questions of type 3, like a|df and a|ef have one answer as an outcome; however a realizable questionnaire starting from a|df has three optimal trichotomic subquestionnaires with respectively 10, 9 and 1 answer(s) and only one question with basis 2. Despite the small rank of the answer isolated at the first question, the routing length of this questionnaire is small.

For the other formable questions, there is still a power of 3 lying between 4 and 10 or 4 and 12, which proves that the questionnaire is not optimal. Since each outcome (i) of these questions corresponds to an even number of answers, there will necessarily be a question with basis 2 in each sub-questionnaire with root (i). Consequently, the best questionnaire we

can hope to realize will have three questions with basis 2.
Its routing length will then be at least equal to $L_H + \frac{2}{20}$
because each new question with basis 2, starting from the
second, generates a proper noise of 1/N.

We note that $20 = 3^2 + 5 \cdot 2 + 1$ and that therefore $L_H = 2 + \frac{17}{20}$ (optimal).

Now we show that it is indeed possible to construct a
questionnaire by merging two lists of 3 ordered numbers having
$L = 2 + \frac{19}{20}$. For this we will give realizable questionnaires using
questions of type 3. (Fig. 26 and 27).

We note that $20 = 2^4 + 4 \cdot 1$,
and therefore the routing
length of the optimal
dichotomic questionnaire
which merges the two
lists of three numbers
(that is 4.4) is clearly
greater than that of the
questionnaires realizable
by means of questions of
type 3 considered here.

Problem 8 Equivalent Questionnaires

1. The questionnaires of
B can be considered as
trichotomic in the broad
sense

$N = a^k + \alpha(a-1) + \beta$

$4 = 3^1 + 0 + 1$

$\alpha = 0$, $\beta = 1$

Semi-homogeneous: rank 0:
1 question with basis 3
(or 2)
rank 1: 1 question with
basis 2 (or 3)

Heterogeneous: 2 questions
with different bases.

We describe four types
of questionnaires to
enumerate:

Questionnaires B_1:

rank 1: 2 answers among
4 (the question of rank 0
has basis 3) 1 question with
basis 2

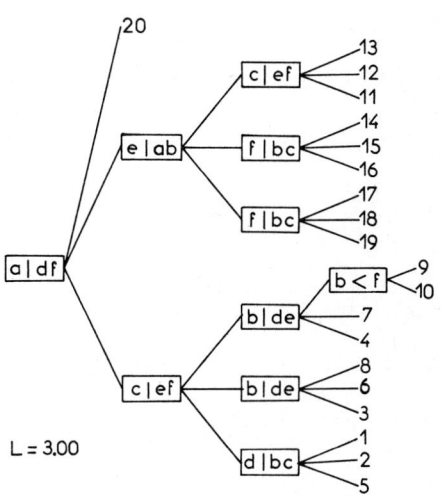

Figure 26

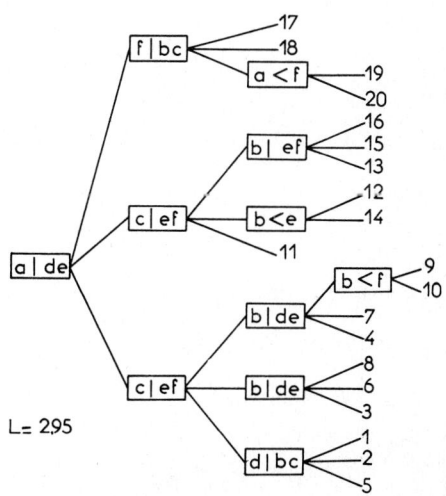

Figure 27

rank 2: 2 answers

There are C_4^2 ways to proceed, that is $C_4^2 = \frac{4 \cdot 3}{2 \cdot 1} = \underline{6}$ classes.

Questionnaires B_2:

rank 1: 1 answer (the question of rank 0 has basis 2)

rank 2: 3 answers

There are 4 ways to proceed. $\underline{4}$ classes.

Thus B has 10 equivalence classes.

Questionnaires C:

$N=4=2^2$, so there are three questions with basis 2.

C_1: all the questions are of rank 0 or 1, the answers of rank 2.

We can group 2 answers among 4 in $C_4^2 (=6)$ distinct ways but here the problem is symmetrical because the simultaneous groupings (p_1, p_2) and (p_3, p_4) are obtained twice and lead to two equivalent questionnaires because all the answers are of rank 2. Thus there are only 3 distinct classes.

C_2: the answers are of rank 1, 2, 3, 3.

If we take p_1 at rank 1, there remain:

$$p_2, p_3 \text{ and } p_4$$
or
$$p_3, p_2 \text{ and } p_4$$
or
$$p_4, p_2 \text{ and } p_3$$

at ranks 2, 3 and 3, and similarly if p_2 or p_3 or p_4 is at rank 1.

That is a total of 12 distinct questionnaires.

Thus C has 15 equivalence classes.

2. Extreme questionnaires

For B , the optimal questionnaire corresponds to type B_1:

2 answers (p_1, p_2) with rank 1

2 answers (p_3, p_4) with rank 2.

Thus
$$L_0 = p_1 + p_2 + 2(p_3 + p_4) = 1 + (p_3 + p_4)$$

the worst questionnaire corresponds to type B_2:

1 answer (p_4) of rank 1

3 answers (p_1, p_2, p_3) of rank 2

$$L_p = p_4 + 2(p_1 + p_2 + p_3) = 2 - p_4 .$$

For C the optimal questionnaire and the worst questionnaire will be distinguished according to type.

For C_1, all the questionnaires have the same routing length:

$$L_1 = 2$$

For C_2, the questionnaire has routing length:

$$L_m = p_1 + 2p_2 + 3(p_3 + p_4) \leq L \leq p_4$$
$$+ 2p_3 + 3(p_1 + p_2) = L_M .$$

Discussion.

We must determine whether L_1 lies between L_m and L_M or not.

$$L_M = 1 + p_3 + 2(p_1 + p_2) = 1 + p_3$$
$$+ 2p_1 + p_2 + p_4 + (p_2 - p_4)$$

$$L_M = 2 + p_1 + (p_2 - p_4) > L_1 \quad \text{because} \quad p_2 > p_4$$

whatever $p_1 > p_2 > p_3 > p_4$.

$$L_m = 1 + p_2 + 2(p_3 + p_4) \geq L_1$$

if and only if

$$p_2 + 2(p_3 + p_4) \geq 1$$

or

$$p_2 + 2(p_3 + p_4) \geq p_1 + p_2 + p_3 + p_4$$

that is

$$p_3 + p_4 \geq p_1 .$$

Figure 28

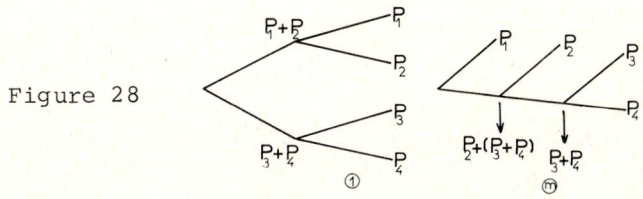

Hence

$$p_3 + p_4 > p_1 \Rightarrow L_1 \text{ is minimal (Fig. 28)}$$
$$p_3 + p_4 = p_1 \Rightarrow L_1 = L_m \text{ are minimal}$$
$$p_3 + p_4 < p_1 \Rightarrow L_m \text{ is minimal.}$$

The worst questionnaire C is always that of type C_2 (Fig. 29), even if the inequalities are taken in the broad sense $(p_1 \geqslant p_2 \geqslant p_3 \geqslant p_4)$.

Comparison of the questionnaires B and C.

For B

$$L_o = 1+(p_3+p_4) \leqslant L \leqslant 2-p_4 = L_p < 2$$

For C:

1. $p_3 + p_4 < p_1$

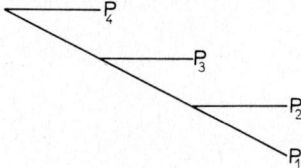

Figure 29

$$L_m = 2 - [p_1 - (p_3 + p_4)] \leqslant L \leqslant 2 + p_1 + (p_2 - p_4)$$

and

$$L_m > L_o$$
$$L_m > L_p, \text{ if } p_1 - (p_3 + p_4) < p_4$$
$$p_1 < p_3 + 2p_4$$
$$L_m < L_p, \text{ if } p_1 > p_3 + 2p_4 \ .$$

2. $p_3 + p_4 \geqslant p_1 \Rightarrow L_p < 2 \leqslant L_m$

$p_1 < p_3+p_4$	$p_1 = p_3+p_4$	$p_3+p_4 < p_1 < p_3+2p_4$	$p_1 = p_3+2p_4$
$L_o < L_p < 2 < L_m < L_M$	$L_o < L_p < L_m = 2 < L_M$	$L_o < L_p < L_m < 2 < L_M$	$L_o < L_m = L_p < 2 < L_M$

$p_1 > p_3+2p_4$
$L_o < L_m < L_p < 2 < L_M$

The questionnaire L_o (type B, optimal) is always the optimal one.

The questionnaire L_M (type C, worst) is always the worst one.

We can have further:

$$L_p < 2 \leqslant L_m \ , \ L_p < L_m < 2 \ , \ L_m \leqslant L_p < 2$$

that is, L_m can admit a certain position relative to interval $[L_p, 2]$, according to the position of p_1 relative to the interval

$$[p_3 + p_4 \ , \ p_3 + 2p_4] \ .$$

3. Now suppose that $p_1 = p_2 > p_3 = p_4$

B:

B_1 There are no longer C_4^2 ways to proceed but only half as many:

rank 1: 1 answer among 4 (2 ways)

rank 2: and either the answer with the same probability (one way) or the other (with repetition) this gives 3 classes:

B_2 rank 0: 1 eventuality: 2 possible ways

rank 1: determined → 2 classes.

Thus B has only 5 equivalence classes.

C:

C_1 We group as outcomes a question of rank 1 two answers with probabilities equal or not → 2 classes

C_2 either rank 1: p_1 (answers)   ⎫
            rank 2: p_1 or p_3 ⎬ → 2 cases

       either rank 1: p_3 ⎫
            rank 2: p_1 or p_3 ⎬ → 2 cases

and 4 classes.

Thus C has only 6 equivalence classes.

Extreme questionnaires

For the case B we have

$$L_o = 1 + 2p_3 \leqslant L \leqslant 2 - p_3 = L_p$$

For the case C we have:

$L_1 = 2$ and directly:

$$L_m = \frac{3}{2} + 3p_3 \leq L < \frac{3}{2} + 3p_1 = L_M .$$

But the relations

$$p_1 = p_2 > p_3 = p_4$$

and

$$p_1 + p_2 + p_3 + p_4 = 1$$

imply

$$p_1 > \frac{1}{4} , \quad p_3 < \frac{1}{4}$$

and hence

$$L_M > 2 + \frac{1}{4} > L_1 = 2 .$$

The result (table) of the discussion of the case $p_1 > p_2 > p_3 > p_4$ remains entirely valid, the position of L_m relative to the interval $[L_p, 2]$ depending on that of p_1 relative to the interval $[2p_3, 3p_3]$

4. Application

$$p_1 = p_2 = \frac{1}{3} > p_3 = p_4 = \frac{1}{6} .$$

We are in the case:

$$p_1 = p_3 + p_4$$

that is

$$L_o < L_p < L_m = 2 < L_M .$$

we find:

$$\begin{aligned}
\text{Type } B_1 &: \quad L_o = 1 + p_1 = 4/3 \\
B_2 &: \quad L_p = 2 - p_3 = 11/6 \\
C_1 &: \quad L_1 = 2 \\
C_2 &: \quad \begin{cases} L_m = 2 \\ L_M = 5/2 \end{cases}
\end{aligned}$$

The routings therefore have the lengths (Fig. 30 to 33)

L:	8/6	9/6	10/6	11/6	12/6	13/6	14/6	15/6
Types:	B_1a	B_1b	B_1c	B_2e	C_1f C_1g C_2h	C_2i	C_2j	C_2k
			B_2d					
Values:	L_o			L_p	$L_1=L_m$			L_M

Information:

$$I_2 = \sum_{i \in E} p_i \cdot \log_2 \frac{1}{p_i},$$

$$I_3 = \sum_{i \in E} p_i \cdot \log_3 \frac{1}{p_i};$$

that is

$$I_2 = 1.91, \quad I_3 = 1.20.$$

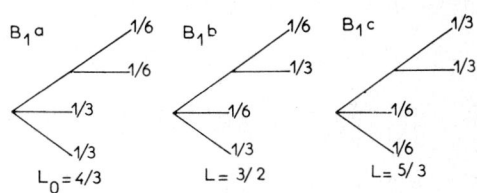

Figure 30

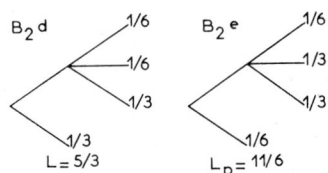

Figure 31

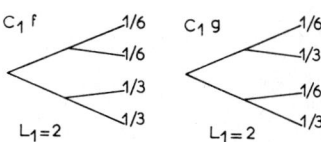

Figure 32

We note that

$I_3 < L_o$ or $(1.20 < 1.33)$

and

$L_p = 1.83 < I_2 < L_1 = 2$.

For questionnaires of type C, it is obvious that I_2 is always less than L.

For questionnaires of type B, it is a fortiori natural that I_3 is less than L because the question with basis 2 (polychotomic

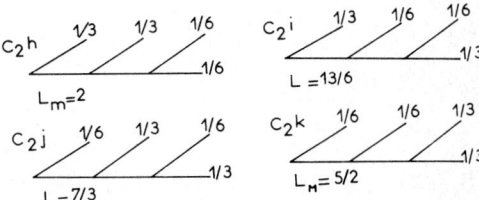

Figure 33

questionnaire in the broad sense) tends to increase the inefficiency.

The fact that I_2 is greater than L_p reveals only that the information (with the least basis as base) is not adapted to the foresight of the routing length of the heterogeneous questionnaire.

The study of I_2 and I_3 is thus a new answer to the initial question: the questionnaires of type B can be considered as homogeneous, semi-homogeneous or heterogeneous.

<u>Problem 9 Reduction of the Circuits of a Flow Chart</u>

1. We take all the triads of 3 elements among M; the ALGOL program ensures that we proceed in order without omission or repetition. Thus there are:

$$X\ MAX = C_M^3 = \frac{M(M-1)(M-2)}{6} \quad \text{questions}$$

and

$$T = \frac{M(M-1)(M-2)}{6} \tau_1$$

A question corresponds to a cycle (commonly called a loop) of the instruction FOR relative to I.

2. I, J, K are fixed.

The questionnaire (Fig. 34) enables us to evaluate C

$A[I,J] \neq 0 \lor A[J,I] \neq 0$.

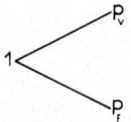

Figure 34

2.1 The double restricted product (Fig. 35) enables us to evaluate B:

There are $N_2 = 4$ answers:

. $A[I,J] = A[J,I] = 0$ with probability p_F

. $(A[I,J] \neq 0 \lor A[J,I] \neq 0)$

$\land (A[J,K] = 0 \lor A[K,J] = 0)$
with probability $p_T p_F$

. $(A[I,J] \neq 0 \lor A[J,I] \neq 0)$

$\land (A[J,K] \neq 0 \lor A[K,J] \neq 0)$

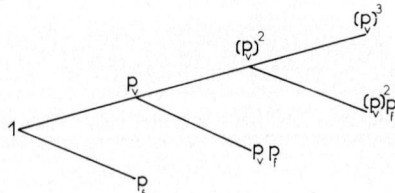

Figure 35

$\land (A[I,K] = 0 \lor A[K,I] = 0)$ with probability $(p_T)^2 p_F$

. B = TRUE with probability $(p_T)^3$.

We will determine

$$L_2 = 1 + p_T + p_T^2$$

and we see that $1 \leq L_2 \leq 3$.

a. $p_T = 1/2 \Rightarrow L_2 = 1 + \frac{1}{2} + \frac{1}{4} = \underline{1.75}$

b. $p_T = 3/4 \Rightarrow L_2 = 1 + \frac{3}{4} + \frac{9}{16} = 2 + \frac{5}{16} = \underline{2.3125}$

2.2 Now we form the usual product with C_M^3 questionnaires as factors.

We then find

$$T = \frac{M(M-1)(M-2)}{6} \tau_2 (1 + p_T + p_T^2) .$$

3 A double restricted product of trichotomic questionnaires (Fig. 36) enables us to decide whether B is TRUE or FALSE.

3.1 As soon as an outcome is p_3 there is no need to go further to evaluate B.

There are 15 answers with probabilities

$p_1^3(1)$, $p_1^2 p_2(3)$, $p_1^2 p_3(1)$,

$p_1 p_2^2(3)$, $p_1 p_2 p_3(2)$, $p_1 p_3(1)$,

$p_2^3(1)$, $p_2^2 p_3(1)$,

$p_2 p_3(1)$, $p_3(1)$.

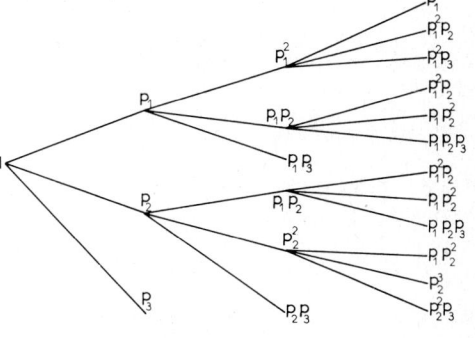

All the answers with rank 1 or 2 contain p_3 as a factor.

Figure 36

The questions determine the routing length:

$$L_3 = 1 + p_1 + p_2 + p_1^2 + 2p_1 p_2 + p_2^2$$

$$L_3 = 1 + p_1 + p_2 + (p_1 + p_2)^2$$

$$L_3 = 1 + (1 - p_3) + (1 - p_3)^2$$

(independent of p_1 and p_2).

This is evident because the first questionnaire (I and J) is certainly applied whereas the other two are applied (restricted products) only with the probabilities $(1-p_3)$ and $(1-p_3)^2$, that is when the first or the second test has not given FALSE.

If we write $L_2 = 1+(1-p_F)+(1-p_F)^2$ and if we identify p_3 and p_F we find:

$$L_2 = L_3 .$$

Furthermore $L_3 = 3-3p_3+p_3^2$ and naturally

$L_3 \to 1$ as $p_3 \to 1$

$L_3 \to 3$ as $p_3 \to 0$

L_3 always lies between 1 and 3.

For $p_3 = \frac{1}{4}$ we obtain

$$L_3 = 2.3125$$

that is the same result as for L_2 when $p_F = \frac{1}{4}$.

This questionnaire is without inefficiency when

$$p_1 = p_2 = p_3 = 1/3$$

and it then has length:

$$L_3 = 2 + \frac{1}{9} = 2.111 ;$$

whereas the preceding questionnaire was without inefficiency for $p_T = p_F = \frac{1}{2}$ which corresponded to $L_2 = 1.75$.

3.2 We then determine:

$$T = \frac{M(M-1)(M-2)}{6} \tau_3 (3 - 3p_3 + p_3^2) .$$

4   The question of time τ_4 is more direct, but we must then make a first restricted product of dichotomic questionnaires to obtain the answer: is C TRUE or FALSE (Fig. 37). Then a double restricted product will give the answer B TRUE or B FALSE. We will therefore have answers of ranks 2 to 6.

We have now to deal with questionnaires without inefficiency in the case where $q_T = 1/2$.

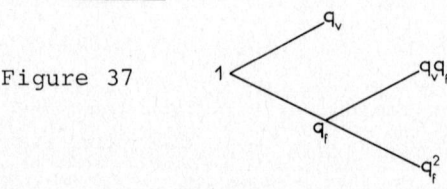

Figure 37

SOLUTIONS TO PROBLEMS

The questionnaire evaluating D has routing length equal to 1.

That evaluating C (case IJ) has length $1+q_F$ and it is necessary to proceed to bring in the following questionnaire with probability $(1-q_F)^2$, or still once more (case IK) with probability $(1-q_F^2)^2$. The questionnaire evaluating B has length:

$$L_4 = (1 + q_F) + (1 + q_F)(1 - q_F^2) + (1 + q_F)(1 - q_F^2)^2$$

that is also

$$L_4 = (1 + q_F)[1 + (1 - q_F^2) + (1 - q_F^2)^2]$$

or

$$L_4 = 3 + 3q_F(1 - q_F - q_F^2) + q_F^4 + q_F^5,$$

this last result also being obtained by applying

$$L = \sum_{x_i \in F_4} p(x_i)$$

to the questionnaire shown in Fig. 38.

For $q_F = \tfrac{1}{2}$ we find

$$L_4 = \frac{3}{2}\left[1 + \frac{3}{4} + \frac{9}{16}\right] = \frac{111}{32}.$$

More generally for $q_F^2 = p_3$ we have:

$$L_4 = (1 + q_F)L_3$$

because the "simulation" of questionnaire C by means of D leads to substituting $(1+q_F)$ question for the 1 question.

We then find

$$T = \frac{M(M-1)(M-2)}{6} \tau_4 L_4 .$$

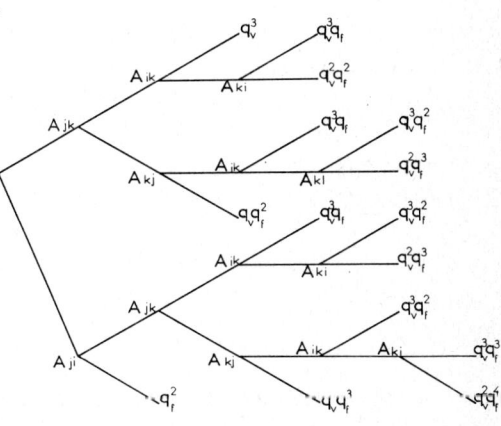

Figure 38

5   When the questionnaire is without inefficiency we find $I_4 = L_4$, where I_4 is evaluated with logarithms to the base 2. But this is the case when $q_F = \tfrac{1}{2}$, that is $q_F^2 = 1/4$.

When $q_F^2 = p_3 = \tfrac{1}{4}$ and $p_1 = \tfrac{1}{2}$ then the questionnaire corresponding to τ_3 processes the same information as the questionnaire working on D: they have the same answers, the same probabilities and we will write them as functions of $p_1 = \tfrac{1}{2}$ by

$$p_2 = p_1^2 \quad , \quad p_3 = p_1^2$$

$$q_F = p_1 \quad , \quad q_T = p_1$$

$$(1)p_1^2, \ (2)p_1^3, \ (5)p_1^4, \ (5)p_1^2, \ (2)p_1^6 \ .$$

Then the information for questionnaire 4 is

$$I_4 = \Sigma p_i \log_2 \frac{1}{p_i} = L_4 = \frac{3}{2} \times 2.3125$$

and for questionnaire 3, it is

$$I_3 = \Sigma p_i \log_3 \frac{1}{p_i} = \frac{1.5}{\log_2 3} 2.3125 \ .$$

Consequently we see that

$$I_3 = \frac{1.5}{1.585} L_3$$

whereas $I_4 = L_4$.

We find $T(M, \tau_3) = T(M, \tau_4)$ for

$$\tau_3 L_3 = \tau_4 L_4$$

or

$$\frac{\tau_3}{\tau_4} = \frac{L_4}{L_3} \ ;$$

but $\ell = \frac{L_4}{L_3}$: thus we must have $\ell = 1.5$.

Since $k = \frac{I_4}{I_3} = 1.585$ this means that $\ell/k < 1$.

This inequality which can also be written as $\frac{L_4}{I_4} / \frac{L_3}{I_3} < 1$ is an interpretation of the inefficiency which is zero for questionnaire 4 (because $L_4 = I_4$) and positive for questionnaire 3. If $\ell > 1.5$ then:

$$T(M, \tau_4) < T(M, \tau_3) \ ,$$

which is the case when the times of the questions vary as the logarithm of the bases. This latter condition will later be assumed to hold.

6   Here we make the hypothesis of trichotomic tests. The restricted product of questionnaires with fixed (I,J,K) will no longer arise because as soon as

$$A[I,J] = A[J,I] = 0$$

we pass to the study of $A[I,J+1]$ or possibly of $A[I+1,I+2]$, while keeping $J<M$. There are therefore C^2_{M-1} possible pairs so that $K<J$.

Consequently the first question

$$A[I,J] \neq 0 \lor A[J,I] \neq 0 \;?$$

is asked $\quad C^2_{M-1} = \dfrac{(M-1)(M-2)}{2} \quad$ times.

The questionnaire relative to (J,K) and (K,I) is then asked in total C^3_n times, if $p_3=0$, and $(1-p_3)C^3_M$ times otherwise, without any necessity to specify the order in the product.

This questionnaire gives a contribution to the routing of

$$((1-p_3)C^3_M)(1+(1-p_3)).$$

Consequently

$$T_6 = \tau_3 \{C^2_{M-1} + (1-p_3)C^3_M(1+(1-p_3))\}.$$

Noting that $C^3_M = C^2_{M-1} \dfrac{M}{3}$, we find:

$$T_6 = \dfrac{(M-1)(M-2)}{2} \times$$

$$\tau_3 \left\{1 + \dfrac{M}{3}(1-p_3)(1+(1-p_3))\right\}$$

which we compare with

$$T_3 = \dfrac{(M-1)(M-2)}{2} \tau_3 \dfrac{M}{3}(3 - 3p_3 + p_3^2).$$

Whence

$$T_3 - T_6 = \dfrac{(M-1)(M-2)}{2} \tau_3 \left(\dfrac{M}{3} - 1\right).$$

Consequently

$$M \geqslant 3 \quad \text{implies} \quad T_3 \geqslant T_6,$$

with equality if and only if $M=3$.

WHATEVER THE PROBABILITY p_3, T_6 therefore stays less than T_3.

We can still try to improve the time by using dichotomic tests of type D as in paragraph 4, which would enable us to obtain a routing length equal to the information, when

$q_F^2 = p_3 = \frac{1}{4}$.

The routing will be analogous, and the new time T_7, will be obtained in the same way by:

$$T_7 = \frac{(M-1)(M-2)}{2} (1 + q_F) \times \tau_4 \left\{ 1 + \frac{M}{3} (1 - q_F^2) [1 + (1 - q_F^2)] \right\}$$

so that

$$\tau_4 - T_7 = \frac{(M-1)(M-2)}{2} \tau_4 (1 + q_F) \left(\frac{M}{3} - 1\right).$$

If $\ell > 1.5$ then the BICYCLE routine can be improved by effective dichotomic tests because then T_7 is the minimal time.

We can now write the new end of the program in question as follows:

```
MONOCYCLE : FOR I := 1      STEP 1 UNTIL M - 2    DO
            FOR J := I + 1  STEP 1 UNTIL M - 1    DO
BEGIN   IF A [I,J] ≠ 0   THEN GO UNTIL UNARC ELSE
        IF A [J,I] = 0   THEN GO UNTIL AUTRARC ;
   UNARC : FOR K := J + 1   STEP 1 UNTIL M    DO
BEGIN   IF A [J,K] ≠ 0   THEN GO TO DEZARC ELSE
        IF A [K,J] = 0   THEN GO UNTIL KAVARI ;
   DEZARC : IF A [I,K] ≠ 0   THEN X := X + 1  ELSE
            IF A [K,I] ≠ 0   THEN X := X + 1 ;
   KAVARI : END ;
AUTRARC : END ; WRITE (M,X)
END
END
```

Although this may take longer to write and compile, it is the quickest program according to the theoretical study of questionnaires and this has been confirmed experimentally on a computer.

TABLES

TABLE 1

p	$\log_2 \frac{1}{p}$	$p \log_2 \frac{1}{p}$	$I_2(p, 1-p)$
0.01	6.64386	0.06644	0.08079
0.02	5.64386	0.11288	0.14144
0.03	5.05889	0.15177	0.19439
0.04	4.64386	0.18575	0.24229
0.05	4.32193	0.21610	0.28640
0.06	4.05889	0.24353	0.32744
0.07	3.83650	0.26856	0.36592
0.08	3.64386	0.29151	0.40218
0.09	3.47393	0.31265	0.43647
0.10	3.32193	0.33219	0.46900
0.11	3.18442	0.35029	0.49992
0.12	3.05889	0.36707	0.52936
0.13	2.94342	0.38264	0.55744
0.14	2.83650	0.39711	0.58424
0.15	2.73697	0.41054	0.60984
0.16	2.64386	0.42302	0.63431
0.17	2.55639	0.43459	0.65770
0.18	2.47393	0.44531	0.68008
0.19	2.39593	0.45523	0.70147
0.20	2.32193	0.46439	0.72193
0.21	2.25154	0.47282	0.74148
0.22	2.18442	0.48057	0.76017
0.23	2.12029	0.48767	0.77801
0.24	2.05889	0.49413	0.79504
0.25	2.00000	0.50000	0.81128
0.26	1.94342	0.50529	0.82675
0.27	1.88897	0.51002	0.84146
0.28	1.83650	0.51422	0.85545
0.29	1.78588	0.51790	0.86872
0.30	1.73697	0.52109	0.88129
0.31	1.68966	0.52397	0.89317
0.32	1.64386	0.52603	0.90438
0.33	1.59946	0.52782	0.91493
0.34	1.55639	0.52917	0.92482
0.35	1.51457	0.53010	0.93407
0.36	1.47393	0.53062	0.94268
0.37	1.43440	0.53073	0.95067
0.38	1.39593	0.53045	0.95804
0.39	1.35845	0.52980	0.96480
0.40	1.32193	0.52877	0.97095
0.41	1.28630	0.52738	0.97650
0.42	1.25154	0.52565	0.98145
0.43	1.21759	0.52356	0.98582
0.44	1.18442	0.52115	0.98959
0.45	1.15200	0.51840	0.99277
0.46	1.12029	0.51534	0.99538
0.47	1.08927	0.51196	0.99740
0.48	1.05889	0.50827	0.99885
0.49	1.02915	0.50428	0.99971
0.50	1.00000	0.50000	1.00000

Table I contd.

p	$\log_2 \frac{1}{p}$	$p \log_2 \frac{1}{p}$	$I_2(p,1-1)$
0.51	0.97143	0.49543	0.99971
0.52	0.94342	0.49058	0.99885
0.53	0.91594	0.48545	0.99740
0.54	0.88897	0.48004	0.99538
0.55	0.86250	0.47437	0.99277
0.56	0.83650	0.46844	0.98959
0.57	0.81097	0.46225	0.98582
0.58	0.78588	0.45581	0.98145
0.59	0.76121	0.44912	0.97650
0.60	0.73697	0.44218	0.97095
0.61	0.71312	0.43500	0.96480
0.62	0.68966	0.42759	0.95804
0.63	0.66658	0.41994	0.95067
0.64	0.64386	0.41207	0.94268
0.65	0.62149	0.40397	0.93407
0.66	0.59946	0.39564	0.92482
0.67	0.57777	0.38710	0.91493
0.68	0.55639	0.37835	0.90438
0.69	0.53533	0.36938	0.83917
0.70	0.51457	0.36020	0.88129
0.71	0.49411	0.35082	0.86872
0.72	0.47393	0.34123	0.85545
0.73	0.45403	0.33144	0.84146
0.74	0.43440	0.32146	0.82675
0.75	0.41504	0.31128	0.81128
0.76	0.39593	0.30091	0.79504
0.77	0.37707	0.29034	0.77801
0.78	0.35845	0.27959	0.76017
0.79	0.34008	0.26866	0.74148
0.80	0.32193	0.25754	0.72193
0.81	0.30401	0.24625	0.70147
0.82	0.28630	0.23477	0.68008
0.83	0.26882	0.22312	0.65770
0.84	0.25154	0.21129	0.63431
0.85	0.23447	0.19930	0.60984
0.86	0.21759	0.18713	0.58424
0.87	0.20091	0.17479	0.55744
0.88	0.18442	0.16229	0.52936
0.89	0.16812	0.14963	0.49992
0.90	0.15200	0.13680	0.46900
0.91	0.13606	0.12382	0.43647
0.92	0.12029	0.11067	0.40218
0.93	0.10470	0.09737	0.36592
0.94	0.08927	0.08391	0.32744
0.95	0.07400	0.07030	0.28640
0.96	0.05889	0.05654	0.24229
0.97	0.04394	0.04263	0.19439
0.98	0.02915	0.02856	0.14144
0.99	0.01450	0.01435	0.08079
1.00	0.00000	0.00000	0.00000

TABLE II

$\frac{1}{p}$	$\log_2 \frac{1}{p}$	$p \log_2 \frac{1}{p}$	$I_2(p, 1-p)$
1	0.00000	0.00000	0.00000
2	1.00000	0.50000	1.00000
3	1.58496	0.52832	0.91830
4	2.00000	0.50000	0.81128
5	2.32193	0.46439	0.72193
6	2.58496	0.43083	0.65002
7	2.80735	0.40105	0.59167
8	3.00000	0.37500	0.54356
9	3.16993	0.35221	0.50326
10	3.32193	0.33219	0.46900
11	3.45943	0.31449	0.43950
12	3.58496	0.29875	0.41382
13	3.70044	0.28465	0.39214
14	3.80735	0.27195	0.37123
15	3.90689	0.26046	0.35336
16	4.00000	0.25000	0.33729
17	4.08746	0.24044	0.32276
18	4.16993	0.23166	0.30954
19	4.2 993	0.22358	0.27947
20	4.32193	0.21610	0.28640
21	4.39232	0.20916	0.27620
22	4.45943	0.20270	0.26676
23	4.52356	0.19668	0.25802
24	4.58496	0.19104	0.24988
25	4.64386	0.18575	0.24229
26	4.70044	0.18079	0.23519
27	4.75489	0.17611	0.22854
28	4.80735	0.17169	0.22228
29	4.85798	0.16752	0.21640
30	4.90689	0.16356	0.21084
31	4.95420	0.15981	0.20559
32	5.00000	0.15625	0.20062
33	5.04439	0.15286	0.19591
34	5.08746	0.14963	0.19143
35	5.12928	0.14655	0.18718
36	5.16993	0.14361	0.18312
37	5.20945	0.14080	0.17926
38	5.24793	0.13810	0.17557
39	5.28540	0.13552	0.17204
40	5.32193	0.13305	0.16866
41	5.35755	0.13067	0.16543
42	5.39232	0.12839	0.16233
43	5.42626	0.12619	0.15935
44	5.45943	0.12408	0.15649
45	5.49185	0.12204	0.15374
46	5.52356	0.12008	0.15110
47	5.55459	0.11818	0.14855
48	5.58496	0.11635	0.14609
49	5.61471	0.11459	0.14373
50	5.64386	0.11288	0.14144

Table II contd.

$\frac{1}{p}$	$\log_2 \frac{1}{p}$	$p \log_2 \frac{1}{p}$	$I_2(p, 1-1)$
51	5.67243	0.11122	0.13923
52	5.70044	0.10962	0.13710
53	5.72792	0.10807	0.13504
54	5.75489	0.10657	0.13304
55	5.78136	0.10512	0.13111
56	5.80735	0.10370	0.12923
57	5.83289	0.10233	0.12742
58	5.85798	0.10100	0.12566
59	5.88264	0.09971	0.12395
60	5.90689	0.09845	0.12229
61	5.93074	0.09723	0.12068
62	5.95420	0.09604	0.11912
63	5.97728	0.09488	0.11759
64	6.00000	0.09375	0.11612
65	6.02237	0.09265	0.11468
66	6.04439	0.09158	0.11327
67	6.06609	0.09054	0.11191
68	6.08746	0.08952	0.11058
69	6.10852	0.08853	0.10929
70	6.12928	0.08756	0.10802
71	6.14975	0.08662	0.10679
72	6.16993	0.08569	0.10559
73	6.18982	0.08479	0.10442
74	6.20945	0.08391	0.10328
75	6.22882	0.08305	0.10216
76	6.24793	0.08221	0.10107
77	6.26679	0.08139	0.10000
78	6.28540	0.08058	0.09896
79	6.30378	0.07979	0.09794
80	6.32193	0.07902	0.09694
81	6.33985	0.07827	0.09597
82	6.35755	0.07753	0.09502
83	6.37504	0.07681	0.09408
84	6.39232	0.07610	0.09317
85	6.40939	0.07540	0.09228
86	6.42626	0.07472	0.09140
87	6.44294	0.07406	0.09054
88	6.45943	0.07340	0.08970
89	6.47573	0.07276	0.08888
90	6.49185	0.07213	0.08807
91	6.50779	0.07151	0.08728
92	6.52356	0.07091	0.08650
93	6.53916	0.07031	0.08574
94	6.55459	0.06973	0.08500
95	6.56986	0.06916	0.08426
96	6.58496	0.06859	0.08354
97	6.59991	0.06804	0.08284
98	6.61471	0.06750	0.08214
99	6.62936	0.06696	0.08146
100	6.64386	0.06644	0.08079

BIBLIOGRAPHY

CHAPTER 1. Fundamental properties of graphs

ANDERSON S.S., Graph theory and finite combinatorics, Markham, Chicago, 1970.
ANDRE D., Des notations mathématiques, énumération, choix et usage, Gauthier-Villars, 1909.
BERGE C., Théorie des graphes et ses applications, Dunod, 1958.
BERGE C., Réflexions sur la place de la Théorie des Graphes, in Rome-66(*), p. XIII-XVIII.
BERGE C., Contribution de la Théorie des graphes à l'étude des relations d'ordre, in Aix-en-Provence-67, p. 143-162.
BERGE C., Principes de Combinatoire, Dunod, 1968.
BERGE C., Graphes et Hypergraphes, Dunod, 1970.
BERGE C. et GHOUILA-HOURI A., Programmes, jeux et réseaux de transport, Dunod, 1962.
BELLMAN R. and COOKE K.L., The Königsberg bridges problem generalized, J. Math. Analysis and Appl. 25 (1969) p.1-7.
BONDY J.A., Cycles in graphs, in Calgary-69, p. 15-18.
BUSACKER R.G. and SAATY T.L., Finite graphs and networks: an introduction with applications, McGraw Hill, 1965.
CAMION P., Quelques propriétes des chemins et des circuits hamiltoniens dans la théorie des graphes, Cahiers Cent. Etudes Rech. Opér. Bruxelles 2 (1960) p. 10-15.
CAMION P., Analyse de la portée explicative, en théorie des graphes, de propriétés de modules de points entiers. In Rome-66, p. 51-56.
CARTIER P. et FOATA D., Problèmes combinatoires de commutation et réarrangement, Lecture notes 85, Springer, 1969.
CESARI Y., Questionnaire, codage et tris, Thèse 3e Cycle, Paris, 1967.
CHATY G., Algorithmes d'orientation progressive des graphes, Thèse 3e Cycle, Paris, 1966.
CHATY G., Graphes fortement connexes c-minimaux, C.R. Acad. Sci. Paris, 266A (1968) p. 907-909.
CHATY G., Unicité de certains chemins dans les graphes fortement connexes C.R. Acad. Sci. Paris 272 A (1971) p. 710-713.
CHATY G., Graphes fortement connexes C-minimaux et graphes sans circuit co-minimaux, J. of Combinatorial Theory B. 10 (1971) p. 237-244.
CHATY G., Cheminements remarquables dans les graphes: existence, obtention, conservation, Thèse Sci. Math., Paris, 1971.
COMTET L., Analyse combinatoire, Tome I, tome II, Presses Universitaires de France, 1970.
ERDOS P. and GALLAI T., On maximal paths and circuits of graphs, Acta Math. Sci. Hung. 10 (1959) p. 337-356.
EULER L., Solutio problematis ad geometriam situs pertinentis, (Commentaires de l'Académie des Sciences de St-Pétersbourg 8 (1736) p. 128-140).
Translated by Coupy :
Solution d'un problème appartenant à géométrie de situation, Nouvelles Annales de Math. 10 (1851) p. 106-118.
FRASNAY C., Quelques problèmes combinatoires concernant les ordres totaux et les relations monomorphes, Ann. Inst. Fourier (Grenoble) 15 (1965) p. 415-524.
GHOUILA-HOURI A., Flots et tensions dans un graphe. Ann. Sci. Ec. Norm. Sup. 3e série 81 (1964) p. 267-339.

(*) *Edited volumes from which individual articles are mentioned are collected separately, starting on page 417.*

GOULD R., The application of graph theory to the synthesis of contact networks. Proc. of an International Symposium on the Theory of Switching (1957). The Annals of the Computation Laboratory of Harvard University, 29, 1959.
HALL, M., Combinatorial Theory, Blaisdell, 1967.
HARARY F., On the consistency of precedence matrices, Journal A.C.M. 7 (1960) p. 255-259.
HARARY F., Graph Theory, Addison-Wesley, 1969.
HARARY F., NORMAN R.Z. and CARTWRIGHT D., Structural Models. An introduction to the theory of directed graphs, Wiley, 1965. French translation: Introduction à la theorie des graphes orientés. Modelès structuraux, Dunod, 1968.
JORDAN C., Sur les assemblages de lignes, J. reine angewtande Math. 70 (1869) p. 185-190.
KONIG D., Theorie der endlichen und unendlichen Graphen (1e ed. Akademie Verlag MBH, Leipzig 1936), Chelsea, 1950.
LAS VERGNAS M., Sur l'existence des cycles hamiltoniens dans un graphe, C.R. Acad. Sci. Paris 270A (1970) p. 1361-1354.
LUCAS E., Récréations mathématiques, nouvelle édition, 4 tomes, Blanchard, 1961.
MAC MAHON P.A., Combinatory analysis (1e ed. Cambridge University Press, 1915, 1916), Chelsea, 1960.
MALGRANGE B., Sous-matrices premières d'une sous-matrice à coefficients binaires, application à certains problèmes de graphes, 2e Congrès de l'AFCALTI, Paris 1961, Guathier-Villars, (1963) p. 231-240.
MARIMONT R.B., A new method of checking the consistency of precedence matrices, Journal A.C.M. 6 (1959) p. 164-171.
MINTY G.J., Monotone networks, Proc. Roy. Soc. A257 (1960) p. 194-212.
MIRSKY L. and PERFECT HAZEL., Applications of the notion of independance to problems of combinatorial analysis, Journal of Combinatorial Theory 2 (1967) p. 327-357.
NASH-WILLIAMS C. St.J.A., A survey of the theory of well-quasi-ordered sets, in Calgary-69, p. 293-299.
NOLIN L., Traitement des données groupées, Publication de l'Institut Blaise Pascal, 1964.
ORE O., Graphs and matching theorems, Duke Math. J. 22 (1955) p. 625-639.
ORE O., Theory of graphs, American Mathematical Society, Colloquium 38, 1962.
PETERSEN J., Die Theorie der regulären Graphs, Acta Math. 15 (1891) p. 193-220.
PICARD C.F., Panorama des liaisons entre graphes et ordinateurs, in Rome-66, p. 301-314.
PICKETT H.E., Subdirected representations of relational systems, Fundamenta Mathematica 56 (1969) p. 223-240.
PONSTEIN J., A generalization of the incidence matrices of a graph, in Rome-66, p. 315-332.
READ C., Teaching graph theory to a computer, in Waterloo-68, p. 161-173.
RIORDAN J., An introduction to combinatorial analysis, Wiley, 1958.
RIORDAN J., Combinatorial identities, Wiley, 1968.
ROTA G.C., The number of partitions of a set, Amer. Math. Monthly 71 (1964) p. 498-504.
ROY B., Cheminement et connexité dans les graphes. Thèse Sci. Math. Paris. 1961.
RYSER H.J., Combinatorial mathematics, Wiley, 1963, traduction : Mathématiques Combinatoires, Dunod 1969.

SACHS H., Construction of non-Hamiltonian planar regular graphs of degrees 3,4 and 5 with highest possible connectivity, in Rome-66, p. 373-382.
SACHS H., Einführung in die Theorie der endlichen Graphen, teil 1, Leipzig, 1970.
SAINTE-LAGUE M.A., Les réseaux ou graphes, Gauthier-Villars, 1926.
SESHU S. and REED M.B., Linear graphs and electrical networks, Addison-Wesley, 1961.
STEEN J.P., Algorithme de recherche d'un isomorphisme entre deux graphes, thèse 3e Cycle, Lille, 1968.
STEEN J.P., Principe d'un algorithme de recherche d'un isomorphisme entre deux graphes, Revue francaise d'Informatique et de Recherche Opérationnelle R-3 (1969) p. 51-69.
SWAN R.G., An application of graph theory to algebra, Proc. Amer. Math. Soc. 14 (1963) p. 367-373.
TUTTE W.T., The factorization of linear graphs, J. London Math. Soc. 22 (1947) p. 107-111.
TUTTE W.T., On the problem of decomposing a graph into n connected factors, J. London Math. Soc. 36 (1961) p. 221-230.
TUTTE W.T., How to draw a graph, Proc. London Math. Soc. 13 (1963) p. 743-767.
UNGER S.H., GIT a heuristic program for testing pairs of directed line graphs for isomorphism, Comm. A.C.M. 7 (1964) p. 26-34.
WATKINS M.E. and MESNER D.M., Cycles and connectivity in graphs, Journal Canadien de Math. 19 (1967) p. 1319.
WELCH J.T. Jr., A mechanical analysis of the cyclic structure of undirected linear graphs, Journal A.C.M. 13 (1966) p. 205-210.
ZYKOV A.A., Théorie des graphes finis (en russe), Nauka (section sibérienne), 1969.

CHAPTER II. Latticoids and arborescences

BALAS E., Finding a minimaximal Path in a disjunctive PERT network, in Rome-66, p. 21-30.
BARBUT M., Vocabulaire des ensembles ordonnés et treillis, in Aix-en-Provence-67, p. 3-23.
BELLMAN R., Dynamic Programming treatment of the travelling salesman problem, Journal A.C.M. 9 (1962) p. 61-63.
BELLMORE M. and NEMHAUSER G.L., The travelling salesman problem, Oper. Res. 16 (1968) p. 538-558.
BERGE C., La fonction de Grundy d'un graphe infini, C.R. Acad. Sci. Paris, 242 (1956) p. 1404-1407.
BERGE C., Sur l'equivalence du problème du transport généralisé et du problème des réseaux, C.R. Acad. Sci. Paris 251 (1960) p. 324-326.
BORILLO M., Quelques remarques sur la détermination de l'algorithme optimal pour la recherche du plus court chemin dans un graphe sans circuits, Revue Francaise d'Informatique et de Recherche Opérationnelle 33 (1964) p. 385-388.
CAYLEY A., On the theory of the analytical forms called trees, Philosophical Magazine 13 (1857) p. 172-176, Math. papers Cambridge 3 (1889-1897) p. 242-246.
CAYLEY A., On the analytical forms called trees, with application to the theory of chemical combinations. Report of the British Asso. for the Advancement of science (1875) p. 257-305, Math. Papers 9, p. 427-460.
CAYLEY A., A theorem on trees, Quart. J. of Math. 23 (1889) p. 376-378, Math. papers 13, p. 26-28.

CHOQUET G., Etude de certains réseaux de route, C.R. Acad. Sci. Paris 206 (1938) p. 310-313.
CROISOT R., Condition suffisante pour l'égalité des longueurs de deux chaînes de même extrémité dans une structure, applications aux relations d'équivalence et aux sous-groupes, C.R. Acad. Sci. Paris 226 (1948) p. 767-768.
ČULIK K., On the ordered trees used in theory of languages, in Rome-66, p. 69-76.
DANTZIG G.B., On the shortest route through a network, Management Sci. 6 (1960) p. 187-190.
DANTZIG G.B., BLATTNER W.O. and RAO M.R., Finding a cycle in a graph with minimum cost to times ration with application to a ship routing problem, in Rome-66, p. 77-83.
DANTZIG G.B., BLATTNER W.O. and RAO M.R., All shortest routes from a fixed origin in a graph, in Rome-66, p. 85-90.
DANTZIG G.B., All shortest routes in a graph, in Rome-66, p. 91-92.
DANTZIG G.B., Linear Programming and extensions, Princeton University Press, 1963.
DEDEKIND R., Uber die drei Moduln erzeugte Dualgruppe, Math. Annalen 53 (1900) p. 371-403, Gesammelte Werke 2 p. 236-271.
DERNIAME J.C. and PAIR C., Problèmes de cheminement dans les graphes, Monographies d'Informatique n° 8 Dunod, 1971.
EDMONDS J. and KARP R.M., Theoretical improvements in algorithmic efficiency for network flow problems, in Calgary-69, p. 93-96.
FAURE R., Eléments de la recherche opérationnelle, Gauthier-Villars, 1968, 2e éd. 1971.
FOATA D. et SCHUTZENBERGER M.P., Théorie géométrique des polynômes eulériens, Lecture notes 158, Springer, 1970.
FORD L.R. and FULKERSON D.R., Network flow and systems of representatives, Journ. Canadien de Math. 10 (1958) p. 78-84.
FORD L.R. and FULKERSON D.R., Flows in networks, Princeton University 1962.
FRANK H. and FRISCH I.T., Communication, transmission and transportation networks, Addison-Wesley, 1971.
FULKERSON D.R., Expected critical path lenghts in PERT networks, Operations Research 10 (1962) p. 808-817.
GHERSI T., Questionnaires latticiels, réseaux de transport et optimalité, Thèse de Docteur-Ingénieur, Paris 1970.
GLICKSMAN S., On the representation and enumeration of trees, Proc. Cambridge philos. Soc., 59 (1963) p. 509-517.
GOOD I.J., The generalization of Lagrange's expansion and the enumeration of trees, Proc. Cambridge Philos. Soc. 61 (1965) p. 499-517.
GROSS M. et LENTIN A., Notions sur les grammaires formelles, Gauthier-Villars, 1967.
GRUNDY P.M., Mathematics and Games, Eureka 2(1939) p. 6-8.
HALL M., Distinct representatives of subsets, Bull. Amer. Math. Soc. 54 (1948) p. 922-926.
HITCHCOCK F.L., The distribution of a product from several sources to numerous localities, J. Math. and Phys. 20 (1941) p. 224-230.
HOANG TUY et NGUYEN QUANG THAI., Sur deux problèmes d'affectation, Studia Scientiarum Mathematicarum Hungarica 4 (1969) p. 13-30.
HU T.C., Revised matrix algorithms for shortest paths, SIAM J. Appl. Math. 15 (1967) p. 207-218.
HU T.C., Integer programming and network flows, Addison-Wesley, 1969.
IVANESCU P.L. and RUDEANU S., A pseudo-boolean approach to matching problems in graphs with applications to assignment and transportation problems in Rome-66, p. 161-175

KNUTH D.E., The art of computer programming, vol. 1, Addison-Wesley, 1968.
KONIG D., Uber Graphen und ihre Anwendung auf Determinantentheorie und Mengenlehre, Mathematische Annalen 77 (1916) p. 453-476.
KREWERAS G., Counting problems in dendroids, in Calgary-69, p. 223-226.
KRUSKAL J.B., On the shortest spanning subtree of a graph and the travelling salesman problem, Proc. Am. Math. Soc. 71 (1956) p. 48-50.
LUDOT J., Contribution à la paramétrisation des graphes d'ordonnancement, Thèse 3e cycle Paris, 1966.
MATULA D.W., On the number of subtrees of a symmetric n-ary tree; SIAM J. Appl. Math. 18 (1970) p. 688-723.
MENON V.V., On the existence of trees with given degrees, Sankhya, 26 (1964) p. 63-68.
MURCHLAND J.D., Construction of a basis of elementary circuits and cocircuits in a directed graph, in Calgary-69, p. 286-288.
NASH WILLIAMS C. St. J.A., Edge disjoint spanning trees of finite graphs, J. London Math. Soc. 36 (1961) p. 445-450.
NIVAT M., Graphes et transducteurs, in Rome-66, p. 267-270.
PERFECT H., Aspects of transversal theory, in Calgary-69, p. 319.
PICARD C.F., Combinatorial problems concerning rooted-trees, in Balatonfüred-69, p. 909-918.
READ C., How to grow trees, in Calgary-69, p. 343-347.
RENYI A., On the enumeration of trees, in Calgary-69, p. 355-360.
RENYI A. and SZEKERES G., On the height of trees, J. Australian Math. Soc. 7 (1967) p. 497-507.
RENYI C. and RENYI A., The Prüfer code for k-trees, in Balatonfüred-69, p. 945-971.
ROSENSTIEHL P., L'arbre minimum d'un graphe, in Rome-66, p. 357-368.
ROY B., Algèbre moderne et théorie des graphes, Dunod, tome 1 1969, tome II, 1970.
RUDEANU S., Notes sur l'existence et l'unicité du noyau d'un graphe, Rev. Fr. Rech. Opér. 33 (1964) p. 345-352. et 41 (1966) p. 301-310.
SIMOES-PEREIRA J.M., Some results on the tree realization of a distance matrix, in Rome-66, p. 383-388.
SZASZ G., Introduction to lattice theory, Academic Press and the Hungarian Academy of Sciences Budapest, 1963.
TARRY G., Le problème des labyrinthes, Nouvelles Annales de Math. (3) 14 (1895) p. 187-190.
TIERNAN J.C., An efficient search algorithm to find the elementary circuits of a graph, Comm. A.C.M. 13 (1970) p. 722-726.
VIZING V.G., A bound on the external stability number of a graph, Soviet Math. Doklady 164 (1965) p. 729-731.
WOSTON J., Algorithme de résolution du problème du voyageur de commerce, METRA 8 (1969) p. 535-542.

CHAPTER III. Operations on graphs

ABERTH O., On the sum of graphs, Revue Fr. Rech. Oper. 33 (1964) p. 353-358.
BENEJAM E., Une méthode de synthèse des circuits séquentiels fiables, Thèse 3e Cycle Paris, 1966.
BENZAKEN C., Contribution des structures algébriques ordonnées à la théorie des réseaux, Thèse Sci. Math. Grenoble, 1968.
BIRKHOFF G., Lattice theory, Amer. Math. Soc. Colloquium, 25, édition révisée 1948.
BOSAK J., On the iteration of a graph transformation, in Calgary-69, p. 19-21.

CAPOBIANCO M.F., Tournaments and tensor products of diagraphs, SIAM J. App. Math. 15 (1967) p. 624-626.
CAPOBIANCO M.F., Tensor products of diagraphs and the structure of groups of pairs, Bulletin of Math. Biophysics 31 (1969) p. 319-326.
CAPOBIANCO M.F., Some properties of tensor composite graphs, in Calgary-69, p. 27-28.
CAPOBIANCO M.F., Some theorems on tensor composite graphs, Journal of research of the National Bureau of Standards, Mathematical Sciences 74B, n° 4, 1970.
CAPOBIANCO M.F., On characterizing tensor composite graphs, in New York-70, Art. 1.
CARVALLO M., Monographie des treillis et algèbre de Boole, Gauthier-Villars, 2e éd. 1966.
CHAO, Y.C., On a problème of Claude Berge, Proc. Amer. Math. Soc. 14 (1963) P. 80.
CHEIN M., Etude des décompositions d'un réseau. Application à l'écriture des fonctions booléennes en sommes et produits, Thèse de Docteur-Ingénieur, Grenoble, 1967.
CHEIN M., Graphe régulièrement décomposable, Revue Francaise d'Informatique et de Recherche Opérationnelle 7 (1968), p. 27-42.
CHEIN M., Sur des problèmes de décomposition d'un graphe, liés à l'implantation. Thèse Sci. Math. Grenoble, 1970.
COOPER D.C., Reduction of programs to a standard form by graph transformation, in Rome-66, p. 57-68.
DRAGOMIRESCU M., L'algorithme de min-addition et les chemins critiques dans un graphe, Rev. Roum. Math. Pures et Appl. XII, 8 (1967) p. 1045-1051.
DUBREIL P., Algèbre, (Cahiers Scientifiques XX) Gauthier-Villars, 1963.
DUBREIL-JACOTIN M.L., LESIEUR L. et CROISOT R., Leçons sur la théorie des treillis géométriques, (Cahiers Scientifiques XXI), Gauthier-Villars, 1953.
FARBEY B.A., LAND A.H. and MURCHLAND J.D., The cascade algorithm for finding all shortest distances in a directed graph. Management Science 14 (1967) p. 19-28.
FAURE R. et HEURGON E., Structures ordonnées et algèbres de Boole (Collection "Programmation"), Gauthier-Villars, 1971.
FLAMENT C., Applications of graph theory to group structure, Prentice Hall, 1963, traduction française : Théorie des graphes et structures sociales, Gauthier-Villars, Mouton, 1968.
HARARY F. and TRAUTH C., Connectedness of products of two directed graphs, SIAM J. Appl. Math. 14 (1966) p. 250-254.
HARARY F. and WILCOX G.W., Boolean operations on graphs, Math. Scand. 20 (1967). p. 41-51.
HAUSDORFF F., Grundzüge der Mengenlehre, Leipzig, 1914, revised by Chelsea, 1949.
HEDRLIN Z., On endomorphisms of graphs and their homomorphic images, in Ann Arbor-68, p. 73-85.
IMRICH W., Kartesisches Produkt von Mengensystemen und Graphen, Studia Sci. Math. Hungar, 2 (1967), p. 285-290.
IMRICH W., Uber das schwache Kartesische Produkt von Graphen, to appear in Journal of Combinatorial Theory.
JORDAN P., The mathematical theory of quasi-order semi-group of idempotent and non commutative lattices - a new field of modern algebra. Aeronautical Res. Lab. office of Aerospace USAF, European office, Final report (1961).
KUNTZMANN J., Algèbre de Boole, Dunod, 1965.

LUNTS A.G., Méthodes algébriques d'analyse et de synthèse des schémas
à contacts, Izv. Acad. Nauk S.S.S.R. Ser. Math. 16 (1952) p. 405-426
(in Russian).
McANDREW M.H., On the product of directed graphs, Proc. Amer. Math. Soc. 14
(1963) p. 600-606.
MILLER D.J., The categorial product of graphs, Journ. Canadien de Math. 20
(1968) p. 1511-1521.
MILLER D.J., Edge equivalence related to cartesian decompositions of graphs,
in Calgary-69, p. 273-276.
MILLER D.J., The automorphism group of graphs, Proc. Amer. Math. Soc. 25
(1970) p. 24-28.
MILLER D.J., Weak Cartesian Product of graphs, Colloq. Math. 21 (1970)
p. 55-74.
ORE O., Chains in partially ordered sets, Bull. Amer. Math. Soc., 49 (1943)
p. 558-566.
PAIR C., Sur des algoritmes pour les problèmes de cheminement dans les
graphes finis, in Rome-66, p. 271-300.
PETEANU V., An algebra of the optimal path in networks, Mathematica 9 (32)
(1967) p. 335-342.
PETEANU V., Optimal paths in networks and generalizations,
    (I) Mathematica (Cluj), 11 (34), (1969) p. 311-327.
    (II) Mathematica (Cluj), 12 (35), (1970) p. 159-186.
PETEANU V. et RADO F., Structures algébriques rattachées aux problèmes
d'ordonnancement. Colloque sur la théorie de l'approximation des
fonctions, Cluj (1967).
PICARD C.F., Latticoid product, sum and product of graphs, in Calgary-69,
p. 321-322.
PICARD C.F., Distributivité d'opérations sur les graphes, C.R. Acad. Sci.
Paris, 270A (1970) p. 1219-1221.
PICARD C.F., Produit et somme de certains graphes quasi-fortement connexes,
C.R. Acad. Sci. Paris 271A (1970) p. 145-148.
PICARD C.F., Distributivity theorems in graph theory, in Ilmenau-70,
p. 229-236.
PICHAT E., Algorithmes pour rechercher les éléments maximaux de structures
algébriques, Séminaire sur les problèmes combinatoires, Institut Henri
Poincaré Paris, 1969.
PICHAT E., Contribution à l'algorithmique non numérique dans les ensembles
ordonnés, Thèse Sci. Math. Grenoble, 1970.
PRUFER H., Neuer Beweis eines Satzes über Permutationen, Archiv der
Mathematik und Physik (3) 27 (1918) p. 142-144.
REID K.B., Connectivity in products of graphs, SIAM J. Appl. Math., 18
(1970) p. 645-651.
ROBERT P. et FERTAND J., Généralisation de l'algorithme de Warshall, Revue
française d'Informatique et de Recherche Opérationnelle 7 (1968)
p. 71-85.
ROY B., Cheminement et connexité dans les graphes, applications aux prob-
lèmes d'ordonnancement METRA, Série spéciale 1962.
SABIDUSSI G., Graph Multiplication, Math. Z. 72 (1960) p. 446-457.
SHANNON C.E., The Zero-error capacity of a noisy channel I.R.E. Trans. 3
(1956), p. 3-15.
SHAPIRO H., The embedding of graphs in cubes and the design of sequential
relay circuits (unpublished). Bell Tel.Lab. Memorandum, July 1953.
SZAMKOLOWICZ, Remarks on the cartesian product of two graphs, Colloq. Math.
9 (1962) p. 43-47.
TOMESCU I., Méthode pour la détermination de la fermeture transitive d'un
graphe fini, Revue Française d'Informatique et de Recherche
Opérationnelle 4 (1967) p. 33-37.

TOMESCU I., Méthode pour la détermination de la fermeture transitive d'un graphe fini, Revue Française d'Informatique et de Recherche Opérationnelle 4 (1967) p. 33-37.
TOMESCU I., Un algorithme pour la détermination des plus petites distances entre les sommets d'un réseau, Revue française d'Informatique et de Recherche Opérationnelle 5 (1967) p. 133-139.
TOMESCU I., Sur l'algorithme matriciel de B. Roy, Revue française d'Informatique et de Recherche opérationnelle 7 (1968) p. 87-91.
VARSHAVSKY V.J., MARAKHOVSKY V.B. and PESCHANSKY V.A., Synchronization of Interacting Automata, Mathematical systems theory 4 (1970) p. 212-230.
VIZING V.G., Cartesian product of graphs,(in Russian) Computer elements and systems 9 (1963) p. 30-43, Israel program for sci. translations, Jerusalem 1966, p. 352-365.
WARSHALL S., A theorem of boolean matrices, Journal A.C.M. 9 (1962) p. 11-13.
WATKINS M.E., Still another product for graphs, in Calgary-69, p. 467-468.
WEICHSEL P.M., The Kronecker product of graphs, Proc. Amer. Math. Soc. 13 (1962),p. 47-52.
YOELI M., A note on a generalization of Boolean Matrix Theory, Amer. Math. Monthly 68 (1961), p. 552-557.
YOELI M., Generalized direct decompositions of transformation graphs, in Rome-66, p. 403-412.

CHAPTER IV. General properties of questionnaires

CHERRY E.C., Generalized concepts of networks, in Brooklyn-54, p. 175-184.
ERDÖS P., Graph theory and probability, Journal Canad. Math. 11 (1959) p. 34-38 and 13(1961) p. 346-352.
FELLER W., An Introduction to probability theory and its applications, John Wiley, vol. 1 (1950), vol. 2 (1966).
FORD L.R. and FULKERSON D.R., Flows in networks, Princeton University, 1962. (French translation: Flots dans les graphes, Gauthier-Villars, 1967).
JORDAN P., Zum Dekekindschen Axiom in der Theorie der Verbände, Abhandl. Hamburg 16 (1949) p. 71-73.
KOLMOGOROV A.N., Grundbegriffe der Wahrscheinlichtskeitsrechnung, Springer, 1933 (English version, Chelsea, 1956).
LOEVE M., Probability Theory, Van Nostrand, 1963 (3e Ed.).
METIVIER M., Notions fondamentales de la théorie des probabilités, Dunod, 1968.
NEVEU J., Bases mathématiques du calcul des probabilités (2e edition) Masson, 1970.
PICARD C.(F)., Théorie des Questionnaires, Thèse Sc. Math. Paris, 1963 (revised and completed in the following work).
PICARD C.(F)., Théorie des Questionnaires, Gauthier-Villars, 1965; translation : Theorie der Fragebogen, Akademie-Verlag, Berlin 1971.
PICARD C.F., Probabilités sur des graphes et information traitée par des questionnaires, in Prague-71.
RENYI A., Foundations of probability, Holden Day, 1970.

CHAPTER V. The construction of questionnaires

CHATY G., Cheminements remarquables dans les graphes : existence, obtention, conservation, Thèse Sc. Math. Paris, 1971.
DEDEKIND R., Uber Zerlegungen von Zahlen durch ihre grössten gemeinsamen Teiler, (published in 1897), Gesammelte Werke 2 (1931) p. 103-148.

GHERSI T., Questionnaires latticiels, réseaux de transport et optimalité, Thèse Docteur Ingénieur Paris, 1970.
JORDAN C., OEuvres, publiées par J. Dieudonné, see note on Jordan's work, Gauthier-Villars, 1961.
KISLITSYNE S.S., Une nouvelle limite inférieure du nombre moyen de comparaisons nécessaires au classement complet de N objets (in Russian), Vestnik Leningrad Univ. Ser. Math. Mech. Astron. (1963), 1 p. 162-163.

CHAPTER VI. Optimal routing

BELLMAN R., Dynamic programming, Princeton University Press, 1957.
BELLMAN R. and DREYFUS S.E., Applied dynamic programming, Princeton University Press, 1957; transl.La programmation dynamique et ses applications, Dunod, 1965.
CESARI Y., Optimisation des questionnaires avec contrainte de rang, in Balatonfüred-69, p. 213-215.
CAMPBELL L.L., Note on the connection between search theory and coding theory, in Debrecen-67, p. 85-88.
DUBAIL F., Algorithmes de questionnaires réalisables optimaux au sens de différents critères, Thèse 3e Cycle, Lyon, 1967.
FANO R.M., Transmission of information, MIT-Press and J. Wiley, 1961.
GILBERT E.N. and MOORE E.F., Variable length binary encodings, Bell syst. Tech. J. 38 (1959) p. 933-967.
HUFFMAN D.A., A method for the construction of minimum redundancy codes, Proc. Inst. Radio Engrs. 9 (1952) p. 1098-1101.
PETOLLA G., Coûts, contraintes, ordres et questionnaires, Thèse 3e cycle, Lyon, 1970.
PETOLLA S., Extension de l'algorithme d'Huffman à une classe de questionn-aires avec coûts, Thèse 3e Cycle, Lyon, 1969.
PICARD C.(F)., Cheminement optimum et théorie de l'information, in Munich-62, p. 720-721.
SHANNON C.E., A mathematical theory of communication, Bell System Tech. J. 27 (1948), p. 379-423, p. 623-656.
Reprinted in C.E. Shannon and W. Weaver: The mathematical theory of communication, University of Illinois Press, Urbana, Ill. 1949.
SCHWARTZ E.S., An optimum encoding with minimum longest code and total number of digits, Information and Control 7 (1964), p. 37-44.
ZIMMERMAN S., An optimal search procedure, Amer. Math. Monthly, 66 (1959) p. 690-693.

CHAPTER VII. Informational study of questionnaires

ACZEL J., Lectures on functional equations and their applications, Academic Press, 1966.
ACZEL J. et DAROCZY Z., Sur la caractérisation axiomatique des entropies d'ordre positif y comprise l'entropie de Shannon, C.R. Acad. Sci. Paris 257 (1963), p. 1581-1584.
ADLER R.L., KONHEIM A.G., McANDREW M.H., Topological entropy, Trans. Amer. Math. Soc. 114 (1965), p. 309-319.
AGGARWAL N.L. and CANONGE J.C., La détermination, mesure de la connaissance sur des ensembles métriques, in Besançon-66, p. 269-275.
ASH R., Information theory, J. Wiley, 1965.
BARNARD G.A., The Theory of Information, J. Roy, Statist. Soc. Ser. B 13 (1951), p. 46-64.
BAIOCCHI C., Su un sistema di equazioni funzionali connesso alla teoria dell'informazione, Boll. Un. Mat. Ital. (2) 22 (1967) p. 236-246.

BERZTISS A.T., Data structures, theory and practice, Academic Press, 1971.
ERDÖS P., On the distribution function of additive functions, Ann. of Math 47 (1946), p. 1-20.
FADDEEV D.K., Zum Begriff der Entropie eines endlichen Wahrscheinlichkeitsschema, in URSS-57, p. 86-90.
FANO R.M., Information theory and generalized networks, in Brooklyn-54, p. 3-10.
FANO R.M., Transmission of information, MIT Press and J. Wiley, 1961.
FEINSTEIN A., Foundations of information theory, McGraw Hill, 1958.
FORTE B., Measures of Inforamtion : The general axiomatic theory : Revue Française d'Informatique et de Recherche Opérationnelle 3 (1969) série R-2, p. 63-84.
GUIASU S. et THEODORESCU R., La théorie mathématique de l'information (translation from Rouman.), Dunod, 1968.
HALMOS P.R., Measure theory, Van Nostrand, 1950.
HARDY G., LITTLEWOOD J. and POLYA G., Inequalities, Cambridge University Press, 1952.
HARTLY R.V., Transmission of Information, Bell System Techn. Journ. 7 (1928), p. 535-563.
HUFFMAN D.A., Information conservation and sequence transducers, in Brooklyn-54, p. 291-307.
INGARDEN R.S., A simplified axiomatic definition of information, Bull. Acad. Pol. Sci. Ser. Sci. Math. Astronom. Phys. 11 (1963), p. 209-212.
INGARDEN R.S. and URBANIK K., Information without probability, Colloquium Math. 9 (1962), p. 131-150.
JAYNES E.T., Probability theory in science and engineering, McGraw Hill, 1961.
JELINEK F., Probabilistic Information Theory, McGraw Hill, 1968.
KAMPÉ DE FERIET J., Mesure de l'information fournie par un évenement, in Clermont-69, p. 191-221.
KAMPÉ DE FERIET J., Mesures de l'information pour un ensemble d'observateurs, C.R. Acad. Sci. Paris 269A (1969), p. 1081-1085 et 271A (1970), p. 1017-1021.
KAMPÉ DE FERIET J. et FORTE B., Information et probabilité, C.R. Acad. Sci. Paris 265A (1967), p. 110-114, 142-146, 350-353.
KAMPÉ DE FERIET J., FORTE B., BENVENUTI P., Forme générale de l'opération de composition continue d'une information, C.R. Acad. Sci. Paris 269A (1969), p. 529-534.
KHINTCHINE A.I., Mathematical foundations of information theory, American edition, Dover, 1957, (Uspehi. Math. Nauk 8 (1953), p. 3-20 and 11 (1956), p. 17-75 and in URSS-57, p. 8-25 and p. 26-85).
KOLMOGOROV A.N., Three approaches to the definition of the concept "quantity of information", Problemy Peredacy Informacii, 1 (1965), p. 3-11.
KOLMOGOROV A.N., Logical basis for information theory and probability theory, IEEE Trans. on Inf. Th. IT 14 (1968), p. 662-664.
PEREZ A., Notions généralisées d'incertitude, d'entropie et d'information du point de vue de la théorie des martingales, in Prague-56, p. 183-208.
PEREZ A., Sur la théorie de l'information dans le cas d'un alphabet abstrait, in Prague-56, p. 209-243.
PEREZ A., Risk estimates in term of generalized f-entropies, in Debrecen-67, p. 299-315.
PICARD C.(F.)., Information, acquisition, précision, in Besançon-66, p. 253-268.
PICARD C.(F.)., Valeur maximale de l'information traitée, C.R. Acad. Sci. Paris 265 A (1967), p. 624-627.

PICARD C.F., Interprétation informationnelle de l'algorithme de Huffman, C.R. Acad. Sci. Paris 266 A (1968), p. 998-1000.
PICARD C.F. et SCHNEIDER M., Information du type M transmise par un questionnaire latticiel, C.R. Acad. Sci. Paris 274 A (1972).
PINTACUDA N., Shannon entropy. A more general derivation, Statistica 26 (1966), p. 509-524.
RENYI A., On a theorem of Erdös and its application in information theory, Mathematica (Cluj) 1 (1959), p. 341-344.
RENYI A., On measures on entropy and information, in Berkeley-60, t.1, p. 547-561.
RENYI A., Dimension, entropy and information, in Prague-62, p. 545-556.
RENYI A., Wahrscheinlichkeitsrechnung mit einem Anhang über Informationstheorie, V.E.B. Deutscher Verlag der Wissenschaften, 1962 (French translation: Calcul des probabilités, Dunod, 1966).
RENYI A., On the foundations of information theory, Rev. Intern. Stat. Inst. 33 (1965), p. 1-14.
RENYI A., On the amount of missing information and the Neyman-Pearson lemma, Festschrift for J. Neyman, Wiley, 1966, p. 281-288.
RENYI A. und BALATONI J., Uber den Begriff der Entropie in URSS-57, p. 117-134.
RIGAL J.L., AGGARWAL N.L. et CANONGE J.C., Incertitude et fonction d'imprécision liées à un questionnaire sur un espace métrique, C.R. Acad. Sci. Paris, 263 A (1966), p. 268-270.
RIGAL J.L., AGGARWAL N.L. et CANONGE J.C., Une mesure de l'interdétermination sur un espace métrique de réponses, in Debrecen-67, p. 359-365.
SCHUTZENBERGER M.P., Contribution aux applications statistiques de la théorie de l'information, Publ. Inst. Statis. Univ. Paris, 3, (1954), p. 5-117.
SHANNON C.E., A mathematical theory of communication, Bell System Techn., J. 27 (1948), p. 379-423, 623-656.
Reprinted in C.E. Shannon and W. Weaver, The Mathematical Theory of Communication, University of Illinois Press, Urbana, Illinois, 1949.
VILLE J.A., Leçons sur quelques aspects nouveaux de la théorie des probabilités. Ann. Inst. Henri Poincaré, 14 (1954), p. 61-143.
VILLE J.A., Théorie de l'Information, cours de la Chaire d'Econométrie, Faculté des Sciences, Paris, 1961.
WATANABE S., A mathematical explication of inductive inference, in Tihany-62, p. 67-107.
WATANABE S., Une explication mathématique du classement d'objets, in Bruxelles-62, p. 39-76.
YAGLOM A.M., YAGLOM I.M., Probabilité et Information (Moscow 1957) (French translation: Dunod 1st ed. 1959, 2nd ed. 1969).

CHAPTER VIII. Information and routing length

BURGE W.H., Sorting, Trees and Measures of order, Information and Control 1 (1958), p. 181-197.
KRAFT L.G., A device for quantizing, grouping and coding amplitude-modulated pulses, Thesis, Massachusets Institute of Technology (1949).
MANDELBROT B., On recurrent noise limiting coding, in Brooklyn-54, p. 205-221.
PICARD C.F., Quasi-questionnaires, codes and Huffman's length, Kybernetika (Academia Praha) 6 (1970), p. 418-435.
RENYI A., On the enumeration of search-codes, Acta Math. Acad. Sci. Hung. 21 (1970), p. 27-33.

SZILLARD L., Uber die Entropieverminderung in einem thermodynamischen System bei Eingriffen intelligenter Weser, Zeitschrift für Physik, 53 (1929), p. 840-856.
TRIBUS M., The use of the maximum entropy estimate in the estimation of reliability, in Lafayette-61, p. 102-140.

CHAPTER IX. Conditioning of the questions and answers

ABRAMSON N., Information theory and coding, McGraw Hill, 1963.
AGGARWAL N.L., Sur la perte d'information dans un questionnaire, C.R. Acad. Sci. Paris 270 A (1970), p. 1190-1193.
BAER J.L., BOVET D.P. and ESTRIN G., Legality and other properties of graph models of computations, Journal A.C.M. 17 (1970), p. 543-554.
BELIS M. and GUIASU S., A quantitative-qualitative measure of information in cybernetic systems, I.E.E.E. Trans. Information Theory IT-14 (1968), p. 593-594.
BERSTEL J., Résolution par un réseau d'automates du problème des arborescences dans un graphe, C.R. Acad. Sc. Paris, 264 (1967), p. 388-390.
BERSTEL J., Quelques applications des réseaux d'automates à des problèmes de la théorie des graphes, Thèse 3e Cycle, Paris, 1967.
BETTMAN J.R., Applying a new methodological approach to a problem of information processing model validation, working paper n° 152, UCLA, Los Angeles, October 1969.
CALINGAERT P.A., System performance evaluation, survey and appraisal, Comm. A.C.M. 10 (1967), p. 12-18.
CAMPBELL L.L., A coding theorem and Renyi's Entropy, Information and Control 8 (1965), p. 423-429.
CESARI Y., Questionnaire, codage et tris, thèse 3e Cycle, Paris, 1968.
CESARI Y., Optimisation des questionnaires avec contrainte de rang, in Balatonfüred-69, p. 213-215.
CHENAIS D., DUBAIL F. et TERRENOIRE M., Questionnaires réalisables optimaux, in Paris-68.
COOPER D.C., Theoretical results concerning programs regarded as directed graphs, Second Machine Intelligence Conference, Edinburgh, 1966.
DOMOTOR Z., Qualitative information and entropy structure, in Inference-70 p. 148-194.
DORFMAN R., The detection of defective members of populations, Ann. of Math. stat., 14 (1943), p. 436-440.
DUBAIL F., Algorithmes de questionnaires réalisables optimaux au sens de différents critères, Thèse de 3e Cycle, Lyon, 1967.
DUNCAN G.T., Information and questionnaires in statistical inference (Thesis) Technical Report 140, School of Statistics, Univ. of Minnesota, Minneapolis, 1970.
DUNCAN G.T., In annals of mathematical statistics, October 1970.
DYNKIN E.B., Theory of Markov Processes, English edition, Pergamon Press, 1960.
FINUCAN H.M., The blood testing problem, Applied Statistics, 13 (1964), p. 43-50.
GORDON P., Théorie des chaînes de Markov finies et ses applications, Dunod, 1965.
GUIASU S., On the most rational algorithm of recognition, Kybernetik 5 (1968), p. 109-113.
GUIASU S., Weighted entropy, Reports on Mathematical Physics 2 (1971), p. 165-179.

GUIASU S. et PICARD C.F., Borne inférieure de la longueur utile de certains codes, C.R. Acad. Sc. Paris, 273 A (1971), p. 248-251.
KNUTH D., Fundamental algorithms, in The art of computer programming, Addison Wesley, Vol. 1, (1968), vol. 2 (1969).
KULLBACK S., Information Theory and statistics, J. Wiley, 1959.
KUMAR S., A generalization of the group testing problem, technical report n° 70, University of Minnesota, 1966.
LUDDE E., Optimierung von logischen Enstcheidungen, V. Internationaler Kongress über Anwendungen der Mathematik in den Ingenieurwissenschaften, Weimar, 1969, p. 265-267.
LUDDE E. und THIELE H., Anhang, in C.F. Picard, Theorie der Fragebogen, Akademie-Verlag, Berlin 1971.
MacCARTHY J., A mathematical theory of computation, in Computer programming and formal systems, North-Holland, 1963.
MacCARTHY J. and SHANNON C.E., Automata studies, Study 34, Princeton 1956.
MASSON Ph., Sur quelques problèmes de pesée, Diplôme d'Etudes Supérieures Paris, 1967.
MATALON B., L'analyse hiérarchique, Gauthier-Villars and Mouton, 1965.
PAGER D., On the efficiency of algorithms, Journal A.C.M. 17 (1970), p. 708-714.
PARKHOMENKO P.P., Optimum questionnaires with questions of unequal value, C.R. Acad. Sc. U.R.S.S. 184 (1969), p. 51-54 (in Russian), transl. in Soviet Physics Doklady 14 (1969), p. 12-14.
PARKHOMENKO P.P., Sur le diagnostic technique (in Russian), Izdatelstvo Znanie, Moscou, 1969.
PATRIS J., Questionnaires avec circuits, Thèse 3e cycle, Paris, 1971.
PETOLLA G., Coûts, contraintes, ordres et questionnaires, Thèse 3e Cycle, Paris 1970.
PETOLLA S., Extension de l'algorithme d'Huffman à une classe de questionnaires avec coûts, Thèse 3e Cycle Lyon, 1969.
PICARD C.F., Automatisation des méthodes mathématiques employées en calcul automatique, 7^a Rassegna Elettronica, Rome, 1960, Congresso per l'elettronica, p. 407-424.
PROST Ch., Séminaire sur les questionnaires, Institut H. Poincaré, Paris 1971.
RAO C.R., Asymptotic efficiency and limiting information, in Berkeley-60 t.1, p. 531-545.
RENYI A., Sur la théorie de la Recherche aléatoire, in Besançon-66, p. 281-287, et Bull. Amer. Math. Soc. 71 (1965), p. 809-828.
SCHNEIDER M., Information généralisée et questionnaires, Thèse 3e Cycle, Lyon, 1970.
SHANNON C.E., Partial ordering for communication channels, Information and control 1 (1958), p. 390-397.
SOBEL M., Group testing to classify all defectures in a binomial sample, in Lafayette-59, p. 127-161.
SOBEL M., Summary of binomial and hypergeometric group-testing, in Besançon-66, p. 277-279.
SOBEL M., Optimal group testing, in Debrecen-67, p. 411-488.
SOBEL M. and GROLL A., Group testing to eliminate efficiently all defectives in a binomial sample, Bell System Tech. J., 38 (1959), p. 1179-1252.
TABOURIER Y., Sur les plans d'inspection. Note de travail 141, SEMA, Paris, 1971.
THIELE H., Wissenschaftstheoretische Untersuchungen in algorithmischen Sprachen, VEB Deutscher Verlag der Wissenschaften, 1966.

UNGAR P., The cut-off point for group testing, Comm. on pure and applied Math. 13 (1960), p. 49-54.

CHAPTER X. Interrogations, comparisons, sortings

CARTER W.C., Mathematical analysis of merge sorting techniques, in Munich-62, p. 62-66.
CESARI Y., Questionnaire, codage et tris, Thèse 3e Cycle, Paris, 1967.
CHEN C.H., A note on sequential decision approach to pattern recognition and machine learning, Information and Control 2 (1966), p. 549-562.
CHENAIS D., Ordre sur un ensemble de questions, C.R. Acad. Sc. Paris 273 A (1971), p. 419-421.
CHENAIS D. et TERRENOIRE M., Détermination de la réalisation d'un processus aléatoire à l'aide de pseudoquestionnaires, in Balatonfüred-69, p. 217-237.
DOOB J.L., Stochastic Processes, John Wiley, 1953.
FALKOFF A.D., Algorithms for parallel-search memoires, Journal A.C.M., 9 (1962), p. 488-511.
FLORES I., Computer sorting, Prentice Hall, 1969.
FORD L.R. and JOHNSON S.M., A tournament problem, Amer. Math. Monthly 66 (1959), p. 387-389.
FRAZER W.D. and McKELLAR A.C., Samplesort: a sampling approach to minimal storage tree sorting, Journal A.C.M. 17 (1970), p. 496-507.
FRIEND E.H., Sorting on electronic computer systems, Journal A.C.M. 3 (1956), p. 134-168.
FRITZCH K., Der Magnet-Band Speicher als binärer Ubertragungskanal, Electronische Informations Verarbeitung und Kybernetik, 5 (1969), p. 331-346.
GLICKSMAN,S.,Concerning the merging of equal length tape files, Journal A.C.M. 12 (1965).
GOETZ M.A., Internal and tape sorting using the remplacement-selection technique, Comm. A.C.M. 6 (1963), p. 201-206.
GROOT (de) M., Optimal stochastical decisions, McGraw Hill, 1970.
GUILBAUD G. et ROSENSTHIEL P., Analyse algébrique d'un scrutin, Mathématiques et Sciences de l'Homme 4 (1963), p. 9-33.
GUTTMAN L., An approach for quantifying paired comparisons and rank order, Ann. Math. Stat. 17 (1946), p. 144-163.
HADIAN A., Optimality properties of various procedures for ranking n different numbers using only binary comparisons, Tech. Rep. 117, Dep. of Stat., Univ. of Minnesota (1968).
HIBBARD Th. N., Some combinatorial properties of certain trees with application to searching and sorting, Journal A.C.M. 9 (1962), p. 13-28.
HIBBARD Th. N., An empirical study of minimal storage sorting, Comm. A.C.M. 6 (1963), p. 206-213.
ISAAC E.J. and SINGLETON R.C., Sorting by address calculations, Journal A.C.M. 3 (1956), p. 169-174.
IVERSON K.E., A programming language, J. Wiley, 1962.
JONES B., Variations on sorting by address calculations, Comm. A.C.M. 13 (1970), p. 105-107.
KARIVCKIY V.V., PARKHOMENKO P.P., SOGOMONIAN E.C., Problèmes sur les diagnostics techniques (in Russian), in Abstraktnay a strukturnaya teoriya releiynir ustroiystv, Institut d'automatique et de télémécanique, Acad. Sci. URSS, Izdatelstvo, "Nauka" Moscou (1966), p. 189-224.

KISLITSYNE S.S., Amélioration de la borne inférieure du nombre de comparaisons nécessaires à l'ordination complète d'un ensemble fini (in Russian), Vestnik Leningrad Univ. Ser. Math. Mech. Astron.(1963), 19, p. 143-145.
LOUIT G., Algorithmes de tri, Dunod, 1971.
LUCAS E., Théorie des nombres, Paris, 1891; new edition: Blanchard, Paris, 1961.
LUSTED L.B., Logical analysis in medical diagnosis, in Berkeley-65, t.4, p. 903-923.
LYNN BEUS H., The use of information in sorting, Journal A.C.M. 17 (1970), p. 482-495.
MARTIN J.J. Bayesien decision problems on Markov chains, J. Wiley, 1967.
MARSCHAK J., Producing storing transporting and using knowledge, Working paper n° 165 UCLA, Los Angeles, September 1970.
MATTEI M., Les ordinateurs à l'aide du diagnostic médical, application en toxicologie, Thèse Doctorat en Médecine, Grenoble, 1969.
MATTEI M., FAURE J. et YACOUB M., Les ordinateurs à l'aide du diagnostic médical, application en toxicologie, in Journées Internationales d'informatique médicale, Toulouse (Mars 1970), I.R.I.A.ed.
MOOERS C.N., Some mathematical fundamentals of the use of symbols in information retrieval, in Information Processing-Congress, UNESCO Paris 1959, p. 315-321.
MOON J.W., Topics on tournaments, Holt, Rinehart and Wiston, 1968.
MOON J.W., The expected strengths of losers in knockout tournaments, in Calgary-69, p. 277-281 (see Collectives, volume 1).
NARAYANA T.V., Contribution to the theory of tournaments, University of Alberta, 1968, (mimeo).
NEMETZ T., Information theory and the testing of a hypothesis, in Debrecen-67, p. 283-294.
OETTINGER A.G., Account identification for automatic data processing, Journal A.C.M. 4 (1957), p. 245-253.
PICARD C.(F)., Quelques idées récentes sur le problème du tri, Revue Française du Traitement de l'Information, 9 (1966), p. 41-46.
PINSKER M.S., Information and information stability of random variables and processes; translated by Amiel Feinstein, Holden Day, 1964.
PRYWES N.S. and GRAY H.J., The Multilist system for real storage and retrieval, in Munich-62, p. 273-278.
SALTON G., Automatic information organization and retrieval, McGraw Hill, 1968.
SANDELIUS M., On an optimal search procedure, American Math. Monthly, 68 (1961), p. 133-134.
SCHUTZENBERGER M.P., On an application of semi-group methods to some problems in coding, I.R.E. Trans. Inf. Theory, IT-2 (1956), p. 47-53.
SIMON J.C. et GUIHO G., Une nouvelle méthode d'établissement et d'utilisation de mémoires ordonnées, C.R. Acad. Sc. Paris, 271 A (1970), p. 1022-1025.
SIMON J.C. and ROCHE C., Application of questionnaire theory to pattern recognition, Second International Joint Conference on artificial intelligence, The British Computer Society, London,Sept. 1971.
SOBEL M., On an optimal search for the t best using only binary errorless comparisons: part 1 the ordering problem, part 2 the selection problem, Tech. Rep. 113 and 114, Depart. of Statistics, Univ. of Minnesota 1968 (to appear in French).
STANFEL L.E., Tree structures for optimal searching, Journal A.C.M. 17 (1970), p. 508-517.

STEINHAUS H., Cent problèmes élémentaires de mathématiques, (translation), Gauthier-Villars, 1966.
SUSSENGUTH., Use of tree structures for processing files, Comm. A.C.M. 6 (1963), p. 272-279.
TABOY J.P., Détermination d'une organisation mémoire permettant la partition d'un fichier par la théorie des questionnaires (Cas des arborescences avec coût, fonction positive croissante de la base), Thèse 3e cycle, Paris, 1971.
TERRENOIRE M., Une généralisation des questionnaires: les pseudoquestionnaires, C.R. Acad. Sc. Paris 270 A (1970), p. 263-265.
TERRENOIRE M., Pseudoquestionnaires et information, C.R. Acad.Sci. Paris 271 A (1970), p. 884-887.
TERRENOIRE M., Un Modèle mathématique de processus d'interrogation : les pseudoquestionnaires, Thèse Sci. Math. Grenoble, 1970.
WALD A., Sequential Analysis, J. Wiley, 1947.
WALD A., Statistical decision functions, J. Wiley, 1950.
WELLS M.B., Application of a language for computing in combinatorics, Proceedings of IFIP Congress 65 t.2, Spartan (1965), p. 497-498.
WOODRUN L.J., Internal sorting with minimal comparing, IBM System 8, (1969), p. 189-203.

COLLECTIVE BOOKS

AIX-EN-PROVENCE-67
Ordres totaux finis. Travaux du séminaire sur les ordres totaux finis, Gauthier-Villars, Mouton 1971.

ANN ARBOR-68
Proof Techniques in graph theory (2nd Ann Arbor Graph Theory Conference, February 1968) edited by F. Harary, Academic Press 1969.

BALATONFURED-69
Proceedings of the Colloquium on Combinatorial theory and its applications, edited by P. Erdős, A. Renyi and Vera T. Sos, North-Holland 1970 (3 vol.).

BERKELEY-50, BERKELEY-54, BERKELEY-60, BERKELEY-65
Proceedings of the 1st (2nd, 3rd, 4th, 5th) Berkeley Symposium on math. statistics and probability theory. Univ. of California Press.

BESANÇON-66
La programmation en analyse numérique, Colloque international du CNRS, éd. CNRS, Paris, 1967.

BROOKLYN-54
Symposium on information networks, Polytechnic Institute of Brooklyn, 1955.

BRUXELLES-62
Information and Prediction in Science, edited by Dockx S. and Bernays P., Academic-Press, 1965.

CALGARY-69
Combinatorial Structures and their applications, Proceedings of the Calgary international conference on combinatorial structures and their applications, held at the University of Calgary (1969), Gordon and Breach 1970.

CHAPEL HILL-67
Conbinatorial Mathematics and its applications (Conference Chapel Hill, 1967), edited by R.C. Bose and T.A. Dowling, The University of North Carolina 1969.

CLERMONT-69
Les probabilités sur les structures algébriques, Colloque International du CNRS, éd. CNRS, Paris, 1970.

DEBRECEN-67
Proceedings of the colloquium on information theory, edited by A. Renyi, J. Bolyai Math. Soc, (Hungary) 1968 (2 vol.).

HAMILTON-68
International symposium on Probability and information theory, Lecture Notes in Math., 89, Springer, 1969.

ILMENAU-70
XV. Internationales Wissenschaftliches Kolloquium Teil Al, Ilmenau (D.D.R.) 1970.

KALAMAZOO-68
The many Facets of graph Theory, Conference held at Western Michigan University, Kalamazoo (1968), Lecture notes in Mathematics, 110, Springer 1970.

LAFAYETTE-59
Information and decision processes, Symposium, Purdue University, edited by R.E. Machol, McMillan, 1960.

LAFAYETTE-61
Recent developments in information and decision processes, Symposium, Purdue University, edited by R.E. Machol and P. Gray, McMillan, 1962.

LONDON-60
Fourth London symposium on Information Theory, edited by Colin Cherry, Butterworths, 1961.

LONDON-63
A Seminar on graph theory, London 1963, edited by F. Harary; Holt, Rinehart and Winston 1967.

MUNICH-62
Information processing 62, proceedings of the IFIP Congress, edited by C.M. Popplewell, North-Holland, 1963.

NEW YORK CITY-70
Recent trends in graph theory, Proceedings of the 1st New-York City Graph Theory Conference (1970), Lecture notes in Mathematics 186, Springer 1971.

OXFORD-69
Combinatorial mathematics and its applications Proc. of a Conference, Oxford, July 69, edited by D.J.A. Welsh, Academic Press 1971.

PARIS-67
Algorithmes de Graphes, Contrat D.G.R.S.T. rédigé par l'Institut de Programmation de Paris, miméographie de l'Institut de Programmation, Université Paris VI.

PARIS-68
Information théorique et questionnaires, Journées d'etudes, édité par C.F. Picard, publication de l'Institut Blaise Pascal, CNRS Paris, 1968 (épuisé).

PRAGUE-56
PRAGUE-59, PRAGUE-62, PRAGUE-65, PRAGUE-71
Information theory, statistical decision functions and random processes, transations of the first (2nd, 3rd, 4th, 6th) Prague Conference, publications de l'Académie des Sciences de Tchécoslovaquie (Institut pour la théorie de l'information et l'automatique).

ROME-66
Theorie des Graphes, Journées internationales d'études, Rome 1966, Dunod 1967.

SMOLENICE-63
Theory of graphs and its applications, Proceedings of the symposium held in Smolenice (1963), Academic Press 1964.

STANFORD-70
Information and inference, edited by J. Hintikka and P. Suppes, Synthese Library D. Reidel, Dordrecht, 1970.

TIHANY-62
Colloque sur les fondements des mathématiques, les machines mathématiques et leurs applications, édité par L. Kalmar, Gauthier-Villars, Paris, Nauwelaerts, Louvain, 1965.

TIHANY-66
Theory of graphs, Proceedings of the Colloquium held at Tihany, Hungary, 1966, edited by P. Erdös and G. Katona, Academic Press 1968.

U.S.S.R.-57
Arbeiten zur Informationstheorie (1), von A.I. Khintchine, D.K. Faddeev, A.N. Kolmogorov, A. Renyi und J. Balatoni (translation from Russian or Hungarian into German), Deutscher Verlag der Wissenschaften, 1957.

U.S.S.R.-70
A survey of progress in graph theory in the Soviet Union by J. Turner and W.H. Kautz, S.I.A.M. Rev. 12 (1970) supplement issue.

WATERLOO-69
Recent Progress in Combinatorics, Proceedings of the 3rd Waterloo Conference on Combinatorics, 1968, edited by W.T. Tutte, Academic Press 1969.

INDEX

absolutely optimal questionnaire	8.1.1
accessory information	10.2.2
acquisition	7.2.2
addition (of arc)	3.3
addition (of vertex)	3.3
additivity	4.2.3
additivity (of information)	7.2.1
adjacency matric	1.5.3
adjacent	1.2
admissible partition	9.6.3
ancestor	1.3
answer	4.2.1
antecedent	1.3
antisymmetric	1.2
antisymmetric graph	1.2
arborescence	2.2
arborescence (of paths)	4.3.1
arborescent procedure	2.3.3
arborescent questionnaire	4.2.1
arc	1.2
associative	1.1
balanced arborescence	5.4.2
basis	4.2.1
binary relation	1.1
binding	9.3.1
bound (cost)	9.3.1
broadly speaking	4.2.1
broad transitive closure	1.3
capacity	2.4
capacity	4.2.2
cartesian operation	3.4
chain	1.2
channel	7.1.2
circuit	1.2
circuitless graph	2.1
class (of vertices)	3.4.2
C-optimal questionnaire	9.3.2
cocircuit	1.6.2
cocycle	1.6.2
cocycle rank	1.6.2
coding	1.5.6
come (out), coming out	1.2
comparison operator	9.7.2
compatibility (condition of)	4.3.2
compatibility (relation of)	4.2.1
compatible (arborescent) questionnaire	4.3.2
compatible ordering	2.1
compatible partition set	9.6.3
compatible set	9.6.3
complete graph	1.4
complete system	4.1

complete system of events	4.4.1
complete system of eventualities	4.4.1
component	1.4
condensation	6.3
conditional information	7.3.2
conditional probability	4.4.2
conditional probability (of arc)	4.4.2
connected	1.4
connected component	1.4
conservation of flow	2.4
conservation (of flow)	4.2.2
constraint	4.4.3
consumed information	7.4
contraction	3.3
contributed heterogeneous information	8.1.4
contributed information	8.1.1
contribution of information	8.1.1
contribution of routing length	5.1.2
convex function	7.3.1
cost (of question)	9.3.1
cost (of questionnaire)	9.3.1
cutset	2.4
cutset	4.2.2
cycle	1.2
cycle rank	1.6.2
Dedekindian	2.1
Dedekindian questionnaire	4.2.1
deduced ordering	2.1
degenerate arborescence	2.2
degenerate subquestionnaire	5.1.1
degenerate tree	2.2
degree	1.4
dependent experiment	7.3.2
destination	7.1.2
detector	10.1.2
diagonal graph	3.2
diagonalisation	3.2
dichotomic arborescence	2.2
dichotomic questionnaire	4.2.1
direct interrogation	10.1.1
direct product	3.4.1
discrete probabilisable space	4.2.4
doubling (of a vertex)	3
downstream procedure	5.1.1
edge	1.2
elementary chain	1.2
elementary circuit	1.2
elementary cocircuit	1.6.2
elementary cocycle	1.6.2
elementary cycle	1.2
elementary event	4.2.4
elementary operation	1.3
elementary path	1.2
elementary questionnaire	5.1.1
emitter	7.1.2

empty graph	3.2
endpoint	2.2
entering, enter (into)	1.2
equiprobable questionnaire	6.6
equivalent questionnaire	problem 8
equivalent rank	5.1.2
establish (a relation)	1.1
event	4.2.4
exchange (of arcs)	3.3
expense	9.3.1
extension (of G by H)	3.4
extension	8.1.2
extremity	1.2
factor	1.1
Fibonacci latticoid	5.4.5
finite graph	1.4
first question	4.2.1
flow	2.2
flow chart	9.5.3
flux	2.2
forest	2.2
formulable question	9.6.3
formulated question	9.6.3
Forte and Kampé de Fériet information	7.5.2
free cost	9.3.1
gain (of information)	7.3.2
graph	1.2
(graph) generated (by)	1.2
H-convergent	10.1.3
Hartley information	7.1.1
height	2.3
heterogeneous questionnaire	4.2.1
homogeneous graph	1.4
homogeneous questionnaire	4.2.1
Huffman questionnaire	6.4
incidence matrix	1.5.4
incident	1.2
included ordering	2.1
indegree	1.4
independent experiment	7.3.2
indirect interrogation	10.1.1
inefficiency	8.1.1
infinite arborescence	2.3.5
infinite graph	2.3.5
infinite latticoid	2.3.5
infinite questionnaire	9.5.1
information (of a questionnaire)	8.1.1
information source	7.1.2
information to transmit	10.1.3
initial extremity	1.2
insertion	10.3.2
instantaneous code	2.3.4

instantaneous code	8.3.2
internal question	4.2.1
internal vertex	4.2.1
interpretable vertex	10.1.2
interpretable vertex to within	10.1.2
interpretative flow chart	9.5.3
interpretative questionnaire	10.7.4
intersection (of graphs)	1.3
inverse graph	3.2
inverse relation	1.2
inversion	3.2
isolated	1.2
isomorphism	1.5.2
K-questionnaire	4.4.4
label	1.2
labelled arc	1.5.6
lattice	2.1
latticoid	2.1
latticoid operation	3.5
latticoid product	3.5
latticoid questionnaire	4.2.1
latticoid sum	3.5
L-convergent	10.1.3
length (of an arc)	1.5.6
length (of a path)	1.3
load	9.4.2
loop	1.2
looped arborescence	3.4.3
looped Dedekindian	3.4.3
looped latticoid	3.4.3
looped root	3.4.2
looped vertex	3.4.3
L-optimal questionnaire	5.3
L_u-optimal questionnaire	9.2.1
lower quasi strongly connected lqsc	1.4
maximal height (arborescence of -)	5.4.4
maximal path	1.6.1
maximal path	5.4.5
maximal subarborescence	4.3.1
maximin	5.4.2
measure	4.2.3
measurable space	4.2.3
merging	problem 7
message	7.1.2
minimal question	6.3
minimal redundancy code	6.5
minimal question of maximal rank	6.3
minimax	5.4.2
monoid	2.3.2
M-questionnaire	7.5.2.3
µ-transitive closure	1.3
mutual information	7.3.2
node	3.4.2
noise	8.1.1

INDEX

non-negativity	7.2.2
non-simple	1.2
non-simple chain	1.2
non-simple circuit	1.2
non-simple cycle	1.2
non-simple path	1.6.1
normality (axiom of)	1.1
operating time	9.5.3
operation	3.1
optimality (L-)	5.3
order graph	2.1
ordination of bases	6.2.1
order (of a graph)	1.4
origin	1.2
outcome	4.2.1
outdegree	1.4
pair	1.1
partial graph	1.2
partial subgraph	1.2
partition operator	9.6.1
path	1.2
(perfectly) reliable	10.1.2
permutation (of answers)	6.2.2
permutation (of subarborescences)	6.2.2
p-graph	1.2
placed (vertex - on)	1.2
polychotomic quasi - questionnaire	8.3.1
polychotomic arborescence	2.2
polychotomic questionnaire	4.2.1
posing (of vertex)	3.3
possibility	4.2.4
possible situation	9.6.3
predecessor	1.3
preferable questionnaire	9.2.1
preferable (-vertex)	9.2.1
preference	9.2.1
preordering	1.4
probabilisable space	4.2.4
probability of an answer	4.2.4
probability of an arc	4.4.2
probability of asking a question	4.2.4
probability of a question	4.4.1
probability of a vertex	4.4.1
probability of transit	4.2.4
probabilized arborescence	8.3.1
processes heterogeneous information	8.1.4
processed information	7.4
product	3.1
product (of graphs)	3.4.1
product (of questionnaires)	5.1.1
product of sets	1.1
projection	1.1
proper ancestor	1.3
proper arc	1.2

proper descendant	1.3
proper subgraph	2.2
proper subquestionnaire	5.1.1
proportionality of flows	4.3.2
Prüfer coding	3.3
pseudoquestionnaire	10.1.2
p-tuple	1.1
push down list	1.5.5
Q-questionnaire	4.2.1
quasi-answer	8.3.1
quasi-question	8.3.1
quasi-questionnaire	8.3.1
quasi-strong connectivity	1.4
quasi strongly connected	1.4
question	4.2.1
questionnaire	4.2.1
questionnaire of type t	9.4.1
questionnaire with circuits	9.5.2
questionnaire with costs	9.3.1
questionnaire with logarithmic costs	9.3.4
questionnaire with utility	9.2.1
rank	2.3
reachable (from)	1.3
realisation	9.6.4
realizable L-optimal questionnaire	9.6.4
realizable questionnaire	9.6.3
realizable subquestionnaire	9.6.3
reallocation (of probabilities)	5.1.1
receiver	7.1.2
reduced graph	3.3
reflexive graph	1.3
regular graph	1.4
relation	1.2
relation operator	9.6.1
reliability	9.4.1
Rényi's information	7.5.1
restricted product (of questionnaires)	5.1.1
restricted relation	1.2
restriction	10.1.2
restriction of loops	3.2
result	3.1
risk	8.1.3
root	2.1
routing	4.3
routing length	4.5
σ-additivity	4.2.3
scope (of questions)	10.2.1
section	4.2.2
semantic	4.4.3
semi-degree	1.4
semi homogeneous questionnaire	8.1.4
semi-strong connectivity	1.4
sequential questionnaire	10.2.3

Shannon's information	7.1.1
simple chain	1.2
simple circuit	1.2
simple cycle	1.2
simple path	1.6.1
sink	2.4
situation	9.6.3
source (of network)	2.4
strict compatibility	4.2.1
strictly speaking	4.2.1
strict transitive closure	1.3
strong additivity	7.2.2
strong class	1.4
strong component	1.4
strong connectivity	1.4
strong graph	1.4
strongly convergent	10.1.3
subarborescence	2.2
subgraph	1.2
sublatticoid	2.2
sub-maximum	10.2.1
subpseudoquestionnaire	10.1.2
subquestionnaire	5.1.1
substitution (of arc)	3.3
subtree	2.2
successor	1.3
sum (of graphs)	3.4.1
support	4.2.1
suppression (of arcs)	3.3
symmetric graph	1.2
terminal	2.1
terminal extremity	1.2
to bind	1.2
to join	1.2
to label	5.2
total expense	9.3.2
transfer (of arborescence)	3.3
transformation	3.1
transitive closure	1.3
transitive graph	1.3
transmitted information	7.4
transmitter	1.2
transportation network	2.4
tree	2.2
triple	1.1
trivial graph	3.2
trivial graph of order 1	2.2
trivialisation	3.2
trivial pseudoquestionnaire	10.1.2
truncation	9.5.1
type (of vertex)	3.4.2
unary operation	3.1
union (of graphs)	1.3
upstream procedure	5.1.1

useful information	9.2.2
useful length	9.2.1
utility	9.2.1
value (of flow)	2.4
valuated graph	1.5.6
valuation	1.5.6
vertex	1.2
(vertex preserving) operation	3.1
walk	1.2
weak class	1.4
weak component	1.4
(weak) connectivity	1.4
weighted length	1.5.6

MAIN SYMBOLS

$G = (X,\Gamma)$; $H = (Y,\Delta)$ graphs
a; b; c; i; j; k; t; x; y; z   vertices
u; v; w   arcs
μ chain
$\omega(A)$ cocycle in A
$\omega^+(A)$, $\omega^-(A)$ parts of a cocycle

a	number of true arcs	1.4
$\bar{a}$	number of edges	1.4
b	number of loops	1.4
k	cycle rank	1.6.2
l	cocycle rank	1.6.2
m	number of arcs	1.4
n	number of vertices	1.4
p	number of components	1.4
F	transitive closure of G	1.3
G_s	symmetric graph	1.2
I_x		3.2
T	μ-transitive closure of G	1.3
(X,Γ,V)	valued graph	1.5.6
Γa; ΓA	set of successors	1.3
$\hat{\Gamma} x$	broad transitive closure of x	1.3
Γ_a; Γ_b		3.2
Γ_s		1.2
$\hat{\Gamma}_s$		1.3
$(a,b) \in \Gamma$; $a\Gamma b$	arc	1.2
[a b]; [u v]	path	1.2
)a,b(	edge	1.2
]a b[;]u v[	chain	1.2
$Q = (X, \Gamma, P)$	questionnaire	4.2.1
$K = (X, \Gamma, \mathcal{P}, u, B)$	pseudoquestionnaire	10.1.2
$\bar{K}$	arborescent quasi-questionnaire	8.3.1
(X, Γ, P_1, P_2)	K questionnaire	4.4.4
$(E, \mathfrak{P}(E), P)$	probabilisable space	4.2.4

MAIN SYMBOLS

$(E_c, \mathfrak{P}(E_c), P_c)$ probabilisable space		4.4.1
a, a_i	bases	4.2.1
b_i	number of arcs entering into i	4.3.1
c_i	number of paths entering into i	4.3.1
$c(i)$	cost of question	9.3.1
e, e_j	answers	4.2.1
h, i	questions	4.2.1
i^s	image of i	4.3.1
j, k	vertices	4.2.1
$p(\alpha), p(e), p(i)$		4.2.2
$p_c(i^s), p_c(i^s, j_u^s)$		4.3.2
$p(i,j), p((i,j)\|i), p((i,j)\|\mu_s)$		4.4.2
$[r, s]$	vertex of rank r	5.4.1
A	arborescence	4.3.1
A_N	acquisition	7.2.2
B, C	arborescent questionnaires compatible with Q	4.3.2
D	detector	10.1.2
E	set of answers of Q	4.2.1
E_c	set of answers of C	4.4.1
$E(i)$	event	4.2.3
F	set of questions of Q	4.2.1
$\bar{F} = F - \{\alpha\}$ set of internal questions		4.2.1
$G_N(\mathcal{P}, \mathcal{U})$	useful information	9.2.2
$H(Q)$	heterogeneous information	8.1.4
$I(i)$	contributed information	8.1.1
$I(\mathcal{P})$	information to transmit	10.1.3
$I(Q)$	transmitted information	7.4
$I(B\|A)$	conditional information	7.3.2
$I_N(p_1, p_2, \ldots, p_N)$	Shannon's information	7.2.1
$I_N^\alpha(p_1, p_2, \ldots, p_N)$	Rényi's information	7.5.1
$J(i), J(Q)$	processed information	7.4
L	routing length	4.5
L_c	cost of questionnaire	9.3.1
L_t	Campbell's length	9.4.1

MAIN SYMBOLS

L_u	useful length	8.2.1
L_H	length of L-optimal questionnaire	6.4
$M = \|\Gamma\|$		4.2.1
$N = \|E\|$		4.2.1
$N = a^k + \alpha(a-1) + \beta$	(polychotomic questionnaire)	5.4.2
P_E	valuation of terminals	5.2
$P(E(i))$	probability of event $E(i)$	4.2.4
Q_H	Huffman questionnaire	6.4
Q_σ	subquestionnaire	5.1.1
$V = \|\mathcal{C}\|$	number of distinct paths	4.3.1
α	root of Q	4.2.1
$\lambda:(\cdot), \lambda_u^S(i)$	parameter	4.3.2
μ_S	path $[\alpha i]$	4.3.1
$\nu_{kj}(i)$		10.1.2
Λ	latticoid	4.3
$\Pi(T\|x)$	vector of probabilities	10.1.2
$\mathcal{A}$	set of bases	5.3
$\mathcal{e}$	routing	4.3
C	set of costs	9.3.1
$\mathcal{F}$	family of questionnaires	5.2
$\mathcal{P} = \{p_1, p_2, \ldots p_N\}$	distribution of probabilities	5.2
$\mathcal{U}$	set of utilities	9.2.1
%	euclidean division	
:=	affectation symbol	
$\triangle, \diamond, \boxplus$	latticoid operations	3.5